	I B	II B	III A	IV A	V A	VI A	VII A	VIII A
								2 $1s^2$ **He** 4
			5 (He) $2s^2$ $2p^1$ **B** 11	6 (He) $2s^2$ $2p^2$ **C** 12	7 (He) $2s^2$ $2p^3$ **N** 14	8 (He) $2s^2$ $2p^4$ **O** 16	9 (He) $2s^2$ $2p^5$ **F** 19	10 (He) $2s^2$ $2p^6$ **Ne** 20
			13 (Ne) $3s^2$ $3p^1$ **Al** 27	14 (Ne) $3s^2$ $3p^2$ **Si** 28	15 (Ne) $3s^2$ $3p^3$ **P** 31	16 (Ne) $3s^2$ $3p^4$ **S** 32	17 (Ne) $3s^2$ $3p^5$ **Cl** 35	18 (Ne) $3s^2$ $3p^6$ **Ar** 40
28 (Ar) $3d^8$ $4s^2$ **Ni** 59	29 (Ar) $3d^{10}$ $4s^1$ **Cu** 64	30 (Ar) $3d^{10}$ $4s^2$ **Zn** 65	31 (Ar) $3d^{10}$ $4s^2$ $4p^1$ **Ga** 70	32 (Ar) $3d^{10}$ $4s^2$ $4p^2$ **Ge** 73	33 (Ar) $3d^{10}$ $4s^2$ $4p^3$ **As** 75	34 (Ar) $3d^{10}$ $4s^2$ $4p^4$ **Se** 79	35 (Ar) $3d^{10}$ $4s^2$ $4p^5$ **Br** 80	36 (Ar) $3d^{10}$ $4s^2$ $4p^6$ **Kr** 84
46 (Kr) $4d^{10}$ **Pd** 106	47 (Kr) $4d^{10}$ $5s^1$ **Ag** 108	48 (Kr) $4d^{10}$ $5s^2$ **Cd** 112	49 (Kr) $4d^{10}$ $5s^2$ $5p^1$ **In** 115	50 (Kr) $4d^{10}$ $5s^2$ $5p^2$ **Sn** 119	51 (Kr) $4d^{10}$ $5s^2$ $5p^3$ **Sb** 122	52 (Kr) $4d^{10}$ $5s^2$ $5p^4$ **Te** 128	53 (Kr) $4d^{10}$ $5s^2$ $5p^5$ **I** 127	54 (Kr) $4d^{10}$ $5s^2$ $5p^6$ **Xe** 131
78 (Xe) $4f^{14}$ $5d^9$ $6s^1$ **Pt** 195	79 (Xe) $4f^{14}$ $5d^{10}$ $6s^1$ **Au** 197	80 (Xe) $4f^{14}$ $5d^{10}$ $6s^2$ **Hg** 201	81 (Xe) $4f^{14}$ $5d^{10}$ $6s^2$ $6p^1$ **Tl** 204	82 (Xe) $4f^{14}$ $5d^{10}$ $6s^2$ $6p^2$ **Pb** 207	83 (Xe) $4f^{14}$ $5d^{10}$ $6s^2$ $6p^3$ **Bi** 209	84 (Xe) $4f^{14}$ $5d^{10}$ $6s^2$ $6p^4$ **Po** (210)	85 (Xe) $4f^{14}$ $5d^{10}$ $6s^2$ $6p^5$ **At** (210)	86 (Xe) $4f^{14}$ $5d^{10}$ $6s^2$ $6p^6$ **Rn** (222)

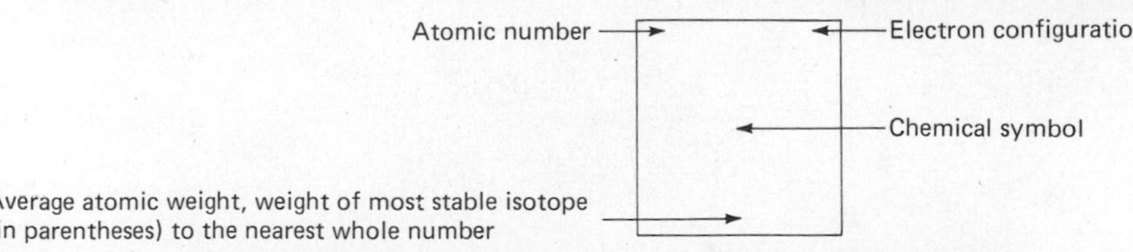

64 (Xe) $4f^7$ $5d^1$ $6s^2$ **Gd** 157	65 (Xe) $4f^9$ $6s^2$ **Tb** 159	66 (Xe) $4f^{10}$ $6s^2$ **Dy** 163	67 (Xe) $4f^{11}$ $6s^2$ **Ho** 165	68 (Xe) $4f^{12}$ $6s^2$ **Er** 167	69 (Xe) $4f^{13}$ $6s^2$ **Tm** 169	70 (Xe) $4f^{14}$ $6s^2$ **Yb** 173	71 (Xe) $4f^{14}$ $5d^1$ $6s^2$ **Lu** 175
96 (Rn) $5f^7$ $6d^1$ $7s^2$ **Cm** (245)	97 (Rn) $5f^8$ $6d^1$ $7s^2$ **Bk** (247)	98 (Rn) $5f^{10}$ $7s^2$ **Cf** (249)	99 (Rn) $5f^{11}$ $7s^2$ **Es** (254)	100 (Rn) $5f^{12}$ $7s^2$ **Fm** (255)	101 (Rn) $5f^{13}$ $7s^2$ **Md** (256)	102 (Rn) $5f^{14}$ $7s^2$ **No** (254)	103 (Rn) $5f^{14}$ $6d^1$ $7s^2$ **Lr** (257)

CHEMISTRY

CHEMISTRY

Harvey A. Yablonsky

Kingsborough Community College
of The City University of New York

THOMAS Y. CROWELL COMPANY
New York Established 1834

ACKNOWLEDGMENTS:

Figure 9–4. Data of V. P. Guinn and H. R. Lukens, Jr., in G. H. Morrison, ed., *Truce Analysis: Physical Methods* (New York: Wiley-Interscience, 1965), p. 345.

Figure 15–3. From John W. Suttie, *Introduction to Biochemistry* (New York: Holt, Rinehart and Winston, Inc., 1972).

Figure 15–8. From W. D. McElroy, *Cell Physiology and Biochemistry,* Second Edition (Englewood Cliffs, N.J.: Prentice-Hall, Inc., 1964).

Figure 17–1. U.S. Department of Health, Education and Welfare, *Air Quality Criteria for Nitrogen Oxide,* p. 23.

U-B-RK

Published simultaneously in Canada by Fitzhenry & Whiteside, Ltd., Toronto.

Library of Congress Cataloging in Publication Data:

Yablonsky, Harvey A.
Chemistry.

Includes bibliographies and index.
1. Chemistry. I. Title.
QD31.2.Y3 1975 540 74-23490
ISBN 0-690-00223-8

Thomas Y. Crowell Company
666 Fifth Avenue
New York, New York 10019

Typography and cover design by Elliot Epstein
Text cartoons by Joe Orlando
Text diagrams and end-paper diagrams by J&R Technical Services
Manufactured in the United States of America

For
Leedra
and
Michael

CONTENTS

PREFACE

More than two decades of teaching have convinced me that few callings are more noble than the humanization of college chemistry. In *Chemistry,* my response to this challenge is the following:

1. In areas where greater than average conceptual difficulty is anticipated, the explanatory sections have been kept as short as practicable to minimize student distress.
2. Concepts introduced in early sections are reexamined in later sections when the student is better able to comprehend and appreciate them. A similar strategy with regard to key reagents helps develop a sense of unity and a feeling for the coherence of the material.
3. Topics are presented in formats that refer to familiar situations. For example, the study of equilibrium kinetics as usually presented is difficult even for the most intrepid. The strain can, however, be lessened if we think in terms of the pairing of partners at a Saturday night dance. Similarly, the gas laws are described in the context of the rules of evidence that prevail at a court trial, and the chemical intricacies of life processes are compared to the operation of an assembly line product.
4. Mathematics beyond basic arithmetic has been relegated to the appendixes.
5. Quantum concepts, described in the most elementary terms, are used very early to give substance to the presentation of chemical principles.
6. Reviews at the end of each section serve to unify the material just presented.
7. Reflexive Quizzes designed to test understanding of concepts occur at the end of each section. The student is provided with short answers to each question. Should he be unable to understand how an answer is arrived at, he may consult the one or two paragraphs in the text whose page reference is given with the answer.

8. The glossary defines important terms, sparing the student from having to search out citations in the text.

The author wishes to express his appreciation to L. Boles, L. Borodkin, and S. Marino for their assistance and to his family for their encouragement and moral support.

H.A.Y.

From caustics and toxins
And foul-smelling vapors
And things that go boom in a flask,
Good Mentor deliver us.

Section 1

The Nuts and Bolts of Matter

Section 1

The Nuts and Bolts of Matter

More than 25 centuries ago, Kanada, the Hindu philosopher, proposed that matter was constructed of small eternal particles. Ancient Greek philosophers as unconfused by the facts as Kanada divided their theories into two schools. Aristotle, as head of one, taught that matter was infinitely divisible. Democritus, founder of the other, taught it was not. Democritus believed that matter was formed of little *atoms* ("uncut" in Greek) that had similar shape and size. It was these non-divisible atoms that produced the properties characteristic of different types of matter. Matter, as a consequence, could not be cut into pieces smaller than these atoms without loss of its characteristic properties.

What Little Atoms Are Made Of

Although two thousand years old, the atomic concept holds true today. An atom is the smallest quantity of an element that will exhibit

the properties that are typical of that element (for example, silver is a *metal;* helium is an *inert gas*). Atoms themselves, however, are composed of smaller, more basic particles. These particles are fundamental to the construction of an atom but do not possess specific atomic characteristics. An analogy can be drawn between an atom and a brick building. The bricks used in constructing a sports stadium do not differ from those used to erect an office building. The number and arrangement of the bricks determine the properties of the building. Similarly, the number and arrangement of *subatomic particles* that are the parts of an atom determine its properties.

The first experimental observation of subatomic particles occurred during the lifetime of people alive today, twenty-two centuries after Democritus set forth his ideas. The discovery of the electron was made in 1897 by J. J. Thomson. This was followed by the discoveries of the proton in 1920 by Ernest Rutherford and the neutron in 1932 by James Chadwick. There is evidence for the existence of other subatomic particles, but their contributions to the chemical properties of elements, if anything, are minimal.

Feathers and Gold Are Not Weighed the Same Way

To describe a man so that he can be recognized, it is necessary to specify his weight, hair and eye coloring, and height. In a similar manner, a subatomic particle can be described in terms of its mass, electronic charge, and size. Before we do this, however, a problem of units must be resolved.

The quantities with which we are dealing are exceedingly small. Were we to use familiar units, we would find ourselves in the same predicament as a man trying to determine the weight in tons of a postage stamp. Since it would take upward of 1,000,000,000,000,000,000,000,000 subatomic particles to weigh as much as a postage stamp, our problem is even more acute. The solution to this difficulty is to weigh particles in terms of each other. This would be equivalent to taking one postage stamp, weighing it, and calling its weight one postal. If we then weighed a sheet of 50 stamps, it would weigh 50 postals. Applying this principle to subatomic particles, scientists determined a unit of mass and called it the *atomic mass unit* (AMU). Happily, the masses of both the proton and the neutron are nearly equal to one AMU. It takes 1837 electrons, however, to equal this mass.

Proton and Electron Charges

Both the proton and the electron are electrically charged. The negative charge of the electron equals exactly the positive charge of the proton and is the smallest quantity ever observed. The amount is so minute that if the entire population of the United States were employed to count the number of electrons necessary to light one small Christmas tree bulb for one second it would take them five hundred years, providing each person counted one a second and no one took time off to eat, sleep, or die.

Bearing this rough but mind-boggling gauge in mind and considering the discussion of atomic masses, we quickly realize that the

TABLE

1–1 Masses and Charges of Subatomic Particles

	Mass	Charge
Electron	$\frac{1}{1837}$ AMU	−1
Proton	1 AMU	+1
Neutron	1 AMU	0

most convenient unit for measuring a subatomic charge is the electron itself. By convention, an electron is assigned a charge of −1, a proton +1, and a neutron 0. The atoms themselves are uncharged because they contain equal numbers of protons and electrons. The number of neutrons in an atom usually varies between one and one and a half times the number of protons. (Hydrogen is an exception to this rule.) Table 1–1 summarizes the data on subatomic particles we have accumulated thus far.

What an Atom Looks Like

To appreciate the size of subatomic particles, we might compare them to the period at the end of this sentence. It would take 50,000,000,000 protons placed side by side in a line to equal the diameter of the period.

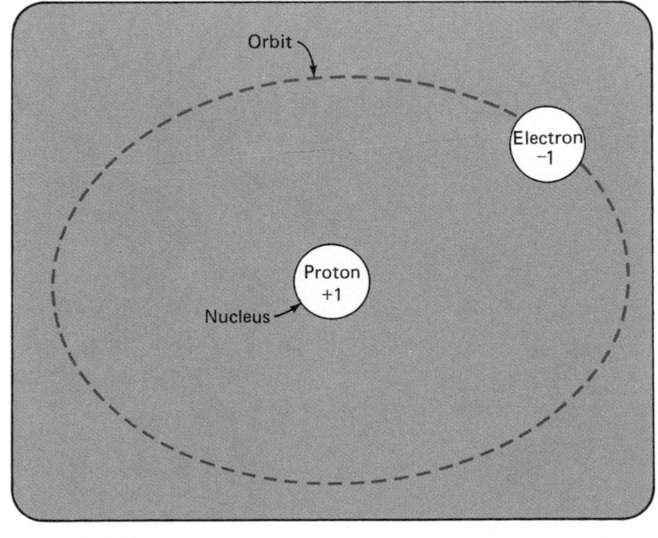

FIGURE

1–1 Bohr Model of a Hydrogen Atom

Obviously, then, if we are to describe this infinitesimal size in any convenient way, we must first modify our system of units and notations. (See Appendix I.) Utilizing the metric system and exponential numbers, we can say that protons, neutrons, and electrons have diameters of approximately 10^{-12} centimeters. Hydrogen, the simplest atom, has a diameter of approximately 10^{-8} centimeters, ten thousand times greater than its constituent particles.

In 1913, Niels Bohr proposed that an atom of hydrogen could be pictured as a negative electron revolving about a positive proton nucleus in a manner similar to the earth revolving about the sun. (Figure 1–1.) This analogy suffers, however, in that it prescribes a fixed orbit for the electron. More recent theories suggest a more complex electron path.

The movement of electrons might more accurately be depicted in terms of a giant egg as tall as the Empire State Building but with a normal size yolk and no white. The protons and neutrons would be confined to the yolk, or nucleus, and the electrons would be restricted to the void left by the missing white. Within this void, we would have to indicate electrons in constant random motion, at speeds so great that they could circle the earth in less than a second. Traveling throughout the space surrounding the nucleus, electrons counteract the positive charge of the nucleus, thus satisfying the requirement that all atoms be effectively neutral.

Such a conceptual model is preferable to the classic Bohr model of electron movement because it accounts for numerous electrons traveling various paths in three dimensions. The orbit that Bohr described is an approximation in two dimensions of the *average* of all the paths that an electron may travel.

REVIEW

1. The smallest quantity of any element still possessing the properties of the element is an atom.
2. Atoms are composed mostly of space and varying amounts of electrons, protons, and neutrons.
3. The properties of an atom are determined by the number and type of its subatomic particles.
4. The nucleus contains the protons and the neutrons.
5. Protons and electrons are charged. Atoms, however, possess equal numbers of protons and electrons and are therefore uncharged.
6. The number of neutrons in a particular atom can vary. In some, it may exceed the number of protons by 50 percent, but only in the case of hydrogen is it ever less than equal.
7. The mass of an atom is mainly the mass of its protons and neutrons. In most calculations, the mass of the electrons may be disregarded.

REFLEXIVE QUIZ

If difficulty is encountered in answering any question, refer to the passages indicated for review by marginal numbers.

1. The mass of an electron is $5 \times 10^{-(\;\;)}$ AMU.
2. The mass of an atom having 10 protons and 10 neutrons is ____ AMU.

3. A particular atom is composed of 3 protons, 4 neutrons, and ___ electrons.

4. The charge on the nucleus of an atom having 9 protons, 9 electrons, and 10 neutrons is ___.

5. Of the following groupings, which can be classified as atoms?

	A	B	C	D
Protons	6	6	6	8
Neutrons	6	8	8	8
Electrons	6	6	8	6

Answers: (1) 4 [3, 5, Appendix I]. (2) 20 [3, 5, 7]. (3) 3 [5, 7]. (4) 9 [5, 7]. (5) A, B [5, 7]. NOTE: Page references for each answer are found in the [square brackets].

SUPPLEMENTARY READINGS

1. "Atomic Theory in the Ancient World." *Chemistry* 44 (1971): 17.

2. R. Moore. *Niels Bohr: The Man, His Science and the World They Changed.* New York: Alfred A. Knopf, Inc., 1966.

3. D. L. Anderson. *The Discovery of the Electron.* New York: D. Van Nostrand Co., Inc., 1964.

4. E. N. da C. Andrade. *Rutherford and the Nature of the Atom.* Garden City, N.Y.: Doubleday & Co., 1964.

Section 2

More Than 100 Ways To Build a Universe

Section 2

More Than 100 Ways To Build a Universe

Basic Rules

One can mix milk, sugar, salt, and eggs and get either a custard or a disaster. Clearly, it is not sufficient to have the proper ingredients. One must also know what to do with them. If a universe is to be built consisting of different kinds of atoms using only electrons, protons, and neutrons, a recipe must be established.

The subatomic-particle relationships common to all atoms were discussed in Section 1. This section will be concerned with the relationships that determine the uniqueness of each type of atom. There are 105 chemically different types of atom, each with a different number of protons in the nucleus. The simplest one—hydrogen—contains 1 proton, the most complex—hahnium—contains 105. *The single property that distinguishes one chemically different atom or element from another is the number of protons in the nucleus.* This number of protons must, of course, be balanced with an equal number of orbital electrons. The manner in which an atom interacts with other atoms is governed by the number of these orbital electrons.

When positioning the paths of electrons in multi-electron atoms, we must (1) consider the mutual repulsion between electrons (like charges repel each other), (2) the forces of attraction between electrons and the nucleus, and (3) the requirement that the net atomic charge be zero. These conditions preclude the possibility of lumping all the electrons together and having them travel as a single entity about the nucleus. Orbits must therefore be assigned so that electrons do *not* collide with each other while they are canceling the positive nuclear charge by moving about in the surrounding atomic space.

Knowledge of the path of any one electron, however, gives little indication of its exact location. The situation is analogous to predicting the location of an elevator in a very large office building at exactly 2:00 P.M. on a given day. The center of the right-hand elevator (Figure 2–1) is (1) at the third floor, (2) 25 feet from the north side of the building, (3) 100 feet from the west side of the building—where the entrance is found—(4) at the given time, 2:00 P.M.

The relationship of elevators and buildings is similar in many ways to that of electrons and atoms:

1. Each elevator will have its own shaft, each electron its own orbit.

2. No two elevators, even if they are on the same floor, will be in exactly the same place at the same time. No two electrons in any atom can be in the same place at the same time. This principle is often referred to as the *Pauli Exclusion Principle* in honor of the man who first proposed it.

3. It is difficult to assign exact locations to either the elevator or the electron because of its continual motion. A *region of space* can be specified for both, however, where the *probability* of finding them is extremely good. The location of the elevator shaft can be described as being x feet in a north-south direction and y feet in an east-west direction from the entrance of the building. The location of an electron orbit, which is a similar though more complex affair, can be specified in terms of four *quantum numbers:* **n**, **l**, **m**, and **s**.

The mathematical foundations of the four quantum numbers are beyond the scope of this book. Fortunately, knowledge of these foundations is not necessary for an understanding of how they *function* in the creation of atoms. The rules governing the allowed values of each quantum number function in a manner similar to the rules governing, for example, the allowed moves of individual checkers or chess pieces. The name of the game is *Aufbau* ("construction" in German).

In this Aufbau game the two quantum numbers of greatest im-

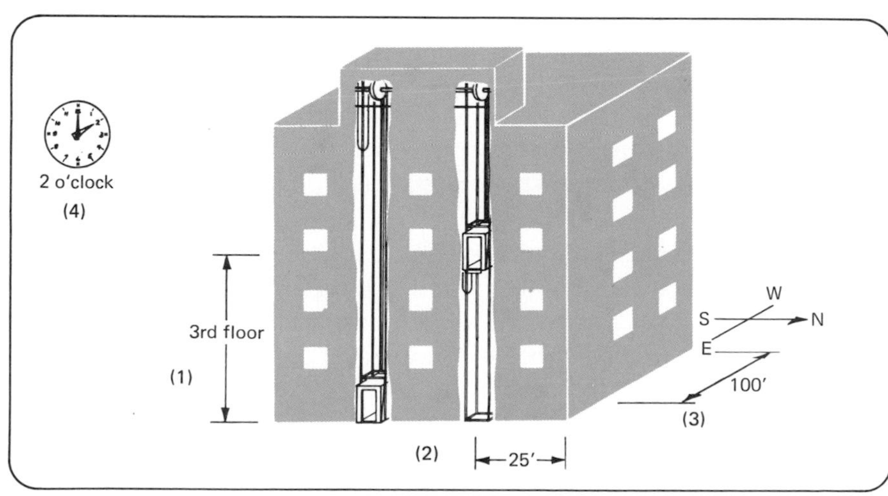

FIGURE 2–1 An Electron Is Like an Elevator.

portance in determining the chemical nature of an atom are the *principal quantum number* **n** and the *azimuthal quantum number* **l.** By analogy to a building with a group of elevators, **n** specifies the floor that the elevator is on and **l** specifies the particular elevator shaft. (The building—atom—must have more floors than elevators.) Just as a municipal building code restricts the placement and number of elevators in a building, the Aufbau Principle similarly restricts the values of the quantum numbers.

Playing the Aufbau Game

If the magnitude of **n** is considered an indication of the *size* of the atom, a pictorial significance can be assigned to the quantum numbers. The larger the value of **n**, if everything else remains constant, the larger the atom. This is because **n** is a measure of the volume of space where the electron *most likely* will be found.

That volume may best be visualized in terms of a balloon with a nucleus at its center. The principal quantum number could be correlated to the number of puffs of air used to inflate the balloon. The volume generated by an electron having 1 as a principal quantum number would correspond to inflating the balloon with a single puff of air. A second puff would increase the volume further and correspond to an **n** of 2. A similar increase in size would occur for successive puffs. However, just as in the case of real balloons, successive puffs or quantum numbers have smaller effects on the volume.

If an atom possesses electrons in more than one principal quantum level (electron orbital volumes with different values of **n**), its structure can be compared either to a balloon within a balloon or to an onion. Each layer of the onion would correspond to a different value of **n**. The farther the layer from the center, the higher the corresponding value of **n**.

Balloons, of course, come in a variety of *shapes,* and so do electron orbitals. The **l** quantum number is a major factor in determining the shape of these orbitals. When the **l** quantum number has a value of 0, the electron orbital it describes is spherical. (Figure 2–2.) This spherical volume is capable of containing as many as *two* of the atom's electrons.

When the **l** quantum number has a value of 1, the volume it describes is composed of three double-sausage-shaped components, or *suborbitals,* that in total are known as a *p* orbital. (Figure 2–2.) A *p* orbital can be formed for each quantum value of **n** the atom possesses in

excess of 1. Each suborbital within the *p* orbital is capable of containing two electrons; it follows that the *p* orbital as a whole can contain a maximum of six electrons.

In the *p* orbital, the three suborbitals are oriented relative to each other much like the axes normally used to describe locations in space—left-right, up-down, front-back—as shown in Figure 2–2. At the point of intersection of these three axes is found the nucleus of the atom. Figure 2–3 shows an *s* orbital and a *p* orbital in an atom. Notice that the center of the sphere describing the *s* orbital coincides with the intersection of the *p* orbital's orientation axes.

There are orbitals of even more complexity than the *p* orbital; they, like the *p* orbitals, are composed of suborbital components, each of which is capable of containing a maximum of two electrons. In summary, with the exception of the *s* orbital, for which l equals 0, all orbitals are composed of more than one suborbital component, each of which, as in the case of the *s* orbital, is capable of containing a maximum of two electrons.

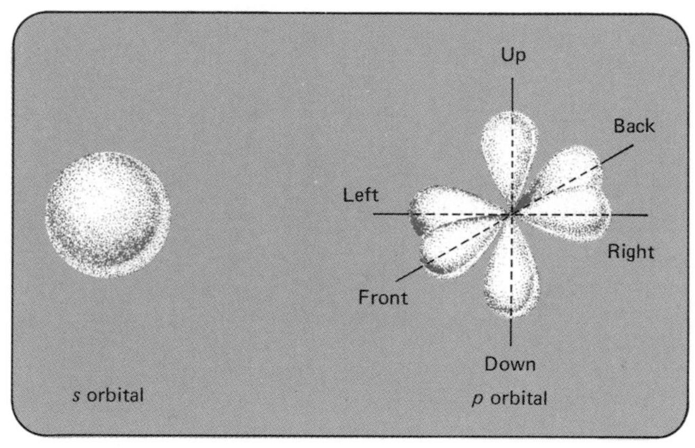

FIGURE 2–2 *s* and *p* Orbitals

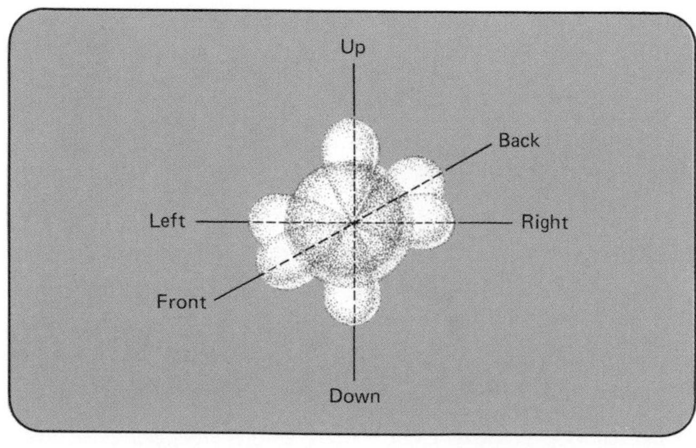

FIGURE 2–3 *s* and *p* Orbitals with a Common Nucleus

As the numerical values of the **l** quantum number increase for a given value of **n**, so too do:

1. The energies of the electrons that these numbers describe.
2. The range of distances from the nucleus where there is an appreciable probability that these electrons can be found.
3. The maximum number of electrons that can be contained in the spaces specified by the higher values of the **l** quantum number.

As can be inferred from Table 2–1, an increase of 1 in the value of **l** produces two additional orbitals, each of which can contain a maximum of two electrons. This results in an increase of 4 in the maximum electron capacity of successively higher **l** orbitals. (The *p* orbital can contain as many as four more electrons than the *s* orbital; the *d* orbital can contain as many as four more electrons than the *p* orbital; and so on.)

It should be understood that the number of electrons contained in an orbital is not necessarily the maximum that that volume can contain. If an atom contains insufficient protons in its nucleus to warrant the complete filling of these volumes, they will not be filled. The situation is analogous to the host or hostess who has a dinnerware service for eight but is expecting only four guests for dinner. The host and his guests will require only five of the available place settings; three will not be used. The host has insufficient guests to warrant using all of his dishes; some of them, therefore, will remain unfilled. Similarly, in describing the quantum numbers of the electrons in an atom, one should not lose sight of the fact that the quantity of these orbiting electrons must equal *the number of protons in the nucleus* and not necessarily the maximum capacity of the orbitals in which they are contained.

With an increase in the value of **n**, the resulting increase in the volume of the atom permits it to contain a greater number of orbitals and consequently a higher maximum number of electrons. The number of orbitals that can be contained in any principal quantum level equals the value of its principal quantum number **n**. Table 2–2 lists the first four primary quantum levels and the orbitals that can be associated with them.

TABLE 2–1 Orbital Electron Capacities

Value of l	Orbital Notation	Maximum Electron Capacity
0	*s*	2
1	*p*	6
2	*d*	10
3	*f*	14

TABLE 2–2 Orbital Capacity of the Principal Quantum Levels

Value of Principal Quantum Number (n)	Possible Associated Orbitals
1	s
2	$s + p$
3	$s + p + d$
4	$s + p + d + f$

Atoms possessing electrons with principal quantum numbers of only 1 can contain at maximum only two electrons. This is because the *s* orbital has a maximum electron capacity of two. If an atom contains electrons with a principal quantum number of 3, however, it can contain not only the electrons that can be associated with the 3*s*, 3*p*, and 3*d* orbitals but also all the electrons that are associated with the lower-energy orbitals, that is, 1*s*, 2*s*, and 2*p*.

The placement of electrons in atomic orbitals is governed by minimum energy considerations. (Electrons will always tend to occupy the orbitals that have the lowest energy first.) The electron configuration of element 1 is $1s^1$. The notation indicates that an atom of element 1 contains one electron in the 1*s* orbital. An atom of element 2 needs two electrons because of the two positively charged protons in its nucleus. Its electron configuration is $1s^2$. The notation indicates that its two electrons are both located in the 1*s* orbital. Because the maximum capacity of any *s* orbital is two electrons, element 3 must place its third electron in the 2*s* orbital. Its electron configuration is therefore $1s^2 2s^1$. The superscripts indicate that of its three electrons, two are located in the 1*s* quantum level and one is located in the 2*s*

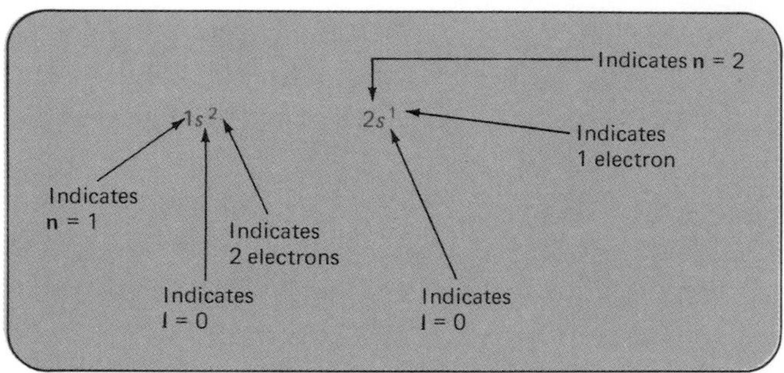

FIGURE 2–4 Electronic Structure of Element 3

quantum level. (Figure 2–4.) ($1s^1 2s^2$ would not be an acceptable electron configuration for element 3 because the electrons in this configuration would be at a higher energy than those of the $1s^2 2s^1$ configuration.)

Just as there is an increase in orbital energy with an increase in the value of **n**, there is also an energy increase with an increase in the value of **l**. This energy increase is not, however, uniform. That is, a *p* suborbital is always at a higher energy than either an *s* orbital of the same principal quantum number or any *p* orbital of a lower principal quantum number, but the energy difference is different in each case. The effect of these differences in energy becomes evident when the first *d* orbital is considered.

This occurs at the third principal quantum level: the energy of the 3*d* suborbital is greater than that of the 4*s*. (For an analogous mental picture, imagine a tall ladder standing on the third floor of a building and projecting through a hole in the ceiling. A person on top of the ladder resting on the third floor would be higher than the level of the fourth floor.) The result of this irregularity is an alteration in what has to this point been a strictly numerical filling of electron quantum levels. Element 19, rather than having its nineteenth electron located in a 3*d* orbital, instead finds it in the lower-energy 4*s* orbital. The electron configuration of element 19 is therefore $1s^2 2s^2 2p^6 3s^2 3p^6 4s^1$ rather than $1s^2 2s^2 2p^6 3s^2 3p^6 3d^1$. (Figure 2–5.)

Figure 2–6 provides a convenient mnemonic (memory aid) for determining the energy levels of each of the common orbitals and the general order of their electron filling. The order of filling commences with the 1*s* orbital and proceeds in the direction indicated by the arrows until the requisite number of electrons have been placed in the atom.

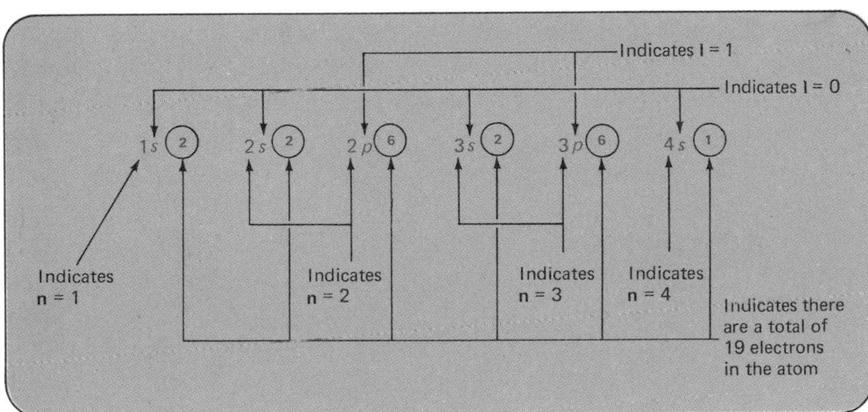

FIGURE 2–5 Electronic Structure of Element 19

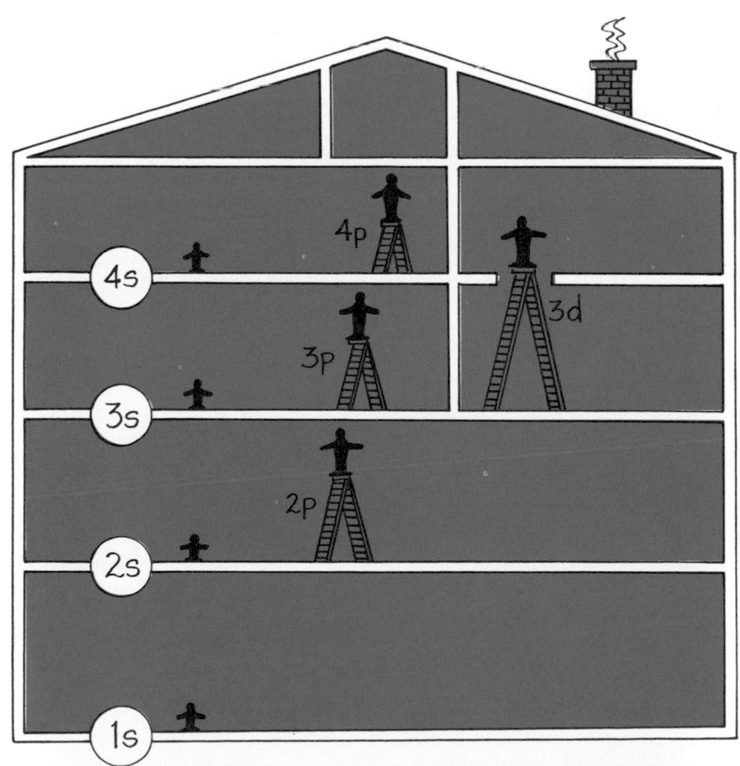

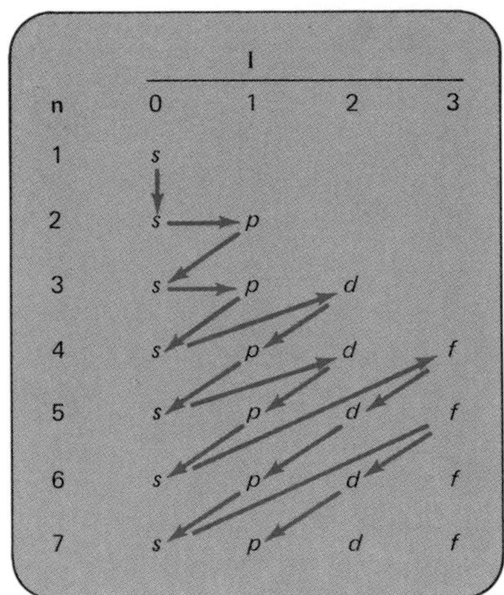

FIGURE

2–6 Energy Levels of Orbitals and Order of Their Electron Filling

TABLE

2–3 Comparison of Theoretical and Observed Electron Capacity

n	Theoretical Electron Capacity	Maximum Observed Electron Capacity
	s p d f g h i	
1	2 = 2	2
2	2 + 6 = 8	8
3	2 + 6 + 10 = 18	18
4	2 + 6 + 10 + 14 = 32	32
5	2 + 6 + 10 + 14 + 18 = 50	32
6	2 + 6 + 10 + 14 + 18 + 22 = 72	11
7	2 + 6 + 10 + 14 + 18 + 22 + 26 = 98	2

Table 2–3 lists the maximum number of electrons that can be contained in each quantum level as compared with what is actually observed. Note that the maximum observed electron capacities of the fifth, sixth, and seventh principal quantum levels are less than their theoretical electron capacities. The complete filling of these principal quantum levels would produce highly unstable atoms. For this reason they are not found in nature. When observed, they will be synthetic products of the atomic age.

These guidelines were utilized in the construction of the periodic chart of the elements. (Figure 2–7.) In this chart each element has an *atomic number* equal to the number of protons in its nucleus.

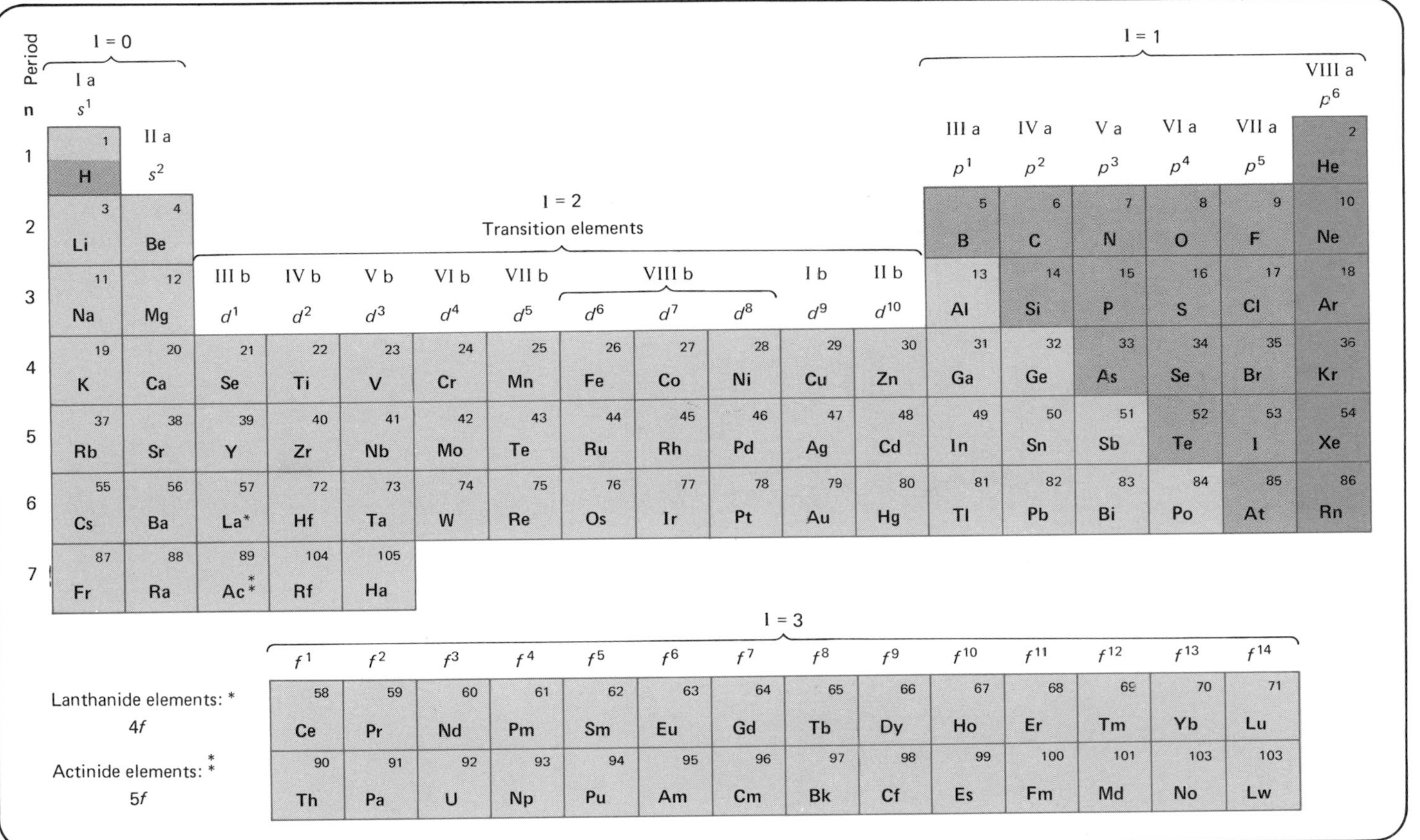

	l = 0		l = 2 Transition elements										l = 1					
Period n	I a s^1	II a s^2	III b d^1	IV b d^2	V b d^3	VI b d^4	VII b d^5	VIII b d^6	d^7	d^8	I b d^9	II b d^{10}	III a p^1	IV a p^2	V a p^3	VI a p^4	VII a p^5	VIII a p^6
1	1 H																	2 He
2	3 Li	4 Be											5 B	6 C	7 N	8 O	9 F	10 Ne
3	11 Na	12 Mg											13 Al	14 Si	15 P	16 S	17 Cl	18 Ar
4	19 K	20 Ca	21 Se	22 Ti	23 V	24 Cr	25 Mn	26 Fe	27 Co	28 Ni	29 Cu	30 Zn	31 Ga	32 Ge	33 As	34 Se	35 Br	36 Kr
5	37 Rb	38 Sr	39 Y	40 Zr	41 Nb	42 Mo	43 Te	44 Ru	45 Rh	46 Pd	47 Ag	48 Cd	49 In	50 Sn	51 Sb	52 Te	53 I	54 Xe
6	55 Cs	56 Ba	57 La*	72 Hf	73 Ta	74 W	75 Re	76 Os	77 Ir	78 Pt	79 Au	80 Hg	81 Tl	82 Pb	83 Bi	84 Po	85 At	86 Rn
7	87 Fr	88 Ra	89 Ac**	104 Rf	105 Ha													

	l = 3													
	f^1	f^2	f^3	f^4	f^5	f^6	f^7	f^8	f^9	f^{10}	f^{11}	f^{12}	f^{13}	f^{14}
Lanthanide elements: * 4f	58 Ce	59 Pr	60 Nd	61 Pm	62 Sm	63 Eu	64 Gd	65 Tb	66 Dy	67 Ho	68 Er	69 Tm	70 Yb	71 Lu
Actinide elements: ** 5f	90 Th	91 Pa	92 U	93 Np	94 Pu	95 Am	96 Cm	97 Bk	98 Cf	99 Es	100 Fm	101 Md	103 No	103 Lw

FIGURE 2–7 Periodic Chart of the Elements

Note: The orbital distribution of electrons presented in this chart is simplistic. When complex energy interactions are considered, the distribution shown here remains strictly valid only for the *regular elements* (groups Ia, IIa, IIIa, IVa, Va, VIa, VIIa, VIIIa). A distribution taking complex energy interactions into consideration is depicted on the inside front cover.

Each principal quantum number **n** *is represented by a separate period* (horizontal row of elements). The period in which an element is listed is the highest value of **n** that must be employed to describe all the electrons in an atom of that particular element. Elements 3 and 5 will both be found in period 2, because 2 is the highest value of **n** necessary to describe their electron structures. The electron configurations are respectively $1s^2 2s^1$ and $1s^2 2s^2 2p^1$. Similarly, elements 12 and 17 will be found in period 3, because 3 is the highest value of **n** that is necessary to describe the arrangement of their electron orbitals. Their electron configurations are respectively $1s^2 2s^2 2p^6 3s^2$ and $1s^2 2s^2 2p^6 3s^2$-$3p^5$.

Each azimuthal quantum number **l** *is represented by a separate group* (vertical column of elements). The group in which an element is listed is determined by the highest value of **l** that must be employed to describe all the electrons in an atom of that element. All s^1 elements are found in the same group; all p^5 elements are found in another. Elements 1, 3, and 11 are all in the s^1 group because their electron configurations are respectively $1s^1$, $1s^2 2s^1$, and $1s^2 2s^2 2p^6 3s^1$. The s^1 orbital configuration is the highest value of **l** in each atom. Similarly, atoms of elements 9 and 17 are found in the same group because their electron configurations are respectively $1s^2 2s^2 2p^5$ and $1s^2 2s^2 2p^6 3s^2 3p^5$ and the p^5 configuration is common to both.

Note. The placement of element 2, helium, in the periodic chart has been shifted from the s^2 column to the p^6 column. This is because the $1s^2$ electron completes the outermost orbital of period 1 in the same manner as the p^6 electron completes the outermost electron orbitals of all the other periods. Consequently, the chemical properties of element 2 are very similar to those of elements 10, 18, 36, 54, and 86, each of which has a p^6 configuration. The elements in the p^6 column are all gases, and they show little tendency to react with other elements. They are known collectively as the *rare* or *noble gases*.

From Figure 2–6 it can be seen that the *f* orbitals as well as the *d* orbitals do not fill until the *s* and *p* of higher numbered periods have already filled. Elements 21–30 comprise the 3*d* (not the 4*d*) elements and have been placed in period 4 only to maintain numbering order. This discrepancy in positioning occurs with all *d* and *f* elements. The *d* shells are positioned one, and the *f* shells two, periods higher than their numbers indicate. This has great significance because it indicates that *d* and *f* orbitals always fill as inner orbitals.

The *s* and *p* orbitals are always at the outermost regions of the atom. They, rather than the *d* and *f* orbitals, are thus in the most suitable location to react with other atoms. Consequently, they

MR. LANTHANIDE AND MR. ACTINIDE ARE A QUEER PAIR!
YES... THEY'RE ALWAYS PUTTING UP A BIG FRONT!
6S
7S
EEEEK!
MY GOSH! THEY'RE MISSING F-ELECTRONS!
4F
5F

2–8 The Sizes of the Elements' Atomic Radii (Angstroms)

Atomic Radii (Angstroms)

Transition elements

Period n	I a	II a	III b	IV b	V b	VI b	VII b	VIII b			I b	II b	III a	IV a	V a	VI a	VII a	VIII a
1	0.37 1 H																	— 2 He
2	1.23 3 Li	0.89 4 Be											0.80 5 B	0.77 6 C	0.74 7 N	0.74 8 O	0.72 9 F	— 10 Ne
3	1.57 11 Na	1.36 12 Mg											1.25 13 Al	1.17 14 Si	1.10 15 P	1.04 16 S	0.99 17 Cl	— 18 Ar
4	2.03 19 K	1.74 20 Ca	1.44 21 Sc	1.32 22 Ti	1.22 23 V	1.17 24 Cr	1.17 25 Mn	1.17 26 Fe	1.16 27 Co	1.15 28 Ni	1.17 29 Cu	1.25 30 Zn	1.25 31 Ga	1.22 32 Ge	1.21 33 As	1.17 34 Se	1.14 35 Br	— 36 Kr
5	2.16 37 Rb	1.41 38 Sr	1.62 39 Y	1.45 40 Zr	1.34 41 Nb	1.29 42 Mo	— 43 Tc	1.24 44 Ru	1.25 45 Rh	1.28 46 Pd	1.34 47 Ag	1.41 48 Cd	1.50 49 In	1.41 50 Sn	1.41 51 Sb	1.37 52 Te	1.33 53 I	— 54 Xe
6	2.35 55 Cs	1.98 56 Ba	1.69 57 La*	1.44 72 Hf	1.34 73 Ta	1.30 74 W	1.28 75 Re	1.26 76 Os	1.26 77 Ir	1.29 78 Pt	1.34 79 Au	1.44 80 Hg	1.55 81 Tl	1.54 82 Pb	1.52 83 Bi	1.53 84 Po	— 85 At	— 86 Rn
7	— 87 Fr	— 88 Ra	— 89 Ac**	— 104 Rf	— 105 Ha													

Lanthanide elements: * 4*f*	1.65 58 Ce	1.65 59 Pr	1.64 60 Nd	— 61 Pm	1.66 62 Sm	1.85 63 Eu	1.61 64 Gd	1.59 65 Tb	1.59 66 Dy	1.58 67 Ho	1.57 68 Er	1.56 69 Tm	1.70 70 Yb	1.56 71 Lu
Actinide elements: ** 5*f*	1.65 90 Th	— 91 Pa	1.42 92 U	— 93 Np	— 94 Pu	— 95 Am	— 96 Cm	— 97 Bk	— 98 Cf	— 99 Es	— 100 Fm	— 101 Md	— 103 No	— 103 Lw

control the chemical characteristics of an element. Since the maximum number of electrons that can be contained in the outer orbitals of any atom is eight (two in the *s* orbital and six in the *p* orbital), the groups of the periodic chart that contain the *regular elements* (elements that do not possess unfilled inner *d* or *f* suborbitals) are numbered from Ia to VIIIa. (Figure 2–7.)

The assignment of any element to a group can therefore be readily made by determining how many of the eight possible electrons are contained in its outer orbitals. Those containing three (s^2p^1) are found in group IIIa. Similarly, those containing a full complement of eight (s^2p^6) are found in group VIIIa. The elements containing unfilled *d* orbitals are known as the *transition elements.* They are found in the groups having the b suffix. Those elements that have unfilled 4*f* and 5*f* orbitals are listed respectively as the *lanthanide* and *actinide elements.*

Of the 105 elements in the periodic chart, 84 are metals, 15 are nonmetals, and 6 are rare gases. Within a vertical group, the greater the proximity of two elements, the greater their chemical similarity. If an element is not a transition, lanthanide, or actinide metal, then the higher the atomic number of that element within a group, the larger its size. Similarly, the lower the atomic number of an element within a horizontal period, the larger its size. These size differences are due to the effect of the attraction between the nucleus and any given electron as the number of protons in the nucleus changes. (Figure 2–8.)

The significance of the third and fourth quantum numbers, **m** and **s**, cannot be evaluated in terms of the periodic chart. Although a discussion of **m** (the magnetic quantum number) is beyond the scope of this book, the significance of **s** (the spin quantum number) can be understood in terms of the formation of orbital component electron pairs.

An electron, like any other charged particle, generates a magnetic field when it moves. If this field is not canceled by the opposing movement of some other electron within the atom, the entire atom will possess magnetic properties. The particular property specified by **s** is the spin of the electron as it moves through the volume of space designated by its orbital. In order for two electrons to fill a particular orbital, their spins must oppose each other. This pairing of opposing spins results in the cancelation of the magnetic fields generated by each, which is the reason orbitals are filled by electron pairs. (The filled *s* orbital contains one pair of electrons; the filled *p* orbital contains three pairs.) In atoms containing an odd number of electrons, complete pairing is not possible, and the magnetic effect generated by the spin of unpaired electrons often is evident in the manner in which these atoms interact with light, which is also electromagnetic in nature.

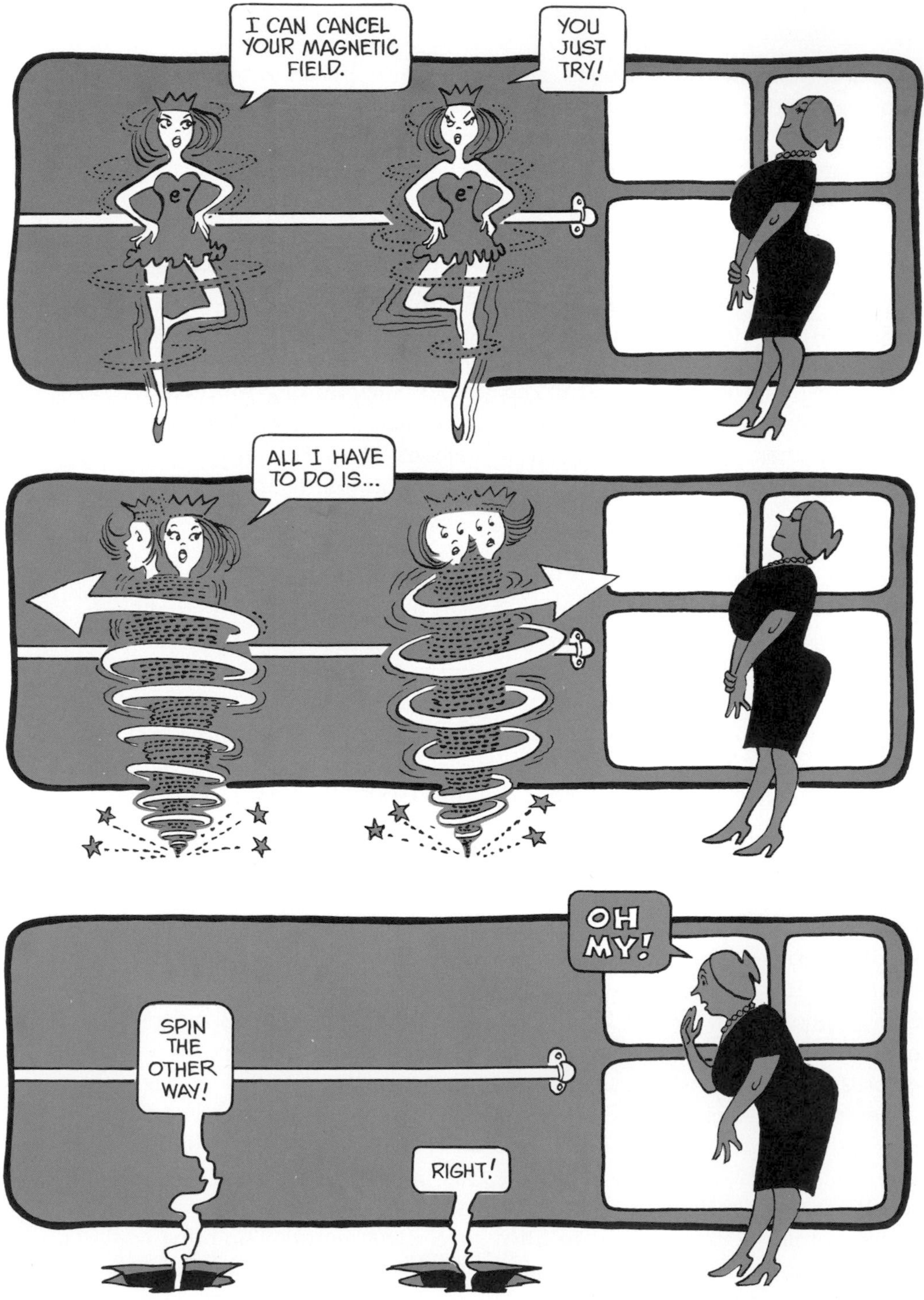
I CAN CANCEL YOUR MAGNETIC FIELD.
YOU JUST TRY!
e⁻
e⁻
ALL I HAVE TO DO IS...
OH MY!
SPIN THE OTHER WAY!
RIGHT!

REVIEW

1. The atomic number of an element is equal to the number of protons in its nucleus.
2. Each element has its own atomic number and characteristic chemical properties.
3. An atom possesses one electron for each proton. The spatial arrangement of these electrons is governed by energy considerations.
4. Elements can be systematically arranged according to their highest quantum numbers. Such an arrangement is often depicted in the form of a periodic chart.
5. The periodic chart is constructed in such a manner that elements in any given vertical group have the same maximum value of **l** and the same number of electrons in their outermost orbits. Such elements possess similar chemical properties.
6. Chemical properties are governed almost entirely by the outermost electrons of an atom.
7. The outermost electrons can be only *s* or *p* electrons, irrespective of the total number of electrons in the atom.
8. The order of filling of electron orbitals is governed by energy considerations. The *s* orbital always fills before the corresponding *p* orbital. However, the *d* and *f* orbitals do not fill until the higher *s* and *p* orbitals are already occupied.
9. Atoms having completely filled principal quantum levels greater than four would be unstable. They have never been observed in nature.

REFLEXIVE QUIZ

1. The number of ____________ in the nucleus determines the nature of an element.
2. "No two electrons can simultaneously occupy the same place" is a statement of the ____________________ Principle.
3. The orbit of an electron can be described in terms of four quantum numbers: ____, ____, ____, and ____.
4. The guidelines for determining permissible values of quantum numbers are often referred to as the ____________ Principle.
5. The quantum number most responsible for describing the size of an atom is ____.
6. The quantum number most responsible for determining the shape of the electron orbit is ____.

7. Element 5 is the first element to utilize a ___ orbital.
8. If elements 1, 3, and 11 were arranged in order of increasing size, their order would be ___, ___, ___.
9. Of the elements 4, 6, 7, 12, and 13, those that would be expected to have similar properties are ___ and ___.
10. The electron configuration of element 9 is $1s^2 2s^2 2p^{(\ )}$.
11. Of the first twenty elements, the chemical properties of element 20 most closely resemble those of elements ___ and ___.
12. The electron configuration of element 22 is

$$1s^2 2s^2 2p^6 3s^{(\ )} 3p^{(\ )} 3d^{(\ )} 4s^{(\ )} 4p^{(\ )}.$$

13. Of the first five principal quantum shells, those that can contain *d* orbitals are ___, ___, and ___.
14. Element 21 is the first element that possesses a partially filled inner orbital. The configuration of the electron in that orbital is ______.
15. Arrange the orbitals 3*s*, 4*s*, 5*s*, 3*p*, 4*p*, and 3*d* in order of increasing energy: ______________.

Answers: (1) Protons [10, 27]. (2) Pauli Exclusion [11]. (3) **n, l, m, s** [11, 12, 25]. (4) Aufbau [11–13]. (5) **n** [13]. (6) **l** [13]. (7) p [16, 17, 20]. (8) 1, 3, 11 [24, 25]. (9) 4, 12 [22, 25, 27]. (10) 5 [17–20]. (11) 4, 12 [20–22, 25, 27]. (12) 2, 6, 2, 2, 0 [17–22]. (13) 3, 4, 5 [20]. (14) $3d^1$ [18–21]. (15) 3*s*, 3*p*, 4*s*, 3*d*, 4*p*, 5*s* [20].

SUPPLEMENTARY READINGS

1. J. E. Huheey. "The Correct Sizes of the Noble Gas Atoms." *Journal of Chemical Education* 45 (1968): 791.
2. H. H. Jaffe. "The Energies of Electrons in Atoms." *Journal of Chemical Education* 33 (1956): 25.
3. "Atomic Theory in the Ancient World." *Chemistry* 44 (1971): 17.
4. H. G. Wallace. "The Atomic Theory, a Conceptual Model." *Chemistry* 40 (1967): 8.
5. H. H. Sisler. *Electronic Structure Properties and Periodic Law.* New York: D. Van Nostrand Co., Inc., 1973.
6. "What Is Matter?" *Scientific American* (September 1953). Reprint 241. San Francisco: W. H. Freeman & Co.

Section 3

The Gaseous Elements of Group VIII

Section 3

The Gaseous Elements of Group VIII

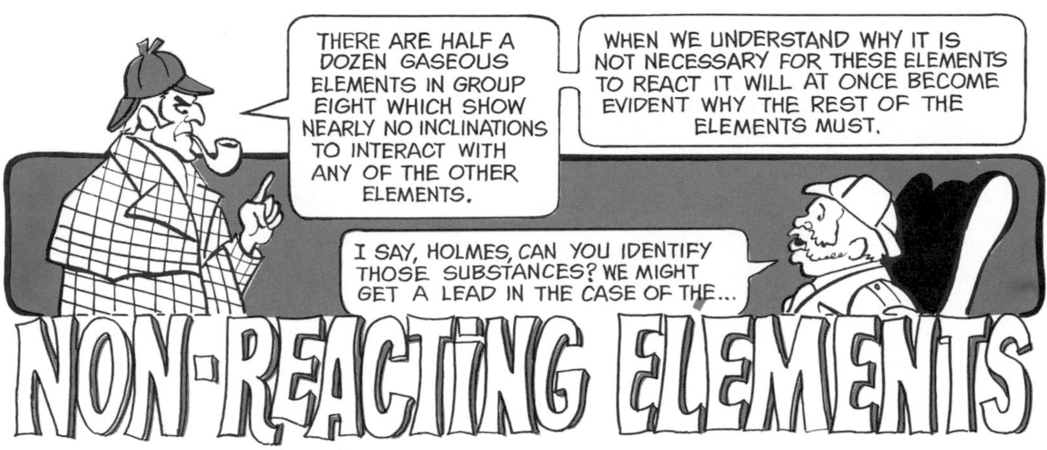

"They are elements 2, 10, 18, 36, 54, and 86, known respectively as helium, neon, argon, krypton, xenon, and radon. The last one, radon, is highly unstable. In one week it will decompose to one quarter its original amount, all the while giving off nasty radiation. It will do so whether or not it is being watched."

"Do you suppose, Holmes, that it is because all of the group VIII elements are colorless, odorless, tasteless gases that they don't react?"

"No, Watson, you have cast the cart before the horse. It is precisely because they do not react that they are colorless, odorless, tasteless gases. I suggest that if you read this section through you will find revealed secrets unknown to the alchemists."

An understanding of the chemistry of rare gases can only be attained by considering the atomic theory developed in previous sections. The behavior of the rare gases is attributable to their p^6 or completed outer shell configuration:

$$_{2}\mathbf{He} = [1s^2]$$

$$_{10}\mathbf{Ne} = 1s^2 2s^2 [2p^6]$$

$$_{18}\mathbf{Ar} = 1s^2 2s^2 2p^6 3s^2 [3p^6]$$

$$_{36}\mathbf{Kr} = 1s^2 2s^2 2p^6 3s^2 3p^6 4s^2 3d^{10} [4p^6]$$

$$_{54}\mathbf{Xe} = 1s^2 2s^2 2p^6 3s^2 3p^6 4s^2 3d^{10} 4p^6 5s^2 4d^{10} [5p^6]$$

$$_{86}\mathbf{Rn} = 1s^2 2s^2 2p^6 3s^2 3p^6 4s^2 3d^{10} 4p^6 5s^2 4d^{10} 5p^6 6s^2 4f^{14} 5d^{10} [6p^6]$$

In the case of helium, which doesn't have a p shield, the $1s^2$ functions as the outer complete shell. It is easily determined from the above electronic descriptions of each of the rare gases that the outer s and p orbitals are in every case completely filled. The situation is one of a place for everything and everything in its place. It is the outermost orbitals that establish the chemical nature of an atom, and it is because these orbitals are completely filled that the rare gases show little tendency to interact with other atoms.

The guiding axiom in chemical reactions is that when two or more atoms react with each other they do so such that all will attain *rare gas configurations.* Each atom completely fills its outer electron orbitals to produce a configuration that is similar to a group VIII element (i.e. s^2p^6). The process is accomplished through donating, acquiring, or sharing electrons. When atoms either lose or gain electrons they are converted to *ions.* An ion has unequal numbers of electrons and protons, and because of this is a charged specie (entity). Ions may be either positively or negatively charged. When charges are produced through chemical reaction, however, they are produced in equal number.

A Give and Take Proposition

The reaction between an atom of lithium ($_{3}\mathbf{Li}^0$)* and an atom of fluorine ($_{9}\mathbf{F}^0$) can be used to illustrate the process of ion formation. The elec-

*The *superscript* associated with each element symbol indicates its respective charge. For example, $\mathbf{F}^{-1}$ has a negative charge of one. The *subscripts* indicate the atomic number of elements. For example, the atomic number of $_{9}\mathbf{F}$ is nine. Superscripts and subscripts are included in the equations to show that only the number of electrons changes. The number of protons remains constant. Later in the text, the numbers in the superscripts are dropped to provide a more compact notation. $\mathbf{Li}^{+1}$, for example, becomes $\mathbf{Li}^{+}$; $\mathbf{F}^{-1}$ becomes $\mathbf{F}^{-}$; and $\mathbf{O}^{-2}$ becomes $\mathbf{O}^{=}$. $\mathbf{Li}^0$, on the other hand, indicates no charge.

tronic configuration of atomic fluorine is $1s^2 2s^2 2p^5$. If it acquires one more electron its configuration becomes $1s^2 2s^2 [2p^6]$. It is important to note that this is also the electronic configuration of neon ($_{10}\mathbf{Ne}^0$). The difference between the two, however, is quite significant. The fluoride ion is an ion containing an acquired electron. It is negatively charged. Although it has ten electrons it has only nine protons and is therefore quite reactive. It is represented by the symbol $_9\mathbf{F}^{-1}$. Neon has no net charge, is unreactive, and is an atom.

If electrons are transferred, the logical question is: Where did they come from? The lithium atom is the answer to this question. Examining the effect of an electron loss on a lithium atom, it is found that it too has converted its electronic structure to one of a rare gas.

$$\text{Atomic lithium} = {}_3\mathbf{Li}^0 = 1s^2 2s^1$$

$$\text{Ionic lithium} = {}_3\mathbf{Li}^{+1} = 1s^2$$

By losing an electron, lithium has acquired the electronic structure of helium. It should be emphasized that lithium has not been converted to helium, but to a lithium ion with a single positive charge ($_3\mathbf{Li}^{+1}$). The three protons that established its character as lithium still remain within its nucleus.

An ion of lithium and an ion of fluorine will combine to form the *ionic compound* lithium fluoride. The smallest unit of a compound composed of more than one atom that still retains the properties of the compound is a *molecule.* The entire sequence of the chemical reaction can be written in a manner quite similar to a mathematical equation:

$$(1) \quad {}_9\mathbf{F}^0 + e^{-1} = {}_9\mathbf{F}^{-1}$$

$$(2) \quad {}_3\mathbf{Li}^0 = {}_3\mathbf{Li}^{+1} + e^{-1}$$

Equation 1 states that a fluoride ion is electrically equivalent to a fluorine atom plus an electron.

Equation 2 states that a lithium atom is electrically equivalent to a lithium ion plus an electron.

Since both equations are equalities, they can, like mathematical equations, be added to produce a third equality:

$$\mathbf{Li}^0 + \mathbf{F}^0 + e^{-1} = \boxed{\mathbf{Li}^{+1} + \mathbf{F}^{-1}}^{0} + e^{-1}$$

Since e^{-1} appears on both sides of the equality it should be apparent that this equality will not be altered when the electrons are removed from each side of the equation. The equation may therefore be written as:

$$\mathbf{Li}^0 + \mathbf{F}^0 = \boxed{\mathbf{Li}^{+1}\mathbf{F}^{-1}}^{\,0}$$

The above equation states that an atom of lithium will combine with an atom of fluorine. The compound **LiF** possesses properties quite dissimilar to those of either lithium or fluorine, which reacted to produce it. Its properties are those of the lithium and fluoride *ions* whose electronic charges hold it together. The molecule, however, has no net charge because of the cancellation of the single positive charge of the lithium ion by the single negative charge of the fluoride ion. The forces that join atoms to form molecules are called *chemical bonds.*

In the production of compounds the elements that *lose* electrons to gain positive ions are referred to as *metals.* Those that *acquire* electrons to form negative ions are *nonmetals.*

Unto Them That Have Shall Be Given.

The rule of thumb as to whether an element will have metallic or nonmetallic characteristics is determined by how many electrons it must gain or lose to acquire a rare gas configuration. Each element can either gain or lose electrons to accomplish this end. The process that will in fact occur is the one requiring the movement of the smallest number of electrons. Reconsidering the example of lithium fluoride, it is seen that lithium could have acquired the rare gas configuration of neon rather than helium:

$$\begin{aligned} {}_3\mathbf{Li}^{+1} &= 1s^2 \\ {}_3\mathbf{Li}^{0} &= 1s^2 2s^1 \\ {}_3\mathbf{Li}^{-7} &= 1s^2 2s^2 2p^6 \end{aligned}$$

(${}_3\mathbf{Li}^{+1}$ to ${}_3\mathbf{Li}^{0}$: Loss of one electron; ${}_3\mathbf{Li}^{0}$ to ${}_3\mathbf{Li}^{-7}$: Gain of seven electrons)

This would, however, have required the transfer of a larger number of electrons, with the associated expenditure of much greater energy. This great discrepancy in both energy and number of electrons makes the formations of a ${}_3\mathbf{Li}^{-7}$ ion much less feasible than a ${}_3\mathbf{Li}^{+1}$ ion.

In the same manner, the formation of a ${}_9\mathbf{F}^{+7}$ ion having the same electron configuration as helium is much less feasible than that of ${}_9\mathbf{F}^{-1}$ with a neon configuration:

$$\begin{aligned} {}_9\mathbf{F}^{+7} &= 1s^2 \\ {}_9\mathbf{F}^{0} &= 1s^2 2s^2 2p^5 \\ {}_9\mathbf{F}^{-1} &= 1s^2 2s^2 2p^6 \end{aligned}$$

(${}_9\mathbf{F}^{+7}$ to ${}_9\mathbf{F}^{0}$: Loss of seven electrons; ${}_9\mathbf{F}^{0}$ to ${}_9\mathbf{F}^{-1}$: Gain of one electron)

The formation of hypothetical fluoro lithium where metallic and nonmetallic characteristics would be reversed is highly improbable. Such a compound would necessitate the transfer of seven electrons rather than one. The charges on the resultant ions would make the energy considerations for the ion formations formidable. In the case of the formation of the $_{3}\mathbf{Li}^{-7}$ ion, three protons would have to counteract the mutual repulsion of ten electrons. $_{9}\mathbf{F}^{+7}$ is an equally unlikely ion. In order for it to be formed seven electrons must be removed from the field of influence of the nine protons in the fluorine nucleus. Because of these energetic considerations, the formation of $_{3}\mathbf{Li}^{-7}$ and $_{9}\mathbf{F}^{+7}$ does not occur.

How Much Should That Ion Be Charged?

The outer orbitals, or *valence shell,* in all atoms can contain a maximum of eight electrons ($s^2 + p^6$). A distinct energetic preference either to gain or to lose electrons, however, will exist only where the number of electrons to be transfered is three or less. If the outer shell contains four electrons, it can theoretically acquire a rare gas configuration, with equal ease, by either gaining or losing four electrons. Because no preferential direction exists, it does neither. Instead, it *shares* its electrons with other atoms to form *covalent bonds.* As a rule those elements with three or less electrons tend to donate, and those with five or more tend to accept electrons when forming ions. Those with three and five electrons, however, do so with great reluctance, often finding the compromise of sharing more acceptable. The number of electrons an atom has available to participate in compound formation is referred to as its *valence.* The valences of atoms in simple ionic compounds never exceed three.

Proper Protocol in Addressing a Reaction

Formulas of chemical compounds are traditionally written so that the more metallic element appears first. It is for this reason a compound of lithium and fluorine is lithium fluoride, not fluoro lithium.

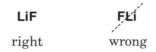

Starting materials are referred to as *reactants.* The process is a *reaction,* and the materials produced at completion are the *products.*

If more than one atom of a reactant reacts with an atom of another reactant, this is indicated by a coefficient such as the 2 preceding the **H** in the reaction below.

$$2\,\mathbf{H} + \mathbf{O} \rightarrow \mathbf{H_2O}$$

The equation indicates that two hydrogen atoms will combine with one oxygen atom to produce one molecule of water. The "ones" are understood and therefore are not written as coefficients. In all reactions the number of atoms of reactants must equal exactly the number of product atoms. During a chemical reaction, atoms are neither created nor destroyed. They are merely rearranged to form new molecules.

Sharing Is the Only Way.

Molecules are sometimes formed by atoms whose aggregate number of electrons is insufficient to convert all atoms into ions with a rare gas configuration. This is accomplished through *sharing* rather than transferring electrons. Electrons are shared between two atoms in order that both may attain a rare gas configuration. Molecules produced in this manner are covalent rather than ionic.

Atoms of hydrogen $_1\mathbf{H}$ combine to form covalent hydrogen molecules. This may be explained in terms of the rare gas rule if we examine the following atomic electronic structures.

$$\text{Hydrogen} = 1s^1$$
$$\text{Helium} = 1s^2$$

Starting with two hydrogen atoms which are identical there is no way of giving *both* atoms $1s^2$ helium structures by transferring electrons. If an electron is removed from one atom and transferred to the other, a hydride ion $\mathbf{H}^{-1}$ and a proton $\mathbf{H}^{+1}$ would be created. This would satisfy the electronic requirements of one atom and at the same time perpetrate a "compound felony" on the other. Because of the $1s^1$ configuration of the hydrogen atom it has no preferential tendency either to gain or to lose electrons. If, however, the atoms are brought into close proximity, the same two electrons can through sharing be utilized to fulfill the rare gas requirements of each without violating electrical neutrality.

The reaction may be written as

$$\mathbf{H}^{\blacksquare}\ (\text{atom}) + \mathbf{H}_{\bullet}\ (\text{atom}) \rightarrow \mathbf{H}^{\blacksquare}_{\bullet}\mathbf{H} = \mathbf{H}^{\times}_{\times}\mathbf{H} = \mathbf{H_2}\ (\text{covalent molecule})$$

The notation signifies that on formation of a covalent bond the electrons of each atom (■, ●) orient themselves so that their influence is extended equally between the two atoms of the molecule. The second and third product notation indicate that once the molecule is formed the electrons become indistinguishable as to the atom of origin (■, ●, → ××).

In a Nutshell:

1. Atoms combine to form molecules.
2. There are two general types of molecules, ionic and covalent. Their distinction is due to their component bonds.
3. Ionic bonds are the result of electronic attraction between oppositely charged ions.
4. Covalent bonds are a result of the mutual utilization of electrons between two atoms so as to provide each with a rare gas configuration.
5. The driving force in bond formation, whether ionic or covalent, is the attainment of a rare gas configuration for each of the combining atoms.
6. Rare gases do not combine with other atoms because they already possess the optimum electronic configuration.
7. When two atoms combine to form an ionic bond, the metallic one donates electrons, the nonmetallic one accepts them. If neither atom donates or accepts electrons, and each shares its electrons with the other, a covalent bond is formed.
8. Many of the characteristics of molecules are a result of the type of bonds utilized in their formation.
9. No matter what the type of bonding, the resultant molecule will possess its own physical and chemical characteristics different from either parent atom.
10. Bonds are formed in a manner such that the number of electrons and the amount of energy transferred is minimized.
11. A molecule is the smallest unit of a compound possessing properties of the compound.

Example. Element 20, calcium, and element 9, fluorine, combine to form a compound. What may be said about this process and its products?

The electronic structures of the reactant atoms are:

$$(a)\quad 1s^2 2s^2 2p^6 3s^2 3p^6 4s^2 \quad \text{(calcium)}$$

The two rare gas configurations closest to this structure are those of argon ($1s^22s^22p^63s^23p^6$) and krypton ($1s^22s^22p^63s^23p^64s^23d^{10}4p^6$). Of these two possible choices it is much simpler for the calcium atom to lose two electrons to acquire an argon configuration than to gain the additional sixteen electrons necessary for a krypton configuration. It should be expected, therefore, that according to the following equation the calcium ion would be double positive.

$$_{20}\mathbf{Ca}^0 \rightarrow {}_{20}\mathbf{Ca}^{+2} + 2\,e^{-1}$$

(The 2 preceding the electron on the righthand side of the equation indicates that each calcium atom, when it becomes an ion, releases two electrons.)

(*b*) $1s^22s^22p^5$ (fluorine)

The closest rare gas configurations are those of helium ($1s^2$) and neon ($1s^22s^22p^6$). Applying the same line of reasoning to fluorine as to calcium, it is evident that the acquisition of one electron has a higher probability of occurrence than the loss of seven. The process may therefore be written as:

$$_9\mathbf{F}^0 + e^- \rightarrow {}_9\mathbf{F}^{-1}$$

Then, if the reactants and products are added, we arrive at:

$$_{20}\mathbf{Ca}^0 + {}_9\mathbf{F}^0 + \cancel{e}^{-1} \rightarrow \boxed{\mathbf{Ca}^{+2}\mathbf{F}^{-1}}^{+1} + 2\,\cancel{e}^{-1}$$

But two things are wrong with this equation:

1. Chemical reactions do not produce an excess or deficiency of free electrons. All the electrons produced by one reactant are utilized by the other. (There are no net free electrons.)
2. Molecules are neutral. They have no electrical charge.

The above equation can be modified so that neither of the two objections exists. This is accomplished by having two fluorine atoms combine with each calcium atom.

$$_{20}\mathbf{Ca}^0 \rightarrow {}_{20}\mathbf{Ca}^{+2} + 2\,e^{-1}$$

$$2\,{}_9\mathbf{F}^0 + 2\,e^{-1} \rightarrow 2\,{}_9\mathbf{F}^{-1}$$

(adding)

$$_{20}\mathbf{Ca}^0 + 2\,{}_9\mathbf{F}^0 \rightarrow \boxed{{}_{20}\mathbf{Ca}^{+2} + 2\,{}_9\mathbf{F}^{-1}} = \mathbf{CaF_2}$$

The subscript 2 to the right of the fluoride portion of the molecule indicates that there are two fluorine atoms associated with each calcium atom in the calcium fluoride molecule.

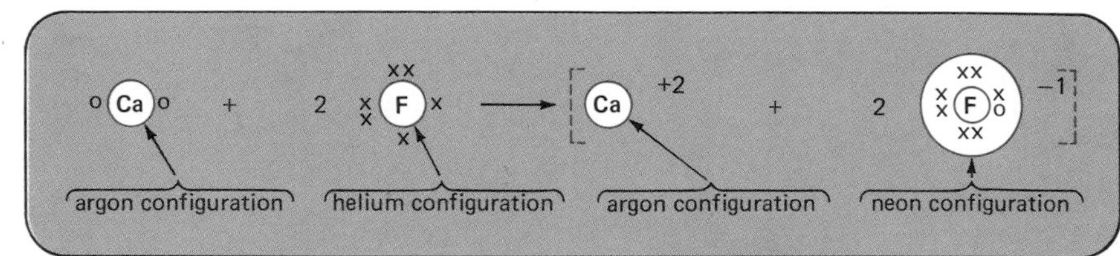

FIGURE 3–1 **Reaction between Calcium and Fluorine**

The reaction between calcium and fluorine may also be represented schematically, as in Figure 3–1, in terms of the valence electrons of the respective atoms. The o's represent the electrons derived from the calcium, whereas the x's indicate that fluorine was the electron source. These distinctions are only for pedagogical purposes. There is no actual distinction. The colored panel indicates that the ions are associated as a compound. Sixteen electrons have been arranged among three atoms to form an ionic compound where each ion has a rare gas configuration.

Example. What type of compound is to be expected from the combinations of carbon and chlorine?

Carbon ($1s^22s^22p^2$) has four electrons in its valence shell, and as a result of this it shares, rather than transfers, electrons. Elements that combine with carbon can, therefore, form only covalent compounds.

Chlorine ($1s^22s^22p^63s^23p^5$), which needs only one electron to form a rare gas chloride ion $\mathbf{Cl^{-1}}$ ($1s^22s^22p^63s^23p^6$), cannot gain one because of carbon's inability to donate. Consequently, the compound resulting from the union of carbon and chlorine atoms must be covalent.

The question arises as to what is the smallest number of carbon and chlorine atoms necessary for the formation of a carbon chlorine compound. This compound must be constructed such that each constituent has a rare gas configuration. From the structure in Figure 3–2 it may be seen that through the sharing of one electron pair with carbon, chlorine could fulfill its rare gas requirements.

Carbon electron requirements are only partially met by this solution. They may, however, be met in full if an additional three chlorine atoms were to associate with the carbon. Through such an association, all five atoms could attain a rare gas configuration, as indicated in Figure 3–3.

This compound, which is formed through the reaction of carbon and chlorine, is carbon tetrachloride ($\mathbf{CCl_4}$). The prefix *tetra* ("four" in

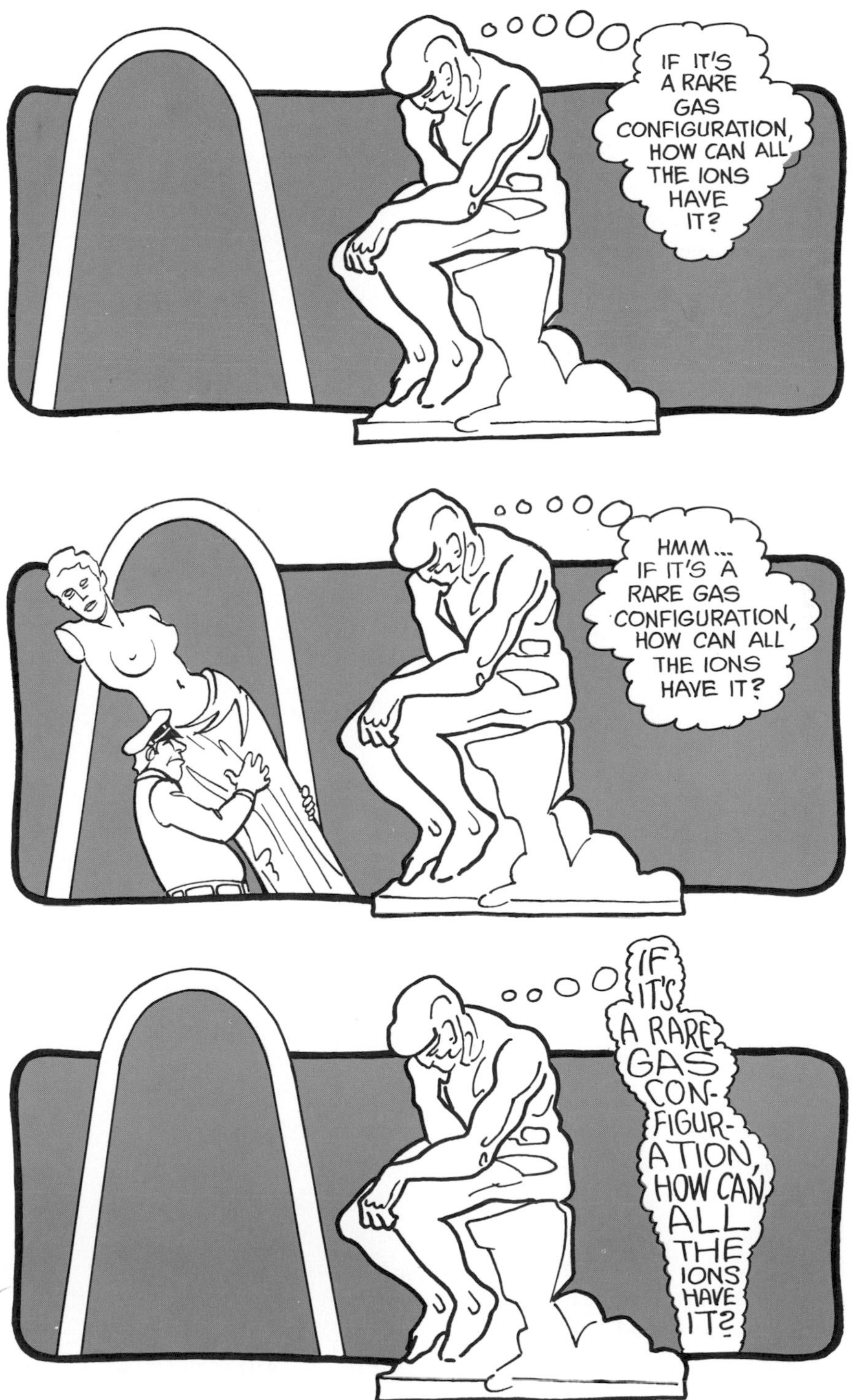
IF IT'S A RARE GAS CONFIGURATION, HOW CAN ALL THE IONS HAVE IT?
HMM... IF IT'S A RARE GAS CONFIGURATION, HOW CAN ALL THE IONS HAVE IT?
IF IT'S A RARE GAS CON-FIGUR-ATION, HOW CAN ALL THE IONS HAVE IT?

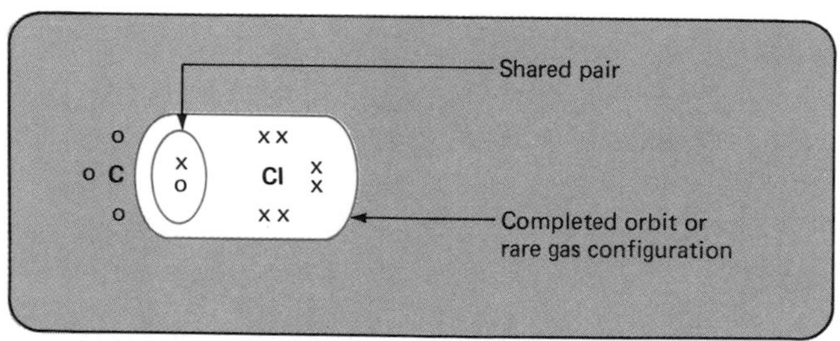

FIGURE 3–2 Carbon Chlorine Compound (One of the Four Carbon Bonds)

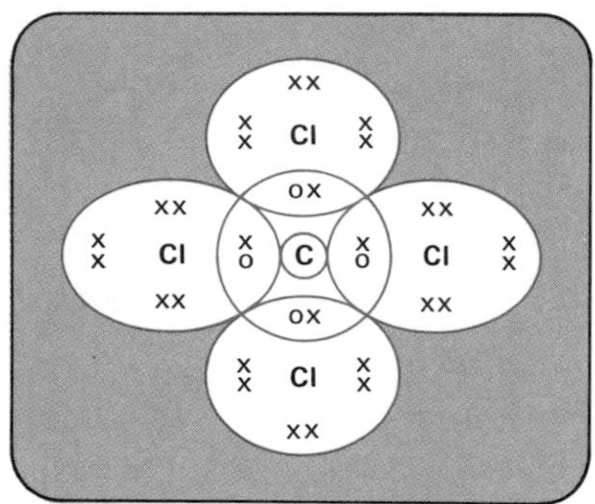

FIGURE 3–3 Rare Gas Configuration of Carbon-Chlorine Compound (Carbon Tetrachloride)

Greek) indicates that there are four chlorine associated with each carbon. The sharing of four electron pairs makes it a decidedly covalent compound.

Metallic Bonds

Metals in their solid or liquid state are held together through a modified covalent bond structure. These metallic bonds are peculiar to the massive (liquid or solid) metal and are the cause of metallic luster and high electrical and thermal conductivity. They are the result of the peregrinations of valence electrons, roaming independent of their atoms of origin. These electrons, rather than being associated with one or two atoms, are shared by the entire metallic commune as a whole. If the metal is heated to a high enough temperature to form a vapor, these bonds are destroyed. This is because the free interchange of electrons between atoms in a vapor is highly restricted due to the extremely large spaces which exist between the atoms.

REVIEW

1. The outermost electron configuration most desirable for an ion of an element is that of the rare gas atom having the closest atomic number.
2. When atoms other than rare gases form ions or covalent bonds they usually acquire rare gas configurations.
3. Ions are electrically charged because of unequal numbers of electrons and protons.
4. Ions may be either positively or negatively charged.
5. The maximum number of electrons that can be contained in the valence shell of an atom is eight.
6. The interactions between atoms to form chemical compounds occurs between the electrons in their valence shells.
7. When bonds are formed, the number of electrons and the amount of energy involved tends to be a minimum.
8. Electrons involved in bond formation are indistinguishable with respect to their atoms of origin.
9. Chemical reactions must be balanced with respect to electrons as well as atoms.
10. Atoms that lose electrons in the process of acquiring rare gas valence shells are metals. Those that gain electrons to accomplish this are nonmetals.
11. The properties of atoms in compounds differ radically from the properties of uncombined atoms. However, atoms that are similarly bonded in more than one type of molecule confer similar properties to those molecules.
12. Ionic compounds are composed of positive and negative ions held together by their opposing electrical charges. They contain both charge types in equal number and therefore have no net charge.
13. Covalent compounds are composed of atoms held together through bonds formed by the sharing of electrons to produce filled valence shells.
14. Metallic bonds are a form of covalent bond found only in metals that are in either solid or liquid state. They are the result of the relative freedom of electrons to associate with several metallic atoms. This freedom produces the high electrical and thermal conductivity associated with metals.

REFLEXIVE QUIZ

1. The guiding principle in chemical reactions is the attainment of a ______ ______ configuration for each reactant atom.

2. The maximum number of electrons that can be contained in a valence shell is ____.

3. Rare gases do not combine with other atoms because they have completely filled _______ shells.

4. The chemical nature of an atom is determined by its __________ orbitals.

5. An atom may acquire a completed valence shell by either gaining, losing, or sharing _________.

6. An atom or group of atoms that contains unequal numbers of protons and electrons is called an ______.

7. Ions are not rare gases, even though they may contain the same number of electrons, because they do not possess the proper number of _________.

8. An ionic compound possesses properties of the _______ rather than the atoms from which it was formed.

9. Elements that lose electrons to form positive ions are known as _______.

10. When ions are formed, atoms transfer the __________ number of electrons that allows them to acquire a complete valence shell.

11. Atoms containing four electrons in their valence shells tend to form __________ compounds.

12. Element 6 (carbon) can combine with element 8 (oxygen). Draw the electron configuration of a possible product.

13. The unique qualities of metallic bonds are due to the ability of an _________ to interact with a large number of atoms.

14. If the same electron is considered in the valence shell of two atoms, the bond between the two atoms is _________.

Answers: (1) Rare gas [30, 31, 41]. (2) 8 [31, 41]. (3) Valence [30, 31, 36]. (4) Outermost [31]. (5) Electrons [31, 41]. (6) Ion [31]. (7) Protons [32]. (8) Ions [33]. (9) Metals [33]. (10) Minimum [33, 41]. (11) Covalent [34]. (12) :O::C::O: or :C:::O: [38, 40]. (13) Electron [40, 41]. (14) Covalent [38, 41].

SUPPLEMENTARY READINGS

1. J. J. Lagowski. *The Chemical Bond.* Boston: Houghton Mifflin Co., 1966.
2. N. Booth. "Chemical Bonds: Ionic Lattices." *Education in Chemistry* 1 (1964): 151.
3. R. T. Sanderson. "Principles of Chemical Bonding." *Journal of Chemical Education* 38 (1961): 382.
4. A. Companion. *Chemical Bonding.* New York: McGraw-Hill Book Co., 1964.

Section 4

Competition for Electrons

Section 4

Competition for Electrons

When more than one atom in a compound vies for a single electron, it is often quite difficult to establish a pecking order. Previous sections have dealt with the correlation between the electron affinity (the desire of an atom to acquire electrons) of an isolated atom and its position in the periodic chart. It was found that excluding the rare gases, the electron affinity within a given period increased with atomic number, being least for elements in groups Ia and IIa and greatest for elements in groups VIa and VIIa. Within a given group this trend is reversed.

In order to evaluate the electron affinity of specific atoms when they are incorporated in compounds, Nobel Laureate Linus Pauling developed a scale of *electronegativities*. According to his system, values for elements may vary from 4.0 for fluorine, the most electronegative element, to 0.7 for cesium and francium. The elements and their respective electronegativities appear in Figure 4–1. It is interesting to note that nitrogen and chlorine have the same electronegativity, as do also carbon, sulfur, and iodine. When comparing elements that are in neither the same period nor the same group, it is found that elements with roughly equivalent electronegativities lie along diagonal lines running from the upper lefthand region to the lower righthand region of the periodic chart.

Because the rare gases do not generally form compounds, insufficient data exist to allow computation of their individual electronegativities. It is safe to predict, however, that when they are computed it will be found that their values are quite small.

The chemical nature of a compound can be predicted on the basis of the electronegativity values of its constituent atoms. A bond formed between two atoms whose difference in electronegativity is 1.7 can be considered as having 50 percent ionic and 50 percent covalent character. Bonds with electronegativity differences greater than 1.7 are considered ionic; those with less than this value covalent. The chemical character of a molecule results from the type and arrangement of bonds that form it. Few molecules are completely ionic or covalent. Most possess properties of both in proportion to the electronegativity differences of their bonds.

Ionic molecules tend to conduct electricity in the molten state, to be soluble in water, and to form comparatively hard solids with high melting points. Sodium chloride is an example of such a molecule.

Molecules composed of atoms with similar electronegativities will be covalent, will usually be insoluble in water, and will not conduct electricity. A typical example is hexane ($\mathbf{C_6H_{14}}$). It is composed of

Transition metals (III b – II b)

	I a	II a	III b	IV b	V b	VI b	VII b	VIII b	VIII b	VIII b	I b	II b	III a	IV a	V a	VI a	VII a	VIII a
1	H 2.1																	He —
2	Li 1.0	Be 1.5											B 2.0	C 2.5	N 3.0	O 3.5	F 4.0	Ne —
3	Na 0.9	Mg 1.2											Al 1.5	Si 1.8	P 2.1	S 2.5	Cl 3.0	A —
4	K 0.8	Ca 1.0	Sc 1.3	Ti 1.5	V 1.6	Cr 1.6	Mn 1.5	Fe 1.8	Co 1.9	Ni 1.9	Cu 1.9	Zn 1.6	Ga 1.6	Ge 1.8	As 2.0	Se 2.4	Br 2.8	Kr —
5	Rb 0.8	Sr 1.0	Y 1.2	Zr 1.4	Nb 1.6	Mo 1.8	Tc 1.9	Ru 2.2	Rh 2.2	Pd 2.2	Ag 1.9	Cd 1.7	In 1.7	Sn 1.8	Sb 1.9	Te 2.1	I 2.5	Xe —
6	Cs 0.7	Ba 0.9	La * 1.0	Hf 1.3	Ta 1.5	W 1.7	Re 1.9	Os 2.2	Ir 2.2	Pt 2.2	Au 2.4	Hg 1.9	Tl 1.8	Pb 1.9	Bi 1.9	Po 2.0	At 2.2	Rn —
7	Fr 0.7	Ra 0.9	Ac ** 1.1	Ku —	Ha —													

Lanthanide series *	Ce 1.1	Pr 1.1	Nd 1.1	Pm 1.1	Sm 1.1	Eu 1.0	Gd 1.1	Tb 1.1	Dy 1.1	Ho 1.1	Er 1.1	Tm 1.1	Yb 1.1	Lu 1.2
Actinide series **	Th 1.3	Pa 1.4	U 1.4	Np 1.4	Pu *	Am *	Cm *	Bk *	Cf *	Es *	Fm *	Md *	No 1.3	Lw —

* between 1.4 and 1.3

FIGURE 4–1 Electronegativities of the Elements

carbon atoms, which have an electronegativity of 2.5 and of hydrogen atoms, which have an electronegativity of 2.1. This compound is covalent, insoluble in water, and does not conduct electricity.

The Reason I Asked You Electrons Here Is. . . .

With compounds of two or more atoms it is often desirable to utilize the Molecular Orbital Theory to determine the placement of electrons. This theory converts the individual atomic orbitals into molecular orbitals (i.e., atomic orbitals of the molecule). It then arranges all the associated atomic electrons so that every atom within the molecule may be considered to have a rare gas configuration. This is more easily done than said, as can be seen from the examples below.

Example. Write the electronic configuration of sodium sulfate (**Na_2SO_4**).

Figure 4–2 represents an arbitrary matrix of atoms that can be used to depict a large number of common chemical substances. The shaded

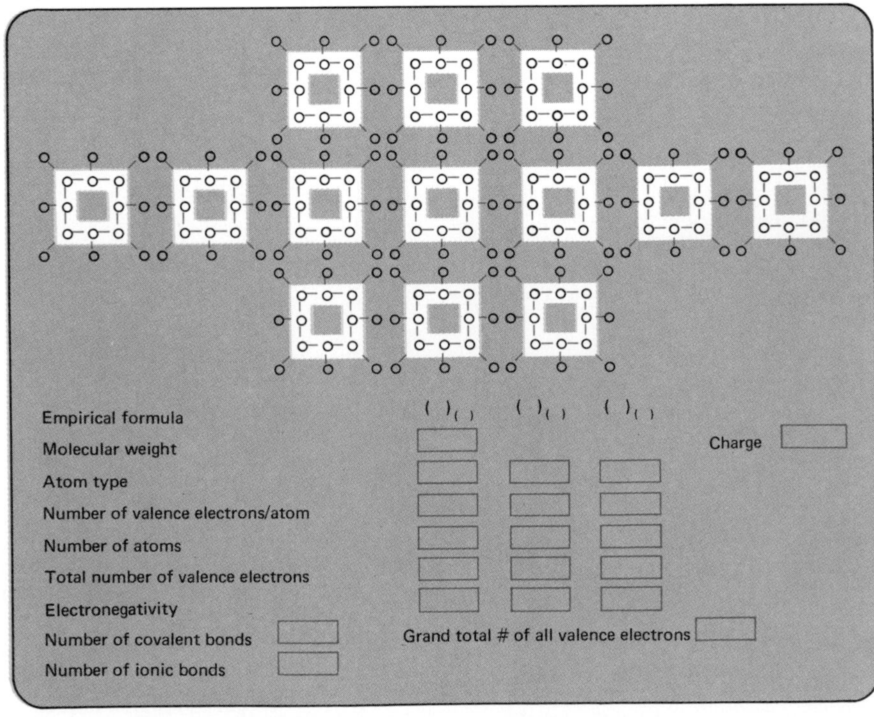

4–2 Matrix for Combining Atoms

square areas each contain eight positions an electron may occupy. They are the atoms' outermost *s* and *p* suborbitals and are the primary valence shells of the atoms they encircle.

The eight positions radiating from these primary valence shells represent the secondary valence shell of covalent electrons that may be associated with each atom in the molecule. Each electron pair that occupies a covalent position can be considered as being shared between the two adjacent atoms to form a covalent bond. The number of electrons involved in satisfying the rare gas (completed valence shell) requirement of each atom is equal to the sum of the electrons in its primary valence shell plus those shared through covalent bonds.

The sulfur atom in sodium sulfate (Figure 4–3) has acquired the eight electrons necessary to complete its valence shell from its four covalent bonds with oxygen. Each oxygen has accomplished the same feat by utilizing the two electrons of the covalent bond between it and sulfur and the six electrons in its own primary valence shell.

The sodium atoms have lost their single electrons to the sulfate complex. Each sodium ion has a complete valence shell because its outer electrons are now the second rather than the third period elec-

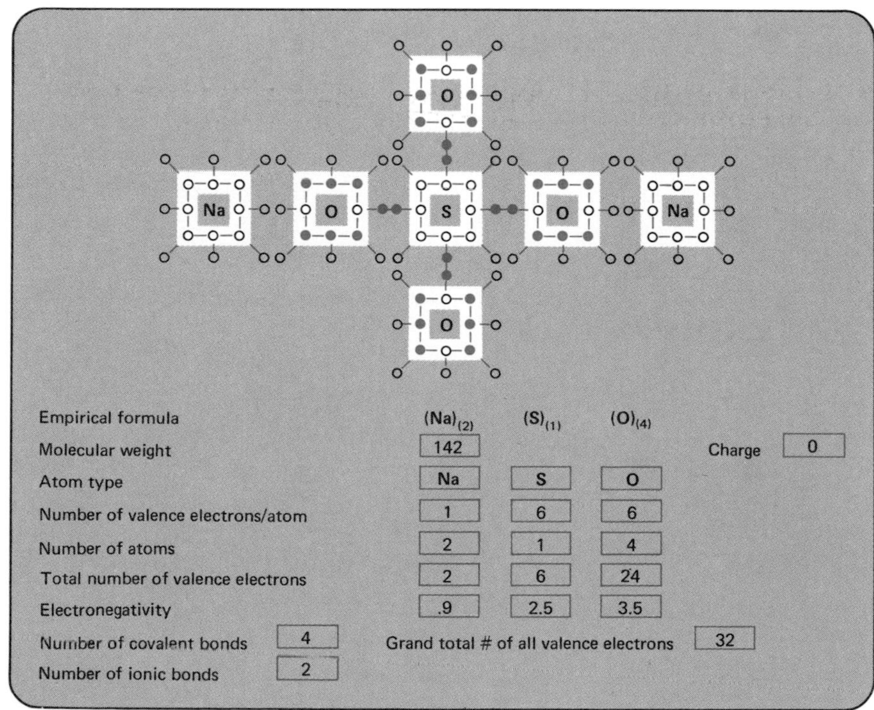

O
Na
O
S
O
Na
O
Empirical formula
(Na)(2)
(S)(1)
(O)(4)
Molecular weight
142
Charge
0
Atom type
Na
S
O
Number of valence electrons/atom
1
6
6
Number of atoms
2
1
4
Total number of valence electrons
2
6
24
Electronegativity
.9
2.5
3.5
Number of covalent bonds
4
Grand total # of all valence electrons
32
Number of ionic bonds
2

FIGURE 4–3 **Matrix of Sodium Sulfate**

trons. (That is, sodium has a complete second valence shell which by virtue of the loss of the single third period electron has become the outer shell.)

The bonds between sodium and the sulfate ion are ionic bonds. Because of this, no electrons are indicated in the adjacent covalent orbits. The ionic nature of the sodium sulfate bonds can be predicted from the 2.6 difference in electronegativity values between the sodium and oxygen atoms. Similarly, the covalent nature of the sulfur-oxygen bond is apparent from the 1.0 difference in their respective electronegativity values. From the data it would be expected that sodium sulfate would ionize in water to give $2\,Na^+ + SO_4^=$. The double negative charge on the sulfate ion is due to the presence of the two electrons donated by each sodium atom.

Note. The unionized sodium sulfate has no net charge. For this reason, a zero has been placed in the box labeled "charge." If the sulfate ion alone were to be considered, a −2 would have been placed in this box.

Example. Write the electron configuration of acetylene (C_2H_2).

Satisfying the electron requirements of the constituent atoms in acetylene is no small task. As can be seen in Figure 4–4, ten electrons must serve to complete the valence shells of four atoms. The electronegativity difference between carbon and hydrogen is 0.4. There is no difference in electronegativity between the two carbon atoms. It should, therefore, be expected that all the bonds in the acetylene molecule be covalent. The covalent bonding allows each of the carbons to count the three pairs of electrons that join them together and the pair that joins them each to hydrogen as part of their respective valence shells.

The three pairs of electrons between the carbon atoms constitute a triple bond between the two atoms. Each carbon is in addition singly bonded to each hydrogen atom.

The lack of any gross differences in electronegativity between carbon and hydrogen as well as the symmetry of the acetylene molecule make acetylene more soluble in petroleum solvents than in water.

Oil and Water Don't Mix.

Most chemical substances can be classified as *polar* or *nonpolar*. Polar materials form solutions with other polar materials. Nonpolar materials similarly form solutions with other nonpolar materials. Most oil and petroleum products and symmetrical covalent compounds are nonpolar. Water, hydrochloric acid, and other asymmetrical covalent

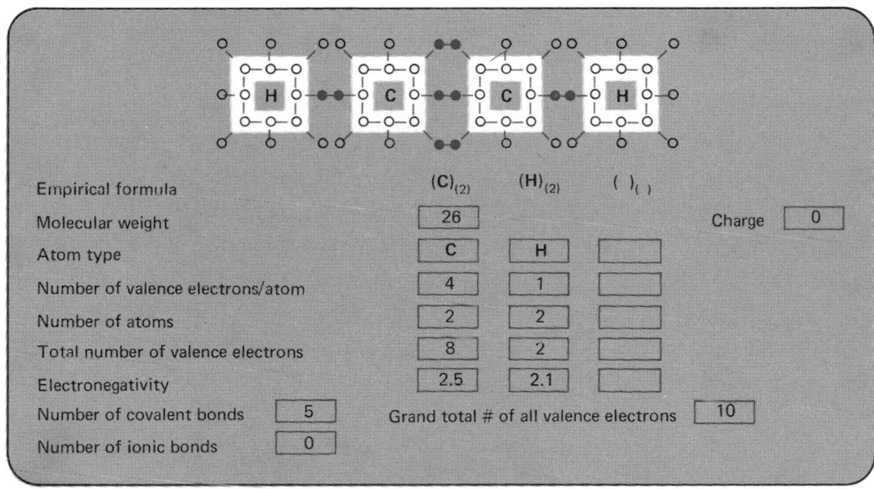

4–4 Matrix of Acetylene **FIGURE**

compounds are usually polar.

If polar and nonpolar materials are mixed, they will separate into two layers. Oil floats on water. Water floats on carbon tetrachloride.

The solubility of one substance in another is readily derivable from its electronic and atomic configurations. The two criteria to be evaluated are molecular charge distribution and molecular symmetry.

1. The charge distribution is directly related to the difference in electronegativity of two atoms held together by a given bond. This may be demonstrated by comparing a molecule of hydrogen with one of hydrogen chloride.

H:H **H⁝Cl⁝**

Both are held together by a single covalent bond. The nature of this bond, however, is quite different for each. In the hydrogen molecule the electron pair is shared equally between the two hydrogen atoms. This is not so, however, in the case of hydrogen chloride, where there is an appreciable difference in the electronegativities of the component atoms. Chlorine has an electronegativity of 3.0, as compared to 2.2 for hydrogen. This results in a molecular *dipole,* where the chlorine side of the molecule is more negatively charged than the hydrogen side producing a polar molecule.

The lowercase Greek deltas (δ) indicate the relative charge distribution. The electron positions indicate that the bonding electron pair is

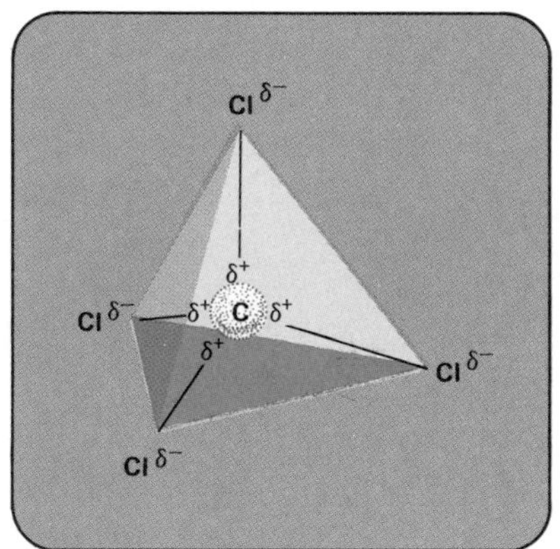

FIGURE

4–5 Carbon Tetrachloride Molecule (Nonpolar)

more closely associated with the chlorine atom than the hydrogen atom, thus producing a polar molecule.

2. The spatial orientation of the constituent atoms is the second factor that determines whether a molecule is polar or not. Large differences in electronegativities do not produce dipolar molecules unless the atoms are properly arranged.

Chlorine combines with carbon to form the compound carbon tetrachloride (CCl_4). On the basis of electronegativity difference alone, the molecule should be polar. (There is a 0.5 electronegativity difference between carbon and chlorine.) The spatial distribution of chlorine atoms about the carbon, however, does not allow the formation of positive or negative poles. Carbon tetrachloride is a symmetrical molecule. The electrical charge measured at any distance from its center is identical to the charge measured at the same distance from the center in the opposite direction. The average center of both the positive and negative charges is located on the central carbon atom. Because of this, the molecule is nonpolar. (Figure 4–5.)

Note. Although it may at first appear that any chlorine atom could provide the CCl_4 molecule with a negative end, on further inspection it will be seen that this is not so. Opposite each chlorine is another equally negative chlorine. The effect of this is to make the midpoint between these two atoms the average center of negative charge. This point, however, is located on the carbon atom, which because of its lower electronegativity is also the center of positive charge.

A molecule of water is composed of two hydrogen atoms (electronegativity 2.1) and one oxygen atom (electronegativity 3.5). This is more than sufficient difference in electronegativity for the formation of a sizable dipole. Thus, the factor that will determine whether or not water will be polar will be its geometry. If the water molecule is linear then, as in the case of carbon tetrachloride, the average centers of positive and negative charge would coincide and no dipole would exist. (Figure 4–6a.)

It is not possible to determine a priori the geometry of the water molecule. Physicochemical measurements have been made, however, which indicate that the angle between the two hydrogens is 104.5° rather than 180°. This deviation from linearity shifts the center of positive charge away from the oxygen atom to a point midway between the hydrogen atoms. Discrete positive and negative poles are produced as a consequence of this shift, making the water molecule polar. (Figure 4–6b.)

Carbon dioxide is an example of a compound similar to water that, because of its linear structure, does not possess a dipole.

$$\overset{\delta^-}{\mathrm{O}}::\overset{\delta^+\ \delta^+}{\mathrm{C}}::\overset{\delta^-}{\mathrm{O}}$$

It is the electrostatic attraction between polar molecules that precludes the formation of stable solutions of polar and nonpolar materials. This may readily be understood if the individual molecules are thought of as little ball bearings with only the polar ones being

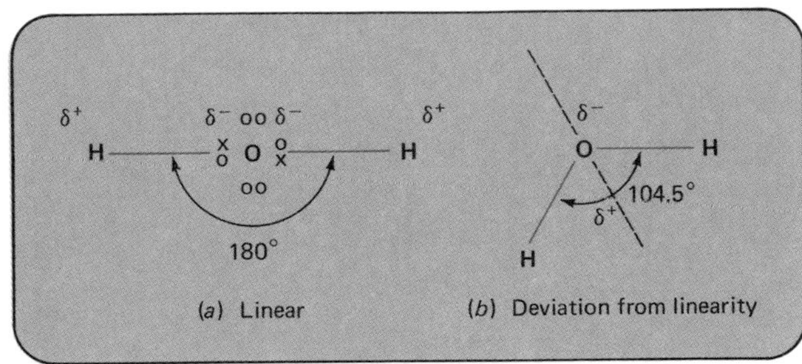

FIGURE 4–6 Geometry of a Water Molecule

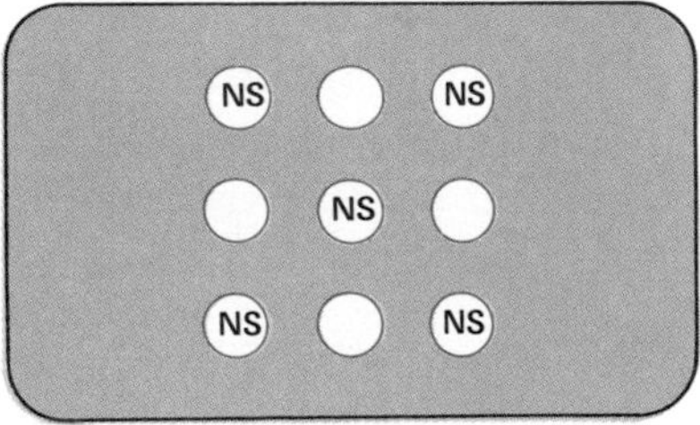

FIGURE 4–7 Solution of Polar and Nonpolar Molecules

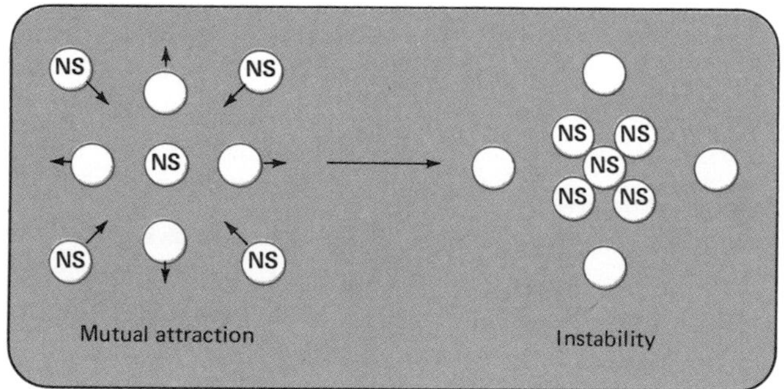

FIGURE 4–8 Electrostatic Attraction of Polar Molecules

magnetic. A solution of both components could be depicted as in Figure 4–7. The north-south magnetic poles are analogous to the positive-negative electrostatic poles of the actual dipole. (Although few readers of this book will be familiar with electrostatics, nearly all will have at some time played with magnets and compasses.) The instability of this solution is due to forces of mutual attraction between the alternating polar molecules. When these molecules come closer together, they squeeze out the intervening nonpolar molecules. (Figure 4–8.) The nonpolar material is expelled from the polar environment just as toothpaste would be expelled from a tube having two powerful magnets on opposing sides.

The instability of a solution composed of both polar and nonpolar materials can be easily demonstrated if a bottle containing oil and water is vigorously shaken and then allowed to stand. Initially the liquid is cloudy, because the molecules of oil and water are interspersed. It doesn't take long, however, before the oil molecules are expelled from those of water, form globules, float to the top, and create a second layer.

Such instability does not occur when solutions are composed entirely of either polar or nonpolar materials. If the mixture is completely polar, then the component atoms will produce stable solutions because there is a dipole attraction between all of the molecules in solution whether they are of the same type or not. If the solution is composed entirely of nonpolar materials, it too will be stable because there are no dipolar forces present to cause some molecules to come together and push aside other molecules. In short, "like dissolves like."

Example. Some typical polar and nonpolar molecules are listed below. Using the molecular matrices furnished with this text as a model, draw the atomic arrangements that could justify their physically observed properties.

Polar: H_2O, NF_3, CH_3COOH, CH_3NH_2
Nonpolar: H_2, CO_2, SiH_4, I_2

Fire Burn and Cauldron Bubble.

Melting and boiling are the result of an increase in the ability of molecules to move relative to their neighbors when the temperature is raised. In a solid crystal the molecules are rigidly held close together. The material offers great resistance to deformation. If the substance is heated until it melts (its melting point), the solid is converted to a liquid. In a liquid the movement of molecules is not so restricted as in solids. This lack of restriction makes the liquid more readily deformable and allows it to be poured. When the liquid is heated until it

DO YOU SEE WHAT I SEE?
I T-THINK THOSE SOLID C-CHECKERS ARE MOVING!
AND IN DIFFERENT DIRECTIONS!
WHY, THOSE CHECKERS ARE MOVING LIKE A **LIQUID**!
NOW THEY'RE BREAKING AWAY!
THAT'S A GAS!

boils (its boiling point), the distance between a molecule and its neighbor becomes very large. This results in an even greater freedom of movement for the molecules.

To visualize the changes in state from solid to liquid to gas think of the molecules as checkers attached by springs to a checkerboard. When the substance is solid, the springs holding each checker to the board are short, and although the molecules are moving, none of them can be displaced beyond the confines of its square. When the temperature is higher, the springs are longer, the substance is liquid, and the molecule can move beyond the boundaries of its square. If the temperature is raised still further, the substance boils, converting the molecules to vapor. This is equivalent to the molecules breaking the springs that bind them to the board. Once these springs are broken, the molecules can move freely, unhampered by the boards restrictions. The removal of these restrictions makes the region in which a molecule can move so great that there is little chance it will come close enough to any other molecule to interact with it.

The checkerboard analogy is equally valid for polar molecules, provided additional allowances are made for the dipole. If a molecule is a dipole, each checker must be thought of as a little magnet with north and south poles. The attractions and repulsions of these poles for the poles of neighboring molecules restrict their movement above and beyond what would be allowed by their anchoring springs. For this reason the temperatures at which polar materials melt and boil are higher than those for nonpolar material.

When a substance changes from solid to liquid to vapor, heat is utilized both to raise the temperature of the material and to change its state. The heat necessary to change a solid to a liquid at the same temperature is known as the *heat of fusion.* The heat necessary to change a liquid to a vapor at the same temperature is the *heat of vaporization.* It is these heats of fusion and vaporization that have the greatest effect on the springs that hold the molecules to the checkerboard.

REVIEW

1. Molecules can be classified as polar or nonpolar.
2. Polar molecules have ends that are charged relatively positive and negative.
3. Nonpolar molecules do not possess ends with discrete charge separation.

4. The polarity of a molecule is the result of two factors: (*a*) a difference in electronegativity of constituent atoms; (*b*) a lack of symmetry in charge distribution.
5. Because of their dipoles, polar molecules tend to associate with themselves and other polar molecules.
6. Because polar molecules in solutions tend to associate in their own layer, mixtures of polar and nonpolar liquids usually are not stable.
7. Polar substances usually have higher melting and boiling points than nonpolar ones because of the intermolecular association of polar molecules.

REFLEXIVE QUIZ

1. The measure of the ability of an atom in a compound to attract electrons is its ________________.
2. Electronegativity values for atoms can be as great as ______ or as small as ______.
3. The atoms that form ionic bonds have differences in electronegativities greater than ______.
4. Atomic orbitals of a molecule are known as __________ __________.
5. If an electron pair is found in the secondary valence shell of two atoms, the bond formed by their presence is __________.
6. Atoms that form ions by losing electrons are ________.
7. When three pairs of electrons bond two atoms together, the bond is a ________ bond.
8. The solubility of one material in another depends upon whether or not they are both ________________.
9. The polarity of a molecule is dependent upon two factors: (*a*) nonuniform molecular charge distribution, and (*b*) ________________.
10. Water is a polar molecule because it is not __________.
11. Of the compounds $\mathbf{C_2H_6}$, **HBr**, $\mathbf{NH_3}$, $\mathbf{CH_3Cl}$, $\mathbf{CH_4}$, and $\mathbf{Cl_2}$, underline those that are polar. (The molecular matrices can be utilized for these determinations.)
12. The temperature at which a substance is converted from a solid to a liquid is its ________________.

13. The heat necessary to convert a material at its melting point from a solid to a liquid is the heat of ________.

Answers: (1) Electronegativity [46]. (2) 4, 0.7 [46, 47]. (3) 1.7 [46]. (4) Molecular orbitals [48]. (5) Covalent [49]. (6) Metals [33, 41]. (7) Triple [50]. (8) Polar or nonpolar [50, 53]. (9) Molecular symmetry [51, 52]. (10) Linear [53]. (11) **HBr, NH_3, CH_3Cl** [54]. (12) Melting point [55–57]. (13) Fusion [57].

SUPPLEMENTARY READINGS

1. L. Pauling. *The Nature of the Chemical Bond,* 3rd ed. Ithaca, N.Y.: Cornell University Press, 1960.
2. H. H. Sisler. *Electronic Structure, Properties and the Periodic Law.* New York: Van Nostrand Reinhold, 1963.
3. P. F. Lynch. *Orbitals and Chemical Bonding.* Boston: Houghton Mifflin Co., 1966.
4. E. J. Little, Jr., and M. M. Jones. "A Complete Table of Electronegativities." *Journal of Chemical Education* 37 (1960): 231.
5. G. E. Ryschkewitsch. *Chemical Bonding and the Geometry of Molecules.* New York: D. Van Nostrand Co., Inc., 1963.

Section 5

Reactions: Substitution

Section 5

Reactions: Substitution

It is impossible to memorize the names and properties of the millions of known chemical substances. Experienced chemists have therefore classified molecules according to specific types. Each type possesses specific identifying characteristics and displays similar chemical properties. Four of the broadest molecular categories (acid-base and oxidant-reductant) can be grouped as pairs because the chemistry of these pairs comprises two of the three most common types of chemical reactions:

1. Substitution reactions
2. Acid-base reactions
3. Oxidation-reduction reactions

The Chemical Square Dance

Substitution reactions can be compared to a chemical square dance. The starting materials or reactants are like couples who come to the dance together. Each compound or couple is composed of two distinct parts or partners. During the dancing, couples change partners many times. If, at the end of the dance, the partners are not the same as at the start, a reaction has occurred.

Example. A solution of potassium chloride (**KCl**) is mixed with a solution of sodium bromide (**NaBr**). The reaction can be written as

$$\mathbf{KBr} + \mathbf{NaCl} \leftrightarrows ?$$

By convention, the reactants are written to the left of an arrow pointing to the right. If a reaction occurs, then the products of this reaction would be written to the right of the arrow. If the products when formed possess the ability to be converted back to the starting material, then an additional arrow, pointing in the opposite direction, is added above the first.

In this reaction, the starting couples are the ions K^+–Br^- and Na^+–Cl^-. The positive potassium and sodium ions (cations) are the male partners. The negative bromide and chloride ions (anions) are the female. If a reaction occurs, the only two possible products are

KCl and **NaBr**. Products such as **KNa** or **ClBr** are not possible as a result of this reaction because like charges repel each other, and each couple needs both a male and a female partner. The reaction may therefore be written as

$$\mathbf{KBr + NaCl \leftrightarrows KCl + NaBr}$$

As written, the reaction is reversible. This means that if we start with the compounds to the left of the arrows, we can produce the compounds to the right, and vice versa. Given a moment's reflection, the student will soon realize that, as a result of this reversibility, there will always be some of each of the four compounds present. This will be so whether the reaction is started with the pair of compounds to the left or the pair to the right of the arrows. Na_2SO_4, $LiNO_3$, **RbCl**, and K_3PO_4 are other examples of ionic compounds that will undergo reversible substitution reactions when mixed in aqueous solutions.

Substitution reactions can be considered reversible unless particular conditions preclude such considerations. A reaction cannot be reversible if the products formed are not capable of further reaction. This is the case when the products leave the scene of the reaction either as gases, precipitates (nonsoluble substances), or nonionizable substances. More will be said about this later.

Was Their Meeting Just Chance?

When the couples arrive at a dance, they might be evenly matched with respect to height. (**KBr** might represent a short couple, **NaCl** a tall.) At the dance, some short girls may decide that they want to pair up with some tall boys. This would leave only short boys for the tall girls to pair up with. The result of this new pairing would produce new couples (called product molecules). When they all leave (at equilibrium), there could be four types of couple: (1) a short boy and a short girl, (2) a tall boy and a tall girl, (3) a short boy and a tall girl, and (4) a tall boy and a short girl.

The number of each type of couple at the end of the dance would, of course, depend on who came at the start. Obviously, no more could leave than came, nor could a tall girl go home with a short boy unless a short girl goes home with a tall boy. (The reaction is a substitution reaction because the products were produced by substitution of one positive or negative ion for another. The ions have exchanged partners.) Because of the attractions of opposite charges, nobody leaves alone.

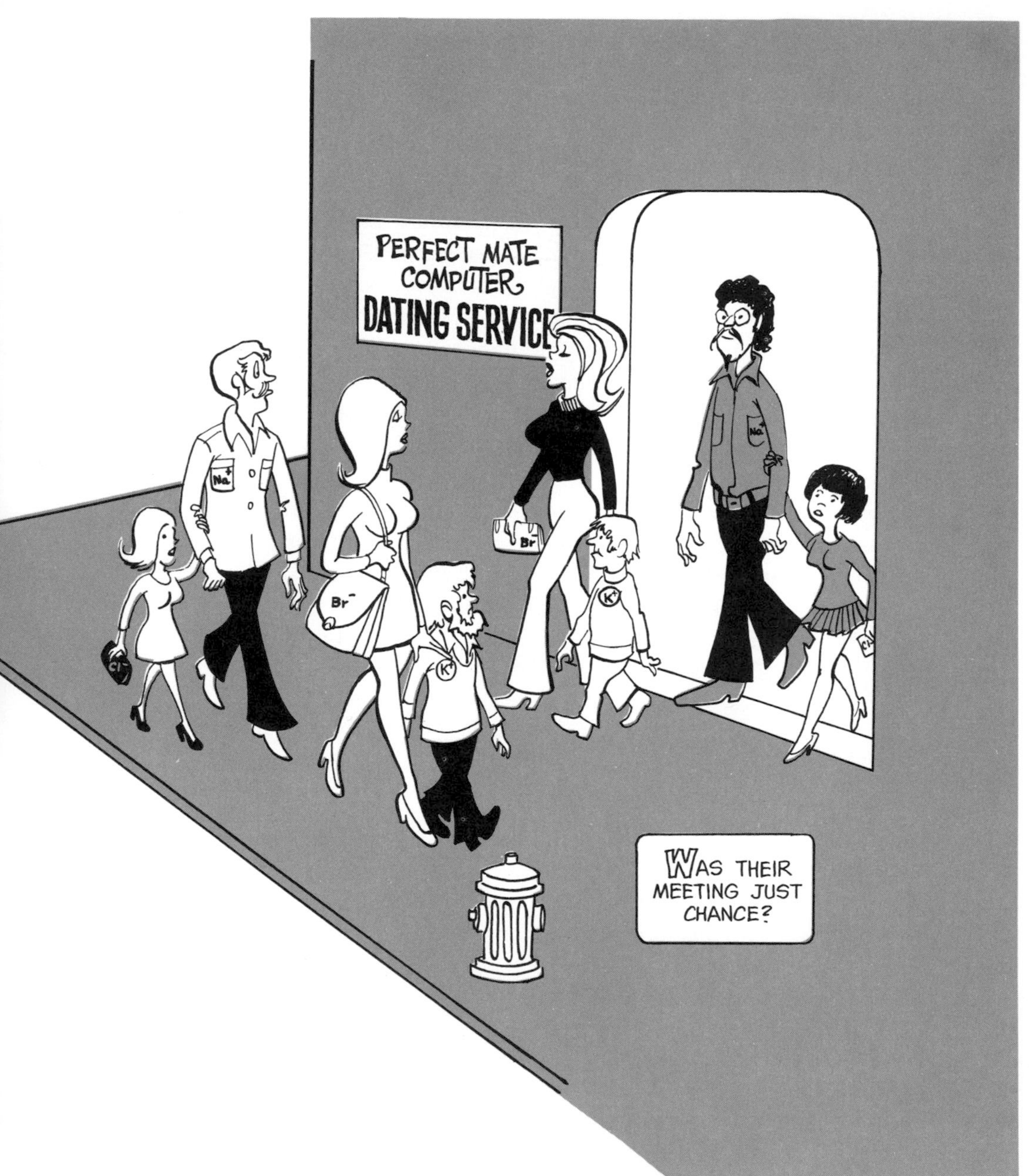

If They Leave, It's All Over.

When a sodium sulfide solution is mixed with hydrochloric acid, hydrogen sulfide gas is produced. The reaction is irreversible because the hydrogen sulfide has left the reaction flask and is therefore incapable of further reaction. An upturned arrow is used to indicate the formation of a gaseous product.

$$Na_2S + 2\,HCl \rightarrow 2\,NaCl + H_2S\uparrow$$

A similar reaction occurs between potassium cyanide and sulfuric acid:

$$2\,KCN + H_2SO_4 \rightarrow K_2SO_4 + 2\,HCN\uparrow$$

It should be kept in mind that the number of atoms of any given type on one side of an equation must equal exactly the number on the other. Atoms are neither created nor destroyed during a chemical reaction. The mathematical operation of equating the atoms on both sides of an equation is known as balancing the equation. When a number precedes a compound, each atom in that compound should be multiplied by that factor. The absence of a number implies that only one molecule is present. When a number appears as a subscript, only the atom or atoms (if parentheses exist) immediately preceding that number should be multiplied. In the reaction between sodium sulfide and hydrochloric acid, there are two sodium, one sulfur, two hydrogen, and two chlorine on each side of the equation. Similarly, in the reaction between potassium cyanide and sulfuric acid there are two potassium, two carbon, two nitrogen, two hydrogen, one sulfur, and four oxygen on each side of the equation.

They Just Dropped Out of Sight.

Substitution reactions rendered irreversible by virtue of product precipitations are common in the formations of mineral deposits.

$$2\,K_3PO_4 + 3\,CaCl_2 \rightarrow 6\,KCl + Ca_3(PO_4)_2\downarrow$$

In the reaction between potassium phosphate and calcium chloride, calcium phosphate is formed. Because of its lack of solubility in water, calcium phosphate is precipitated out of solution. This is indicated by a downturned arrow. Since it is obvious that potassium chloride alone cannot be converted to both potassium phosphate and calcium chloride, the reaction is irreversible. There are six potassium, two phosphorus, eight oxygen, three calcium, and six chlorine atoms represented on each side of the equation. The subscript 2 in the calcium phosphate molecule denotes that everything in the parentheses immediately preceding the two should be multiplied by two.

The reaction between barium chloride and sodium sulfate is another example of a substitution reaction that is irreversible because of precipitate formation:

$$BaCl_2 + Na_2SO_4 \rightarrow NaCl + BaSO_4\downarrow$$

Precipitation reactions, like those involving gas evolution, are irreversible. This is because the product couples when formed discontinue their dancing and leave the ballroom. The direction of the arrows placed next to each signifies the process by which each couple exited. The particular method of exiting each couple may have chosen, however, is of no consequence in determining the irreversibility of the reaction. What is important is that each couple that has left can no longer dance with any of the other ballroom occupants.

The preceding examples contained the polyatomic ions $\mathbf{SO_4^{=}}$ and $\mathbf{PO_4^{\equiv}}$. These are two of the more common polyatomic ions. Polyatomic ions involved in substitution reactions maintain their integrity when going from one side of an equation to the other. A listing of some uni- and polyatomic ions, their charges, and the solubilities of some of their compounds appears in Figure 5–1.

SOLUBILITIES IN WATER												
i – nearly insoluble ss – slightly soluble s – soluble d – decomposes		acetate CH_3COO^-	bromide Br^-	carbonate $CO_3^{=}$	chloride Cl^-	hydroxide OH^-	iodide I^-	nitrate NO_3^-	oxide $O^{=}$	phosphate $PO_4^{\equiv}$	sulfate $SO_4^{=}$	sulfide $S^{=}$
Aluminum	Al^{+++}	s	s		s	i	s	s	i	i	s	d
Ammonium	NH_4^+	s	s	s	s	s	s	s		s	s	s
Barium	Ba^{++}	s	s	i	s	s	s	s	s	i	i	d
Calcium	Ca^{++}	s	s	i	s	ss	s	s	ss	i	ss	d
Copper II	Cu^{++}	s	s	i	s	i	d	s	i	i	s	i
Iron II	Fe^{++}	s	s	i	s	i	s	s	i	i	s	i
Iron III	Fe^{+++}	s	s		s	i	s	s	i	i	ss	d
Lead	Pb^{++}	s	ss	i	ss	i	ss	s	i	i	i	i
Magnesium	Mg^{++}	s	s	i	s	i	s	s	i	i	s	d
Mercury I	Hg^+	ss	i	i	i		i	s	i	i	ss	i
Mercury II	Hg^{++}	s	ss	i	s	i	i	s	i	i	d	i
Potassium	K^+	s	s	s	s	s	s	s	s	s	s	s
Silver	Ag^+	ss	i	i	i		i	s	i	i	ss	i
Sodium	Na^+	s	s	s	s	s	s	s	d	s	s	s
Zinc	Zn^{++}	s	s	i	s	i	s	s	i	i	s	i

FIGURE 5–1 Solubilities in Water of Common Ions

They Act As If They're Not Here at All.

The third type of nonreversible substitution reaction occurs because of the formation of nonionizable products. It is equivalent to two dancers meeting and immediately deciding "to go steady." Nonionizable products differ from precipitates or gases in that they do not necessarily leave the reaction site after formation. The most common example of a nonionizable product is water. Water ionizes only very slightly to give hydrogen and hydroxide ions. At room temperature only 1/10,000,000 of the water molecules formed will ionize.

$$H_2O \leftrightarrows H^+ + OH^-$$

(9,999,999 molecules) (1 ion) (1 ion)

Since only one in ten million product molecules revert to reactants, nonionized water may be said to be the cause of the reaction going to completion.

$$HClO_4 + LiOH \rightarrow LiClO_4 + H_2O$$

Acid-base reactions such as the preceding are the most common examples of nonreversible water-producing substitution reactions. They will be discussed in greater detail in a later section.

In a Nutshell.

1. Substitution reactions mostly involve the interchange of positive and negative ions.
2. The starting materials are reactants.
3. The materials produced as a result of the reactions are products.
4. When writing a reaction, the reactants are written to the left of an arrow pointing right; products are written to the right of the arrow.
5. If the reaction is reversible, a second arrow pointing from right to left is written above the first.

6. A reaction is nonreversible if the reaction products are incapable of further reaction.
7. The three most common processes that render products incapable of further reaction are gas evolution, precipitation, and the formation of a nonionizable product.

That's the Equilibrium Offer—Not a Molecule More!

It is quite simple to determine, when a reaction is complete, if it is irreversible. The starting materials have disappeared and the products remain. In a reversible reaction, however, this evaluation is more complex. Because of its reversible nature, both product and reactant molecules are present at all times. The question is how long it takes before the maximum amount of reactant molecules have reacted to produce the maximum amount of product. This amount of time is the equilibrium time. Before it, less than the maximum amount of reactants have reacted. After it, no further amounts will react. The amounts of both reactants and products at equilibrium are the equilibrium amounts. They are unique for each reaction type under its set of conditions. They are the arrangement of couples after the last dance.*

When everyone leaves at the end of the dance, the chemical reaction can be considered established. The partner pairing is the optimum for the type of people who attended. The numbers of each type of couple are the most workable, or equilibrium, amounts. If there were twice as many tall girls as tall boys at the dance, it should be suspected that, at equilibrium, more tall girls would leave with short partners than if their numbers were equal.

Example. At a dance there are four tall girls, two tall boys, six short girls, and eight short boys. (The number of tall girls is twice as great as tall boys.) This assortment of dancers puts definite restrictions on the type of couple that can leave at equilibrium. Because of the small number of tall boys and the abundance of short ones, there can never be less than four short couples nor more than two tall ones. The three possible distributions of couples are illustrated in the cartoon on page 70.

During the dance each dancer or ion had the opportunity to dance with more than one other ion. At equilibrium the partner pairing that

*In truth, none of the reactions ever cease. After equilibrium is reached, however, no change is discernible because the amount of reactants converted to products equals exactly the amount of products converted back to reactants at any given time.

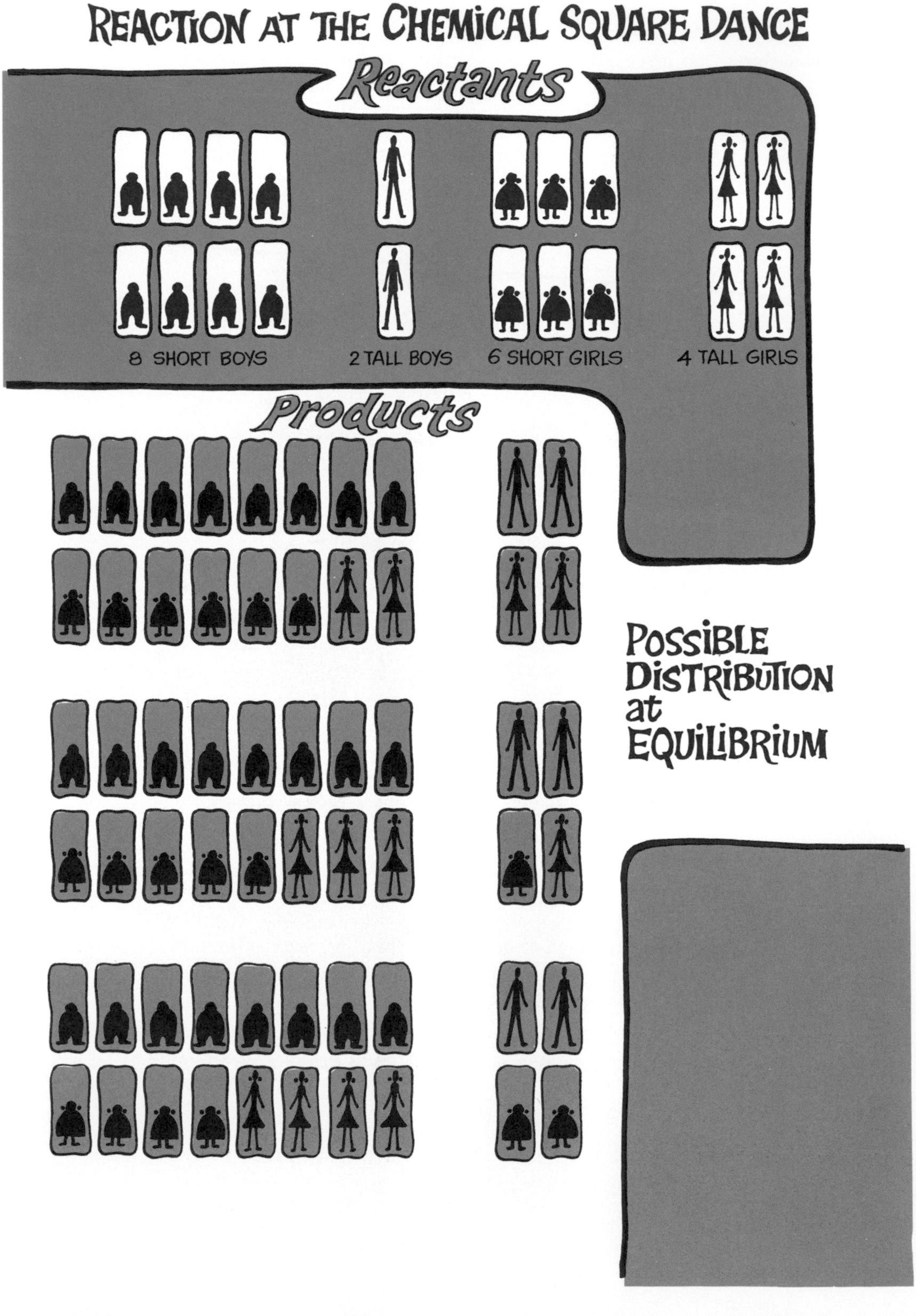

REACTION AT THE CHEMICAL SQUARE DANCE

Reactants

8 SHORT BOYS 2 TALL BOYS 6 SHORT GIRLS 4 TALL GIRLS

Products

POSSIBLE DISTRIBUTION at EQUILIBRIUM

has been established can be considered as optimum for the number and types of dancers or ions that attended. As can be seen in the cartoon the composition of the equilibrium couples is to a great degree dependent upon the composition of the couples at the start of the dance.

It should be noted that when the couples came to the dance, they had to be in one of the three possible arrangements. When they leave, they must leave arranged in one of the three possibilities. There are no other arrangements. If, on leaving, the arrangement is different than when they came, a reaction has taken place. While they are dancing, there is no way to determine the arrangement in which they came. Since they are all fickle, faceless chemicals, you should not be concerned over the possible differences that may exist between two boys or girls of the same height. It matters not which particular ion of a given type comprises the pair as long as the required number of types of pairs are formed. The requirements are determined by the laws of probability in conjunction with the chemical affinities of the particular ions. They are expressed in the form of an *equilibrium constant* (K).

Considering the three possible distributions of couples, the following statements can be made.

1. At the beginning of the dance the couples (or compounds) arrived at the ballroom (reaction site) in one of the three possible combinations, as shown in the cartoon at the left.

Note: In order to maintain electrical neutrality there are no singles at the dance. Each dancer always has a partner.

2. The equilibrium constant when it takes into consideration the personal preferences of each dancer (chemistry of each reactant or product) specifies that of the three possible distributions shown in the cartoon one is more probable than the other two.

3. The most probable distribution is the one that will exist at equilibrium when all the dancers (chemical species) have had sufficient time to dance with a large enough number of other dancers that they can properly choose the most suitable partner. (Chemical species combine to form the most probable compound.)

4. If the choice of partners at the close of the dance is the same as that which existed when the couples entered the ballroom (if the equilibrium mixture is the same composition as the starting reactant mixture) then there will be no evidence that the dance did anything to alter the makeup of the couples (there will be no evidence of a reaction having taken place).

An Equilibrium Is Nothing without a Constant.

We have seen that, on the basis of chance alone, three possible equilibrium mixtures exist for four tall female, two tall male, six short female, and eight short male dancing ions. Three choices rather than one exist because we did not consider the chemical personalities of the ions. It is not reasonable, however, to expect in an actual chemical experiment that the nature of the reaction products will depend solely on chance. When specific chemicals react, specific products are produced. The same products are produced regardless of the number of times the experiment is performed. The invariance of these products is often expressed mathematically by the equilibrium constant of the reaction.

The equilibrium constant for simple reactions is defined as the product of the product amounts divided by the product of the reactant amounts.

$$\underset{\text{reactants}}{A + B} \leftrightarrows \underset{\text{products}}{C + D}$$

$$\text{Equilibrium constant} = K = \frac{(C) \times (D)}{(A) \times (B)}$$

Note. If, as in the reaction below, more than one molecule of a given product is produced

$$E + F \leftrightarrows H + 2I = H + I + I$$

the equilibrium constant is written as

$$K = \frac{(H) \times (I) \times (I)}{(E) \times (F)} = \frac{(H) \times (I)^2}{(E) \times (F)}$$

At a given temperature, it is a constant for any reversible reaction. Its value is determined by the nature of the specific chemicals as well as the amounts in which they are present. Once the nature of the chemicals is considered, the value of the equilibrium constant specifies that one equilibrium distribution of Figure 5–1 is far more probable than the others.

If tall girls, rather than being indifferent to their partners' height, strongly prefer tall boys, this will have a marked effect on the type of couples who leave when the dance is over. This does not mean that every tall girl will leave with a tall boy: that is not possible, because there are more tall girls than tall boys. It also does not mean that every tall boy will leave with a tall girl. This is because there are six short girls as compared with only four tall girls competing for a tall boy. These restrictions make distribution 2 far more probable than

1 or 3. Knowledge of this probability can be used to calculate the equilibrium constant for this and any other dance, with any number of people. The generalized reaction can be written as follows (the letters in color indicate male partners):

$$\mathrm{S\ S + T\ T \leftrightarrows S\ T + S\ T}$$

As written, the dance is pictured as starting with short and tall couples and producing mixed couples as it proceeds. It should be kept in mind, however, that the format is merely a convenient notation and, as we have seen, has no effect on the equilibrium products. The equilibrium constant may then be represented by

$$\frac{(\mathrm{S\ T}) \times (\mathrm{S\ T})}{(\mathrm{S\ S}) \times (\mathrm{T\ T})} = \mathrm{K\ (constant)}$$

When the parentheses are filled with the equilibrium numbers of each type of couple, the numerical constant for this reaction may be determined

$$\frac{(3) \times (1)}{(5) \times (1)} = \frac{3}{5}$$

This constant can be used to calculate the equilibrium distribution of couples for other dances with different numbers of dancers. The equilibrium constant takes into account the nature of the participating chemicals and is a constant for all reactions that incorporate the same chemicals, regardless of the amounts in which they are present.

Example. If the equilibrium constant is 3/5, what will be the couple composition when the dance is over if six tall and eight short couples come to the dance?

With a little algebra, it can be shown that the answer is three tall couples, five short couples, and three of each type of mixed couples.

$$\frac{(\mathrm{S\ T}) \times (\mathrm{S\ T})}{(\mathrm{S\ S}) \times (\mathrm{T\ T})} = \frac{(3) \times (3)}{(5) \times (3)} = \frac{3}{5} = \mathrm{K}$$

If a new couple were to arrive late, or one leave in the middle, the equilibrium would be upset. This would cause the other couples to rearrange themselves so that the value of the equilibrium constant would be maintained. The chemist Le Chatlier was the first to note that the addition of reactants or the removal of products from an equilibrium mixture resulted in the further production of products. The reverse effect occurs when products are added or reactants removed. In this way, the reaction is altered to accommodate the equilibrium constant.

This principle of shifting of products and reactants is known as the Le Chatlier Principle. It is the chemical equivalent of Newton's Law, which requires every action to have a reaction.

Did You Ever See a Mole Dance?

When considering chemicals or dancers it is often more important to know their concentrations than their absolute amounts. Twenty dancers in a 4 × 5-foot room would have a concentration of one dancer per square foot of floor space. A hoedown in this room would be difficult, to say the least. If, however, the same twenty dancers were in a room that measured 20 × 100 feet, each dancer would have an average of 100 square feet of floor space in which to dance—an entirely different situation.

Concentration is equally significant when dealing with chemicals. One teaspoon of salt in a pot of soup—mmmm, good. One teaspoon of salt in a tablespoon of soup—blggh! Chemical concentration is expressed in units of moles per liter. If a solution has a concentration of one gram molecular weight (mole) per liter it is said to be a one molar solution. A liter is slightly more than a quart. A gram molecular weight is equal to the weight in grams of the molecular weight of the substance.

Example. The formula of table salt is **NaCl**. This means that each molecule of salt contains one atom of sodium and one atom of chlorine. The atomic weight of sodium is 23. The atomic weight of chlorine is 35. One gram molecular weight of salt therefore weighs $23 + 35 = 58$ grams. If 58 grams of salt were mixed with sufficient water to give a liter of solution, the concentration of the solution would be one molar. If it were mixed so as to produce ten liters of solution, the resulting solution concentration would be one tenth molar.

To facilitate the exchange of scientific knowledge the units of moles per liter, to express concentration, have been adopted universally by chemists. When molar concentrations of chemical participants are substituted for the absolute number of dancing couples, equilibrium constants become chemically meaningful. The use of concentration rather than absolute amounts removes the restrictions placed upon a reaction by one specific set of conditions and allows the outcome of all similar reactions to be predicted independent of the amounts of chemicals or container size. It doesn't make any difference whether soup is cooked in a small pot to serve four or in a large cauldron to serve forty.

If the same relative proportions (concentrations) of ingredients are employed, the soup will taste the same. The equilibrium constant specifies that the products of a given reaction will be the same whether it is run in a test tube or in a thousand-gallon reaction kettle as long as the concentrations of the participating chemicals remain the same.

The equilibrium constant for the reactions

$$KBr + NaCl \rightleftarrows KCl + NaBr$$

can be expressed as

$$\frac{[KCl][NaBr]}{[KBr][NaCl]} = K \text{ (constant)}$$

The square brackets are used to indicate that the reactant and product amounts are expressed as mole per liter concentrations.

Once the equilibrium constant is evaluated by substituting the value of the equilibrium concentrations for one set of conditions, it can be used for any reaction of these four chemicals in any amounts. At any temperature it will be a constant for this reaction irrespective of the amounts of chemical reacting.

Although the value of the constant for this particular reaction does not differ much from unity, equilibrium constants may be as great as 10^{25} or, when written for the reverse reaction, as small as 10^{-25}. The size of the constant determines whether there will be more products or more reactants present at equilibrium. When it is large there will be almost exclusively products at equilibrium, if comparable quantities of reactants were there at the start. If it is small, then at equilibrium there will be practically no indications that a reaction has taken place. If A and B are mixed with the expectation that C and D will be produced it should be done with the understanding that the equilibrium constant should be greater than 1. If the constant is much smaller than 1, the reaction will not take place.

When equilibrium constants are either extremely large or extremely small, the reactions they govern may be considered irreversible. An extremely large number indicates that there is virtually no chance of the reaction going in the opposite direction. An extremely small number also indicates that the reaction can go only one way. This is so because if the reaction has been written so as to interchange products and reactants, the equilibrium constant for the reaction would be extremely large.

$$A + B \rightleftarrows C + D \qquad K_1 \text{ (very large)}$$

$$C + D \rightleftarrows A + B \qquad K_2 \text{ (very small)}$$

REVIEW

1. When a reversible reaction progresses until there is no change in concentration of either reactant or products, equilibrium has been reached.
2. The concentration of reactants and products at equilibrium are the equilibrium concentrations.
3. The equilibrium concentrations of all the components of a reaction mixture are specified by the equilibrium constant of that reaction.
4. The equilibrium constant is equal to the ratio of the product of the product concentrations divided by the product of the reactant concentrations.
5. The units of concentrations used to calculate equilibrium constants are gram molecular weights (moles) per liter.
6. The equilibrium constant is a constant for all concentrations of chemicals involved in a specific reaction at a particular temperature.
7. When an equilibrium constant for a reaction is numerically greater than 1 the amount of products at equilibrium will be proportionately greater than the amount of reactants. When it is less than 1 the opposite is true. This is so for all reactions occurring between chemicals having comparable concentrations.
8. Virtually no reaction will take place between the reactants of a reaction as written if the equilibrium constant is extremely small.
9. If a reaction is reversible there will be present at all times some of each chemical specie.
10. Substitution reaction will be irreversible if gases, insoluble precipitates, or nonionizable products are formed. In addition, they will be effectively irreversible if their equilibrium constants are either very large or very small.

REFLEXIVE QUIZ

1. ____________ reactions occur when molecules interchange ions.
2. A substitution reaction is nonreversible. Three causes are __________, __________________, ____________________.
3. The units usually employed to express chemical concentrations are ____________.
4. Three solutions whose volumes are 2, 4, and 8 liters each contain 4 moles of lithium acetate. The concentrations of these solutions are respectively ____, ____, and _____ moles/liter.

5. When a reversible reaction reacts until no further change occurs, it can be said that ____________ has been reached.

6. At equilibrium the ratio of the product of the product concentrations divided by the product of the reactant concentrations is referred to as the ____________ constant.

7. In a reaction where two molecules of a product are produced, the product concentration must be __________ when writing the equilibrium constant.

8. An equilibrium mixture when disturbed reacts so as to minimize the disturbance. This is in accordance with the ____________ Principle.

9. If a reversible reaction is to produce a significant amount of products, the value of its equilibrium constant must be greater than ___.

10. An extremely large value of the equilibrium constant indicates that a reaction is ____________.

Answers: (1) Substitution [62, 63]. (2) Gas formation, precipitate formation, nonionizable product formations [63–67]. (3) Moles/liter [74, 76]. (4) 2, 1, and 0.5 [74]. (5) Equilibrium [68–75]. (6) Equilibrium [72–76]. (7) Squared [72]. (8) Le Chatlier [73–74]. (9) 1 [74, 76]. (10) Non-reversible [75–76].

SUPPLEMENTARY READINGS

1. P. G. Ashmore. "Reaction Kinetics and the Law of Mass Action." *Education in Chemistry* 2 (1965): 160.

2. S. W. Benson. "Some Aspects of Chemical Kinetics for Elementary Chemistry." *Journal of Chemical Education* 39 (1962): 321.

3. S. Berline and C. Bicker. "The Law of Mass Action." *Journal of Chemical Education* 46 (1969): 499.

4. E. L. King. *How Chemical Reactions Occur*. New York: W. A. Benjamin, Inc., 1963.

5. C. J. Nyman and R. E. Hamm. *Chemical Equilibrium.* Boston: D. C. Heath & Co., 1967.

6. J. O. Edwards, E. F. Greene, and J. Ross. "From Stoichiometry and Rate Law to Mechanism." *Journal of Chemical Education* 45 (1968): 381.

Section 6

Reactions: Acid-Base

Section 6

Reactions: Acid-Base

Any drinking man can tell you that lemons are sour and quinine is bitter. The ingredients of tonic, which have kept the British content for so many years, contain appreciable quantities of two acids (citric and ascorbic) and the base quinine. It is the acids that taste sour and the base that tastes bitter.

When acids react with bases, a salt and water are produced. Because of the toxicity of many chemicals, it is often unwise to use taste as the criterion for determining whether the substance is either acidic or basic. An alternative procedure is the use of indicator dyes. The most common of these is an extract of lichens called litmus. Litmus turns blue in the presence of bases and red in the presence of acids. Dyes possessing similar properties can also be prepared from cabbage, radishes, and other plant extracts.

The acid and basic qualities that produce the changes in color and difference in taste are the result of the relative hydrogen ion and hydroxide ion concentrations. Water when it ionizes produces equal quantities of each. This occurs, however, in less than one in every five hundred million molecules of pure water (5.5×10^8).

$$H_2O \rightleftarrows H^+ + OH^-$$

At room temperature the equilibrium constant for this reaction is 1.8×10^{-16}. The concentration of either hydrogen or hydroxide ions in water can, however, be increased by the addition, respectively, of either an acid or a base. By definition an acid is any substance that when added to water increases its hydrogen ion concentration. A base is any substance that has a similar effect on the hydroxide ion concentrations.

Peter Piper Pickled His Peppers with Acid.

Hydrogen chloride is an acid because it ionizes in water to give hydrogen and chloride ions

$$HCl + H_2O \rightarrow H_3O^+ + Cl^-$$

H_3O^+, ($H^+ + H_2O$) signifies a hydrogen ion that is associated with water molecules. Such an ion is often called a hydronium ion. The hydronium ion notation is deficient, however, in that it specifies only one water

PREPARE A STIMULATING DRINK QUICKLY, CARROTHERS! LADY ATWATER HAS FAINTED!
ACIDS AND BASES, A SPLASH OF ALCOHOL, AND A TOUCH OF THE OLD JUNIPER!
HURRY, MAN, HURRY!
MMMM...
...DELIGHTFULLY REFRESHING, CARROTHERS! BURP
NOW TELL ME... WHAT DO YOU THINK WE SHOULD DO ABOUT THE OLD GIRL?

molecule as being associated with each hydrated hydrogen ion when in fact the number is often greater.

Sulfur trioxide can be considered an acid although it itself does not contain any hydrogen because it will increase the concentration of hydrogen ions in the solutions to which it is added.

$$SO_3 + H_2O \rightleftarrows H^+ + HSO_4^- \rightleftarrows 2\,H^+ + SO_4^=$$

For clarity in the above equations the hydrogen ions are written unhydrated. If two additional waters are added to the left side of the equation, both hydrogen ions can be written as H_3O^+.

Two theories justify the classification of these two substances as acids. The first theory is attributed to Bronsted and Lowry. It classifies acids as substances capable of donating protons, and bases as species capable of combining with protons. This definition, although applicable to hydrochloric acid (**HCl**), leaves much to be desired when classifying sulfur trioxide, the anhydride of sulfuric acid. (An acid anhydride is a substance that is converted to that acid upon additions of water.) The second and more recent theory was proposed by G. N. Lewis in 1923. He classified acids as substances capable of accepting electron pairs and bases as substances capable of donating them. His theory, by far the more inclusive of the two, can be used to explain the acidic character of both **HCl** and SO_3.

To understand the Lewis theory, let's first examine, as we have done in earlier sections, the electronic structure of the molecules under consideration. The reaction between water and hydrogen chloride can be written as

$$\underset{\text{Water}}{\mathrm{H{:}\ddot{\underset{\cdot\cdot}{O}}{:}\atop H}} \quad + \quad \underset{\substack{\text{Hydrogen chloride}\\ \text{(Acid)}}}{\mathrm{H{:}\ddot{\underset{\cdot\cdot}{Cl}}{:}}} \quad \rightarrow \quad \underset{\text{Hydronium ion}}{\left[\mathrm{H{:}\ddot{\underset{\cdot\cdot}{O}}{:}H\atop H}\right]^+} \quad + \quad \underset{\text{Chloride ion}}{\left[\mathrm{{:}\ddot{\underset{\cdot\cdot}{Cl}}{:}}\right]^-}$$

Thus, we see that the oxygen atom of the water molecule donates an electron pair for the formation of a coordinate covalent bond (a covalent bond in which both electrons of the shared pair have been supplied by a single atom) with the hydrogen of the hydrogen chloride molecule. This results in the formation of a singly hydrated hydrogen ion (hydronium ion). Further, we see that the water molecule is acting as a base.

Similarly, the reaction between sulfur trioxide and water can be written as

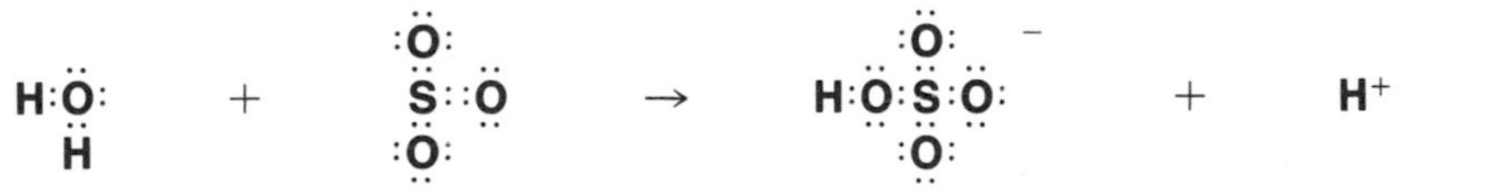

```
                  :Ö:                 :Ö:  −
H:Ö:      +       S::Ö      →     H:Ö:S:Ö:        +       H⁺
  H               :Ö:                 :Ö:
```

Water		Sulfur trioxide (Acid)		Bisulfate ion		Hydrogen ion (hydrated)

In this reaction a covalent bond between the oxygen and hydrogen in water has been broken and a new covalent bond between sulfur and the water oxygen has been formed. The Lewis theory has the advantage over the Bronsted theory in that it is capable of predicting the long-recognized experimental fact that metallic oxides are basic whereas nonmetallic oxides are acidic.

It should be noted that the bisulfate ion is also an acid by both theories because

1. it contains a proton it can donate

$$HSO_4^- \rightarrow H^+ + SO_4^=$$

2. it can accept an electron pair from a water molecule, resulting in the breaking and forming of hydrogen covalent bonds to produce a sulfate ion and a hydrated hydrogen ion.

```
                  :Ö:  −                             :Ö:  =
H:Ö:      +    H:Ö:S:Ö:      →     H:Ö:H  +    +   :Ö:S:Ö:
  H               :Ö:                  H              :Ö:
```

Water		Bisulfate ion (Acid)		Hydronium ion		Sulfate ion

Table 6–1 sets forth some of the common acids, their chemical formulas, and their reactions with water. Note that the two chemicals that make up each of the three pairs of compounds marked by brackets produce the same solution when added to water. The second of each pair is the nonmetallic oxide that is the anhydride of the acid in each pair.

In writing the equations of ionic reactions that take place in water it should be noted that hydrogen ion, by virtue of being hydrated, is not unique. Depending upon the temperature, their concentrations, or their charges, all ions in aqueous solutions (water solutions) are hydrated to varying degrees. Often, either because of the lack of specific knowledge or because of the desire to portray a given reaction in the least confusing manner, these associated waters are not explicitly indicated when writing the reaction equation. The student should be

TABLE

6–1 Some Common Acids, Their Formulas and Reactions with Water

Compound	Formula	Hydrolysis Reaction
Acetic acid	$HC_2H_3O_2$	$HC_2H_3O_2 + H_2O \rightarrow H_3O^+ + C_2H_3O_2^-$
Carbonic acid	H_2CO_3	$H_2CO_3 + H_2O \rightarrow H_3O^+ + HCO_3^-$
Carbon dioxide	CO_2	$CO_2 + 2\,H_2O \rightarrow H_3O^+ + HCO_3^-$
Citric acid	$H_3C_6H_5O_7$	$H_3C_6H_5O_7 + H_2O \rightarrow H_3O^+ + H_2C_6H_5O_7^-$ $H_2C_6H_5O_7^- + H_2O \rightarrow H_3O^+ + HC_6H_5O_7^=$ $HC_6H_5O_7^= + H_2O \rightarrow H_3O^+ + C_6H_5O_7^{\equiv}$
Formic acid	$HCHO_2$	$HCHO_2 + H_2O \rightarrow H_3O^+ + CHO_2^-$
Hydrochloric acid	HCl	$HCl + H_2O \rightarrow H_3O^+ + Cl^-$
Nitric acid	HNO_3	$HNO_3 + H_2O \rightarrow H_3O^+ + NO_3^-$
Phosphoric acid	H_3PO_4	$H_3PO_4 + H_2O \rightarrow H_3O^+ + H_2PO_4^-$ $H_2PO_4^- + H_2O \rightarrow H_3O^+ + HPO_4^=$
Phosphorus pentoxide	$(P_2O_5)_2$	$(P_2O_5)_2 + 6\,H_2O \rightarrow 4\,H_3O^+ + 4\,H_2PO_4^-$ $H_2PO_4^- + H_2O \rightarrow H_3O^+ + HPO_4^=$
Sulfuric acid	H_2SO_4	$H_2SO_4 + H_2O \rightarrow H_3O^+ + HSO_4^-$ $HSO_4^- + H_2O \rightarrow H_3O^+ + SO_4^=$
Sulfur trioxide	SO_3	$SO_3 + 2\,H_2O \rightarrow H_3O^+ + HSO_4^-$ $HSO_4^- + H_2O \rightarrow H_3O^+ + SO_4^=$

aware, however, when reading any equation that no matter how it is written there are always water molecules associated with each ion in solution. Ions in solution are always *all wet.*

Hydroxide Ions and Hydrogen Ions in Water

The addition of bases to water increases hydroxide ion concentration beyond what would be produced by the ionization of water alone. At room temperature the concentration of hydroxide ion in pure water is only one ten-millionth of a mole per liter (1×10^{-7} moles/liter). This is also the value of the hydrogen ion concentration, since both were produced by the splitting of water molecules. Neutral water has an excess of neither hydrogen nor hydroxide ion. The concentration of hydroxide ion in lye solutions typically employed as drain cleanser is often two million times greater than the amount found in neutral water.

The compound soda lye (**NaOH**) is a typical strong base or alkali. In water it dissolves to give sodium and hydroxide ions.

$$NaOH \rightarrow Na^+ + OH^-$$

The **OH⁻** ion classifies as a base according to the Bronsted theory because it is capable of combining with a proton to yield water. It also satisfies the requirements of the Lewis theory because the oxygen atom provides the necessary electron pair to combine with the proton.

$$:\ddot{\underset{..}{O}}:H^- + H^+ \rightarrow H:\ddot{\underset{..}{O}}: \atop \quad\quad\quad\quad H$$

This reaction should appear familiar as it is merely the reverse of the water dissociation reaction.

Ammonia is a basic molecule that does not contain a hydroxide group. It will, however, increase the hydroxide concentration when it is added to pure water.

$$\begin{array}{c}\mathrm{H}\\ \mathrm{H{:}\ddot{N}{:}}\\ \mathrm{H}\end{array} + \begin{array}{c}\\ \mathrm{H{:}\ddot{O}{:}}\\ \mathrm{H}\end{array} \rightleftharpoons \left[\begin{array}{c}\mathrm{H}\\ \mathrm{H{:}\ddot{N}{:}H}\\ \mathrm{H}\end{array}\right]^{+} + \left[\mathrm{{:}\ddot{\underset{\cdot\cdot}{O}}{:}H}\right]^{-}$$

The nitrogen atom provides the electron pairs to form the bond with the proton accepted from water. The reaction satisfies the requirements of both the Bronsted and the Lewis theories.

Table 6–2 lists some typical basic compounds, their chemical formulas, and their hydrolysis products. Note that aqueous solutions of metallic oxides and hydroxide are usually basic.

Dry Acids and Bases

All the examples of acids and bases presented thus far can be discussed in terms of either the Bronsted or the Lewis theory with equal facility. There are compounds, however, that classify as acids and bases only in terms of the Lewis theory. The reaction between fluoride ion and boron trifluoride, for example, can be written as

$$\underset{\text{Base}}{\left[\mathrm{{:}\ddot{\underset{\cdot\cdot}{F}}{:}}\right]^{-}} + \underset{\text{Acid}}{\begin{array}{c}\mathrm{{:}\ddot{F}{:}}\\ \mathrm{B{:}\ddot{\underset{\cdot\cdot}{F}}{:}}\\ \mathrm{{:}\underset{\cdot\cdot}{F}{:}}\end{array}} \rightarrow \left[\begin{array}{c}\mathrm{{:}\ddot{F}{:}}\\ \mathrm{{:}\ddot{\underset{\cdot\cdot}{F}}{:}B{:}\ddot{\underset{\cdot\cdot}{F}}{:}}\\ \mathrm{{:}\underset{\cdot\cdot}{F}{:}}\end{array}\right]^{-}$$

TABLE 6–2 Typical Basic Compounds, Their Formulas and Hydrolysis Products

Compound	Formula	Hydrolysis Reaction*
Ammonium hydroxide	NH_4OH	$NH_4OH \rightarrow NH_4^+ + OH^-$
Sodium borate	$Na_2B_4O_7$	$B_4O_7^{=} + H_2O \rightarrow HB_4O_7 + OH^-$
Sodium carbonate	Na_2CO_3	$CO_3^{=} + H_2O \rightarrow HCO_3^- + OH^-$
Sodium hydroxide	$NaOH$	$NaOH \rightarrow Na^+ + OH^-$
Trisodium phosphate	Na_3PO_4	$\begin{cases} PO_4^{\equiv} + H_2O \rightarrow HPO_4^{=} + OH^- \\ HPO_4^{=} + H_2O \rightarrow H_2PO_4^- + OH^- \end{cases}$

*The positive sodium ions (**Na^+**) and ion-associated water molecules have deliberately been omitted from a number of these equations, since the purpose is to illustrate the manner in which compounds increase the hydroxide ion concentration of water. Neither the sodium ions nor the associated waters affect this process.

Here the fluoride ion behaves as a base by donating an electron pair to the boron trifluoride acceptor acid.

The Lewis theory is especially useful when describing reactions in nonaqueous mediums where there may be a complete lack of hydrogen and hydroxide groups. It should be noted that, although a salt is always produced as the result of an acid-base neutralization, water usually is formed only when the reaction is conducted in an aqueous medium. Compare, for example, these two reactions:

Aqueous Neutralization

$$\boxed{Ca^{++} + 2\,(OH)^-} + \boxed{2\,H^+ + SO_4^=} \rightarrow \boxed{Ca^{++} + SO_4^=} + 2\,H_2O$$

Base	Acid	Salt	Water
(Calcium hydroxide)	(Sulfuric acid)	(Calcium sulfate)	

Nonaqueous Neutralization

$$\boxed{Ca^{++} + :\ddot{\underset{\cdot\cdot}{O}}:^{=}} + \begin{matrix} :\ddot{O}: \\ \ddot{S}::\ddot{O}: \\ :\underset{\cdot\cdot}{O}: \end{matrix} \rightarrow \boxed{Ca^{++} + \begin{matrix} :\ddot{O}: \\ :\ddot{\underset{\cdot\cdot}{O}}:\ddot{\underset{\cdot\cdot}{S}}:\ddot{\underset{\cdot\cdot}{O}}: \\ :\underset{\cdot\cdot}{O}: \end{matrix}^{=}}$$

Base	Acid	Salt
(Calcium oxide)	(Sulfur trioxide)	(Calcium sulfate)

Aside from the presence of water, these two reactions are identical. They both produce the salt calcium sulfate. This is not surprising since calcium hydroxide is the hydrated form (the form to which water has been added) of calcium oxide and sulfur trioxide is the acid anhydride of sulfuric acid.

$$CaO + H_2O \rightarrow Ca(OH)_2$$

$$SO_3 + H_2O \rightarrow H_2SO_4$$

In spite of this identity, however, the Bronsted theory cannot adequately describe the nonaqueous neutralization.

*p*H Is a Phunny Way To Write Hydrogen Ion Concentration.

An aqueous acid-base reaction can be monitored by measuring its *p*H. *p*H is a convenient numbering system by which large ranges of hydrogen ion concentration can be evaluated. An explanation of this system appears in Appendix I. A similar and equally convenient system also exists for measuring the hydroxide ion concentration. It is called *p*OH. If an aqueous solution is neither acidic nor basic, but neutral, then both its *p*H and *p*OH will equal 7. This means that both the hydrogen ion concentration and the hydroxide ion concentration of that solution equal 1×10^{-7} moles per liter. If the hydrogen ion concentration is greater than this value, the solution will be acidic. If the hydroxide ion concentration is greater than this value, it will be alkaline. In either case the respective *p*H or *p*OH will be less than 7. In all cases, however, the sum of the *p*H and the *p*OH of a solution will be 14. This means that if either the *p*H or the *p*OH of a solution is known, the other can be readily calculated by subtracting the known value from 14.

Note. It should be remembered that the *lower* the numerical value of the *p*H or *p*OH, the higher the concentration of the respective ions. A hydrochloric acid solution with a *p*H of 1 contains 100 times more hydrogen ions than a hydrochloric acid solution with a *p*H of 3.

Normality

In the discussion of acid-base reactions no distinction was made between reactions of acids and bases having one hydrogen or hydroxyl group and those having several. It should be obvious that one mole of acid that has two hydrogens in its structure can neutralize two moles of a base that possesses only one hydroxyl group per molecule.

$$H_2SO_4 + 2\,NaOH \rightarrow Na_2SO_4 + 2\,H_2O$$

Similarly, a mole of base having two hydroxyl groups would need either two moles of monobasic acid or one mole of dibasic acid for neutralization.*

$$2\,HCl + Ca(OH)_2 \rightarrow CaCl_2 + 2\,H_2O$$

$$H_2SO_4 + Ca(OH)_2 \rightarrow CaSO_4 + 2\,H_2O$$

Normality was devised to allow for these discrepancies. It is equal to the molarity multiplied by the number of hydrogen or hydroxyl groups in the molecule.

*The terms *monobasic* and *dibasic* indicate the number of hydrogens in the acid molecule. *Mono, di,* and *tri* signify one, two, and three, respectively.

Compound	Molarity	Normality
HCl	1	1
H_2SO_4	1	2
NaOH	1	1
$Ca(OH)_2$	1	2

The normality of a solution is defined in terms of the number of gram equivalent weights of material dissolved in a liter. A gram equivalent weight is equal to the gram molecular weight divided by the number of hydrogen or hydroxyl ions in the molecule.

Compound	Molecular weight	Equivalent weight
HCl	36.5	36.5
H_2SO_4	98	49
NaOH	40	40
$Ca(OH)_2$	74	37

The nice part of equivalent weights is that irrespective of the chemicals involved, one gram equivalent weight of one reactant will react with exactly one equivalent weight of another. For example, 40 grams of sodium hydroxide will neutralize 36.5 grams of hydrochloric acid or 49 grams of sulfuric acid. One liter of a one normal sodium hydroxide solution will neutralize one liter of a one normal solution of any acid. This is so because by definition there is one gram equivalent weight in every liter of a one normal solution.

$$\text{Liters} \times \text{Normality} = \text{Equivalents}$$

Example. How many liters of a four normal solution of acid are necessary to neutralize two liters of a two normal solution of base?

$$\text{Liters} \times \text{Normality (acid)} = \text{Equivalents Acid}$$

$$\text{Liters} \times \text{Normality (base)} = \text{Equivalents Base}$$

In order to neutralize the base completely, it is necessary to have the same number of equivalent of acid.

$$\text{Liters} \times \text{Normality (acid)} = \text{Liters} \times \text{Normality (base)}$$
$$A \times 4 = 2 \times 2$$
$$4A = 4$$
$$A = 1 \text{ Liter Acid}$$

Since the concentration of acid is twice as great as the base, it is only necessary to use one liter of acid, which is half the base volume, for complete neutralization.

A further discussion of other aspects of chemical stoichiometry (the amount of product produced by a given amount of reactants) appears in Appendix II.

REVIEW

1. Acids and bases react chemically to neutralize each other.
2. The product of the neutralization process is a salt.
3. If the neutralization occurs in an aqueous medium, water is produced as an additional product.
4. Acids characteristically have a sour taste. They increase the hydrogen ion and decrease the hydroxide ion concentration in an aqueous solution to which they have been added. They are capable of accepting an electron pair.
5. Bases characteristically have a bitter taste. They decrease the hydrogen ion concentration and increase the hydroxide ion concentration in an aqueous solution to which they have been added. They are capable of donating an electron pair.
6. In aqueous solutions at room temperature, the product of the hydrogen and hydroxide ion concentrations is a constant.
7. *p*H and *p*OH are logarithmic scales for measuring the hydrogen and hydroxide ion concentration in a solution. They equal the negative log of the respective ionic concentration.
8. The sum of the *p*H and the *p*OH of any aqueous solution at room temperature is 14.
9. The hydrogen ion concentration of a solution cannot be altered without also affecting the hydroxide ion concentration.
10. Solutions having *p*H values less than 7 are acid. Those with values greater than 7 are basic. At *p*H 7, the *p*H equals the *p*OH, the solution is neutral, and the hydrogen and hydroxyl ion concentrations are the same.
11. The normality of a solution is equal to the number of gram equivalent

weights of material dissolved in one liter. It is also equal to the molarity divided by the number of hydrogen or hydroxyl groups in the molecule.

12. For acid-base reactions, the gram equivalent weight is equal to the molecular weight divided by the number of hydrogen or hydroxyl groups per molecule.
13. In any chemical reaction, one gram equivalent weight of one reactant will react with exactly one gram equivalent weight of another to produce products.

REFLEXIVE QUIZ

1. The prime cause of basic properties of an aqueous solution is the ________ ion.
2. **HBr** is an acid because it ionizes in water to produce ________ ions.
3. In the reaction between **SrO** and **SO_3**, the sulfur trioxide is an acid according to the Lewis theory because it accepts an ________ ________ from the oxide ion.
4. Nitric acid (**HNO_3**) is a ______ basic acid.
5. The *p*H of an acid solution is less than ____.
6. A solution with a *p*H of 10 is ______.
7. The equivalent weight of a dibasic acid is equal to ________ its molecular weight.
8. The number of gram equivalent weights of a substance in solution can be found by multiplying the volume of the solution in liters by its ________.
9. The *p*OH of a solution with a *p*H of 13 is ____.
10. The molarity of a dibasic acid is ________ its normality.

Answers: (1) Hydroxide [80, 84, 90]. (2) Hydrogen [80]. (3) Electron pair [80, 81, 86]. (4) Mono [88]. (5) 7 [87–90, Appendix I]. (6) Basic [87–90, Appendix I]. (7) One-half [88, 90]. (8) Normality [88, 90]. (9) 1 [87–90]. (10) One-half [87–90].

SUPPLEMENTARY READINGS

1. D. A. MacInnes. "*p*H." *Scientific American* Ja 51 (1951): 40.
2. J. Waser. "Acid-Base Titration and Distribution Curves." *Journal of Chemical Education* 44 (1967): 274.
3. J. N. Butler. *Solubility and* pH *Calculations*. Reading, Mass.: Addison-Wesley, 1964.

4. W. F. Luder. "Contemporary Acid-Base Theory." *Journal of Chemical Education* 25 (1948): 555.

5. C. A. Van der Werf. *Acids, Bases and the Chemistry of the Covalent Bond.* New York: D. Van Nostrand Co., 1966.

Section 7

Reactions: Redox

Section 7

Reactions: Redox

Substitution reactions owe their uniqueness to their ability to exchange ions; acid-base reactions will produce a change in the pH of a solution. The claim to fame of redox reactions (reduction-oxidation reactions) is their ability to transfer electrons.

When an atom *loses* electrons, the process involved is *oxidation*. The extent of the oxidation can be determined from the charge on the resulting ion. If the atom loses one electron it will have an oxidation number of +1. Before the electron loss, it's oxidation number is 0. All elemental uncombined atoms have zero oxidation numbers. Potassium, a group I metal, can be used to illustrate this point.

$$\underset{\text{Atom}}{\mathbf{K^0}} \rightarrow \underset{\text{Ion}}{\mathbf{K^+}} + e^-$$

The superscripts indicate the oxidation number of the two potassium species. If group II elements are considered then the oxidation might be illustrated by the following example.

$$\underset{\text{Atom}}{\mathbf{Ca^0}} \rightarrow \underset{\text{Ion}}{\mathbf{Ca^{++}}} + 2\,e^-$$

In this case the calcium ion has an oxidation number of +2.

As has been pointed out earlier, an atom cannot lose electrons unless another atom is willing to accept them. If an atom *gains* electrons to become an ion, it is said to be *reduced*. Bromine, a group VII atom, can be used to illustrate the process of reduction.

$$\underset{\text{Atom}}{\mathbf{Br^0}} + e^- \rightarrow \underset{\text{Ion}}{\mathbf{Br^-}}$$

The oxidation number of the bromine ion resulting from this reduction is −1. Similarly, group VI atoms, when they are reduced, tend to form ions with an oxidation number of −2. The compound water is composed of two atoms of hydrogen, each of which has an oxidation number of +1, and one atom of oxygen, which has an oxidation number of −2.

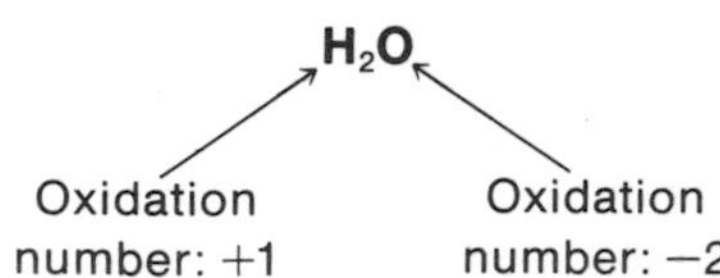

Note that, just as in the case of ionic compounds, where the net charge of the molecule is zero, the total of all the oxidation numbers of atoms in a molecule is also zero. The difference between ionic charges and atomic oxidation numbers becomes apparent when complex ions are considered. The compound potassium perchlorate, for example, is composed of a potassium ion and a chlorate ion.

$$KClO_4 \rightarrow K^+ + ClO_4^-$$

As indicated, the potassium ion has a charge and oxidation number of +1. The perchlorate ion, however, presents a more difficult problem. Although the ion has a charge of −1, the oxidation numbers of the oxygen and chlorine atoms in the perchlorate ion are not immediately apparent. In this particular example the oxidation number of the chlorine is +7 and that of each oxygen is −2. Because of this, the perchlorate ion has a charge of −1 [(+7) + 4(−2) = −1].

There exist several other compounds of potassium chlorine and oxygen. One of these is potassium chlorate. It ionizes in the following manner:

$$KClO_3 \rightarrow K^+ + ClO_3^-$$

Here too, just as in the case of potassium perchlorate, the molecule ionizes to give one singly charged positive ion and one singly charged negative ion. In this case, however, the chlorate ion contains one less oxygen atom than the perchlorate ion.

ClO_3^-	ClO_4^-
Chlorate	Perchlorate

If an oxidation number of −2 is assigned to each oxygen in both of these compounds, the oxidation number of chlorine in chlorate ion must be +5 rather than +7.

ClO_3^-	ClO_4^-
$(+5) + 3(-2) = -1$	$(+7) + 4(-2) = -1$

If there were a reaction in which the chlorate ion is converted to a perchlorate ion, it would be an oxidation reaction.

$$\underset{(+5)}{ClO_3^-} + O \xrightarrow{\text{oxidation}} \underset{(+7)}{ClO_4}$$

This is so because in going from chlorate to perchlorate, the oxidation number of the chlorine atom has become more positive. The increase in positive charge is equivalent to the loss of electrons. In this reaction the chlorate ion is oxidized by the oxygen to perchlorate, and the oxygen atom, which originally had an oxidation number of 0, is reduced to an oxidation number of −2.

$$Cl^{+5} \xrightarrow{-2\ e^-} Cl^{+7}$$

Oxidation

$$O^{0} \xrightarrow{+2\ e^-} O^{-2}$$

Reduction

Note. For every oxidation there is a complementary reduction. If something loses electrons there must also be something to gain them. It is for this reason that these reactions are never considered individually but always as *redox* pairs.

It should not be construed, because chlorate was oxidized to perchlorate by the addition of an oxygen atom, that this is the sole means by which chemicals can be oxidized. Oxidation can occur in the complete absence of oxygen. It is the loss of electrons that is the deciding factor.

For a partial listing of atoms and some of their more common oxidation numbers, see Table 7–1. For a listing of common complex ions and their net charges, see Table 7–2.

Unbalanced Redox Equations Have Deep Electron Problems.

The balancing of redox equations is appreciably more difficult than the balancing of either substitution or acid-base equations. This is because electrons as well as atoms must be balanced. The process is most effectively performed if each partial equation is considered separately. The results of the oxidation part are then combined with the reduction part to provide a complete redox equation.

Example. Potassium permanganate reacts with ferrous (iron II, which is iron with an oxidation number of 2) chloride in the presence of hydrochloric acid to produce ferric (iron III) and manganous (manganese II) chlorides:

$$KMnO_4 + FeCl_2 \xrightarrow{HCl} MnCl_2 + FeCl_3 \qquad \text{(unbalanced)}$$

In this reaction the manganese in the $KMnO_4$ is reduced from a +7 to a +2 oxidation state. This can be determined by using Tables 7–1 and 7–2 to establish the ions and charges produced in solution of these reactants. The net charge of the permanganate ion is −1. The oxida-

7–1 Some Atoms and Their Common Oxidation Numbers

TABLE

Element	Oxidation Number
H	+1 (except −1 in hydrides; for example, **LiH**)
Li	+1
Na	+1
K	+1
Mg	+2
Ca	+2
Zn	+2
Al	+3
Cl	−1 } (−1 in compounds with only two elements; variable oxidation numbers in compounds containing more than two elements, for example, $KClO_3$ and $LiClO_4$)
Br	−1
I	−1
O	−2 (except −1 in peroxides; for example, H_2O_2)
Hg	+1 and +2
Fe	+2 and +3
Sn	+2 and +4

These numbers are often helpful in determining the oxidation numbers of individual atoms in complex ions. (The value of chlorine in both chlorate and perchlorate ion, for example, was derived from the knowledge that oxygen has an oxidation number of −2.)

7–2 Common Complex Ions and Their Net Charges

TABLE

Formula	Name	Charge	Formula	Name	Charge
NH_4^+	Ammonium	+1	$CO_3^=$	Carbonate	−2
$C_2H_3O_2^-$	Acetate	−1	$CrO_4^=$	Chromate	−2
ClO_4^-	Perchlorate	−1	$Cr_2O_7^=$	Dichromate	−2
ClO_3^-	Chlorate	−1	$O_2^=$	Peroxide	−2
ClO_2^-	Chlorite	−1	$SiO_3^=$	Silicate	−2
ClO^-	Hypochlorite	−1	$SO_4^=$	Sulfate	−2
MnO_4^-	Permanganate	−1	$SO_3^=$	Sulfite	−2
NO_3^-	Nitrate	−1	$BO_3^{\equiv}$	Borate	−3
NO_2^-	Nitrite	−1	$PO_4^{\equiv}$	Phosphate	−3
OH^-	Hydroxide (Hydroxyl)	−1	$AsO_4^{\equiv}$	Arsenate	−3

tion number of oxygen in most compounds is −2. The oxidation number (x) of the manganese atom in the permanganate ion can therefore be determined algebraically:

$$-1 = x + 4(-2)$$

$$-1 = x + (-8)$$

The oxidation number of manganese in permanganate is +7. **MnCl** ionizes to give $\mathbf{Mn^{++}}$ and 2 $\mathbf{Cl^-}$. The oxidation number of the manganese atom in this compound is therefore +2. This reduction of oxidation state from +7 to +2 is equivalent to having the manganese atom acquire five electrons:

$$(+7) + 5(e^-) = +2$$

$$\mathbf{MnO_4^-} + 5(e^-) \rightarrow \mathbf{Mn^{++}}$$

The conversion of ferrous (iron II) to ferric (iron III) chloride is an oxidation process. The iron goes from an oxidation state of +2 to one of +3. The process is equivalent to the loss of one electron:

$$\mathbf{Fe^{++}} \rightarrow \mathbf{Fe^{+++}} + e^-$$

Since each manganese atom when reduced gains five electrons, and each iron atom when oxidized only loses one, it is necessary to have five times as many $\mathbf{Fe^{++}}$ ions as $\mathbf{MnO_4^-}$ ions in order to maintain electrical neutrality. As a first step in balancing this equation therefore, we can write

$$\mathbf{KMnO_4} + 5\,\mathbf{FeCl_2} \rightarrow \mathbf{MnCl_2} + 5\,\mathbf{FeCl_3} \qquad \text{(unbalanced)}$$

This equation, although balanced electrically, leaves much to be desired in terms of atoms. The potassium, oxygen, and chlorine must still be balanced. The oxygen can be balanced by adding four water to the righthand side. The hydrogen is balanced by adding eight hydrochloric acids to the left. The eighteenth chloride ion on the righthand side associates with the single potassium to give a molecule of potassium chloride. The addition of these ions is possible because the reaction was run in an aqueous solution of hydrochloric acid. Their addition results in the production of the completely balanced molecular equation:

$$\mathbf{KMnO_4} + 5\,\mathbf{FeCl_2} + 8\,\mathbf{HCl} \rightarrow \mathbf{MnCl_2} + 5\,\mathbf{FeCl_3} + 4\,\mathbf{H_2O} + 3\,\mathbf{KCl}$$

Thus, it is balanced with respect to both electrons and atoms.

A "Whiz Bang" Method for Balancing Redox Equations

As just demonstrated, the balancing of redox equations can be lengthy and tedious, notwithstanding darn near impossible if the oxidation states of the individual atoms are not known. By making it necessary to know only the net ionic charge rather than the individual oxidation states of each atom in a complex ion, the "Whiz Bang" method can ease the task of balancing redox reactions in aqueous mediums. In many cases the "Whiz Bang" method will, in addition, automatically specify the number of nonredox participating solvent molecules.

A "Whiz Bang" Method for Balancing in Acid Solutions

1. Write all reactant ions as they appear in solution.
2. Balance all atoms with the exception of hydrogen and oxygen.
3. For each unbalanced oxygen atom add one water molecule to the opposite side of the equation.
4. For each unbalanced hydrogen atom add a hydrogen ion (H^+) to the opposite side of the equation.
5. Balance all electrons.

Example. Write a balanced equation for the oxidation of ferrous chloride by potassium permanganate. (Note: This is the same as the previous example; this time, however, the "Whiz Bang" method will be used to solve it.)

Step 1. Write all reactant ions as they appear in solution.

$$MnO_4^- \rightarrow Mn^{++}$$

$$Fe^{++} \rightarrow Fe^{+++}$$

Step 2. Balance all atoms with the exception of hydrogen and oxygen.

$$MnO_4^- \rightarrow Mn^{++}$$

$$Fe^{++} \rightarrow Fe^{+++}$$

Since all atoms with the exception of hydrogen and oxygen are already balanced, there is nothing that must be done for this step.

Step 3. For each unbalanced oxygen atom add one water molecule to the opposite side of the equation.

$$MnO_4^- \rightarrow Mn^{++} + 4\,H_2O$$

$$Fe^{++} \rightarrow Fe^{+++}$$

This step is applicable only to the permanganate reduction.

Step 4. For each unbalanced hydrogen atom add a hydrogen ion to the opposite side of the equation.

$$8\,H^+ + MnO_4^- \rightarrow Mn^{++} + 4\,H_2O$$

$$Fe^{++} \rightarrow Fe^{+++}$$

This step too is applicable only to the permanganate reaction.

Step 5. Balance all electrons.

$$5\,e^- + 8\,H^+ + MnO_4^- \rightarrow Mn^{++} + 4\,H_2O$$

$$Fe^{++} \rightarrow Fe^{+++} + e^-$$

Without the addition of electrons, the left side of the permanganate reaction has a charge of +7. This is the sum of the eight positive charges from the hydrogen ions and the one negative charge from the permanganate ion. The right side, by contrast, has only a +2 charge. In order to make the charges on both sides equal, it is necessary to add five negative charges in the form of electrons to the lefthand side. This produces a balanced ionic equation for the permanganate reduction. From the electron balancing, it should be evident that this is a reduction reaction. In the production of products, the reactants have gained electrons.

The loss of an electron by the ferrous ion (Fe^{++}) to produce a ferric ion (Fe^{+++}) is the necessary oxidation to complete the redox reaction.

Although both the oxidation and the reduction equations are balanced with regard to atoms and electrons, no allowance was made for the fact that the oxidation reaction produces only one electron, whereas the reduction reaction utilizes five. To compensate for this disparity, it is necessary to multiply the entire oxidation equation by 5. In order to reduce one permanganate ion, five ferrous ions must be oxidized.

$$5\,e^- + 8\,H^+ + MnO_4^- \rightarrow Mn^{++} + 4\,H_2O$$

$$5\,(Fe^{++} \rightarrow Fe^{+++} + e^-)$$

$$\cancel{5\,e^-} + 8\,H^+ + MnO_4^- + 5\,Fe^{++} \rightarrow Mn^{++} + 4\,H_2O + 5\,Fe^{+++} + \cancel{5\,e^-}$$

Adding the two equations produces the total ionic redox reaction. Since the number of product electrons equals the number of reactant electrons, they can be canceled on both sides.

The total ionic equation, although complete, does not account for the nonreactant ions. This is the province of the molecular equation. A molecular equation can be produced by writing the ion partners of each of the ions in the balanced equation. Because the H^+ came from **HCl**, the MnO_4 came from $KMnO_4$, and the Fe^{++} came from $FeCl_2$, these molecules can be substituted for their respective ions on the lefthand side of the equation.

$$8\,HCl + KMnO_4 + 5\,FeCl_2 \rightarrow$$

In doing this, eighteen chloride (Cl^-) and one potassium ion have been added to the lefthand side. In order to maintain a balanced equation, they must also be added to the righthand side. It does not matter how they are balanced so long as electrical neutrality is maintained.

$$8\,HCl + KMnO_4 + 5\,FeCl_2 \rightarrow MnCl_2 + 4\,H_2O + 5\,FeCl_3 + KCl$$

The results of the "Whiz Bang" method are the same as those obtained using the oxidation numbers of the individual atoms; however, it was not necessary to know the oxidation states of the individual atoms.

A "Whiz Bang" Method for Balancing in Basic Solutions

1. Write all ions as they appear in solution.
2. Balance all atoms with the exception of hydrogen and oxygen.
3. For each unbalanced oxygen atom add a hydroxide ion to the opposite side of the equation.
4. For each unbalanced hydrogen atom add a water molecule to the opposite side of the equation.
5. For each water added to one side of the equation add a hydroxide ion to the opposite side.
6. Balance all charges.

It should be noted that the first two rules are identical to those which were presented for acid solutions.

Example. In a basic solution, ferrous hydroxide ($Fe(OH)_2$) is oxidized by potassium permanganate to ferric hydroxide ($Fe(OH)_3$). In the process, potassium permanganate ($KMnO_4$) is reduced to manganese dioxide (MnO_2). Write a balanced equation for the reaction.

Step 1. Write all ions as they appear in solution.

$$MnO_4^- \rightarrow MnO_2$$

$$Fe^{++} \rightarrow Fe^{+++}$$

The ionic product of the oxidation reaction is unchanged from what it was in acid medium; however, the product of the permanganate reduction is now MnO_2 rather than Mn^{++}. It will be seen that this change in reaction product will have a marked effect upon the number of ferrous ions that the permanganate is capable of oxidizing.

Step 2. Balance all atoms with the exception of hydrogen and oxygen. As in the case of the acid medium, this step can be skipped.

Step 3. For each unbalanced oxygen atom add a hydroxide ion to the opposite side of the equation.

$$\mathbf{MnO_4^-} \rightarrow \mathbf{MnO_2} + 2\,\mathbf{OH^-}$$

$$\mathbf{Fe^{++}} \rightarrow \mathbf{Fe^{+++}}$$

Step 4. For each unbalanced hydrogen atom add a water molecule to the opposite side of the equation.

$$2\,\mathbf{H_2O} + \mathbf{MnO_4} \rightarrow \mathbf{MnO_2} + 2\,\mathbf{OH^-}$$

$$\mathbf{Fe^{++}} \rightarrow \mathbf{Fe^{+++}}$$

Step 5. For each water added to one side of the equation add a hydroxide ion to the opposite side.

$$2\,\mathbf{H_2O} + \mathbf{MnO_4^-} \rightarrow \mathbf{MnO_2} + 4\,\mathbf{OH^-}$$

$$\mathbf{Fe^{++}} \rightarrow \mathbf{Fe^{+++}}$$

Step 6. Balance all electrons.

$$3\,e^- + 2\,\mathbf{H_2O} + \mathbf{MnO_4^-} \rightarrow \mathbf{MnO_2} + 4\,\mathbf{OH^-}$$

$$\mathbf{Fe^{++}} \rightarrow \mathbf{Fe^{+++}} + e^-$$

In order to assure that sufficient electrons exist to reduce the permanganate ion, it is necessary to multiply the entire ferrous oxidation reaction by 3. Once this has been done, the electrons will cancel when the two half reactions have been added.

$$3\,e^- + 2\,\mathbf{H_2O} + \mathbf{MnO_4^-} \rightarrow \mathbf{MnO_2} + 4\,\mathbf{OH^-}$$

$$3\,(\mathbf{Fe^{++}} \rightarrow \mathbf{Fe^{+++}} + e^-)$$

$$\cancel{3\,e^-} + 2\,\mathbf{H_2O} + \mathbf{MnO_4^-} + 3\,\mathbf{Fe^{++}} \rightarrow \mathbf{MnO_2} + 4\,\mathbf{OH^-} + 3\,\mathbf{Fe^{+++}} + \cancel{3\,e^-}$$

The balanced total ionic equation can be converted to a molecular equation by substituting the appropriate molecules for their respective ions. Since the reactants are potassium permanganate and ferrous hydroxide, the lefthand side of the equation can be written:

$$2\,\mathbf{H_2O} + \mathbf{KMnO_4} + 3\,\mathbf{Fe(OH)_2} \rightarrow$$

The addition of one potassium and six hydroxyl nonreactant ions to the left side must be equalized by adding them to the right. This produces the final molecular equation:

$$2\,H_2O + KMnO_4 + 3\,Fe(OH)_2 \rightarrow MnO_2 + 3\,Fe(OH)_3 + KOH$$

"Normality" Is Relative.

With a little reflection, it should be realized that the oxidation of ferrous ion by permanganate has been used as the example of a redox reaction in both acid and basic solutions. The fact that ferrous chloride was utilized in one and ferrous hydroxide in the other is of no consequence, since neither the chloride nor the hydroxyl ions are active participants in the redox process. The point to be gleaned from these reactions is that permanganate ion in a basic solution can oxidize only 60 percent of the ferrous ion that it can oxidize in an acid solution. The situation is quite similar to that encountered in the acid-base reactions where one mole of acid or base can neutralize more than one mole of base or acid. Because of this, in redox reactions, just as in acid-base neutralizations, it is expedient to use equivalent weights and normalities.

The equivalent weight of a molecule in a redox reaction is equal to its molecular weight divided by the number of electrons involved in its half reaction. Only one electron is involved in the ferrous reaction, therefore its equivalent weight, whether in the chloride or hydroxide form, is equal to its molecular weight. Potassium permanganate, however, presents a different situation. Although its molecular weight is always equal to 158 ($K + Mn + 4\,O = 39 + 55 + 64 = 158$), its equivalent weight varies from 31.6, which is (158/5) in acid medium:

$$\boxed{5\,e^-} + 8\,H^+ + MnO_4^- \rightarrow Mn^{++} + 4\,H_2O$$

to 52.7, which is (158/3) when the medium is basic:

$$\boxed{3\,e^-} + 2\,H_2O + MnO_4^- \rightarrow MnO_2 + 4\,OH^-$$

A one normal solution will contain either 31.6 or 52.7 grams of potassium permanganate per liter, depending on whether it is acid or basic. In all cases, however, one equivalent weight of potassium permanganate will react with one equivalent weight of ferrous ion or any other reactant.

Table 7–3 lists some typical oxidizing and reducing agents, their reaction products and electron changes.

7–3 Typical Oxidizing and Reducing Agents, Their Reaction Products and Electron Changes

TABLE

Oxidizing Agents (Oxidized State)	Usual Reduced State	Change in Oxidation Number
ClO_3^-	Cl^-	**Cl** from +5 to −1
MnO_4^- (in acid)	Mn^{++}	**Mn** from +7 to +2
MnO_4^- (neutral or basic)	MnO_2	**Mn** from +7 to +4
$Cr_2O_7^=$	Cr^{+++}	**Cr** from +6 to +3
HNO_3 (dilute)	NO	**N** from +5 to +2
HNO_3 (concentrated)	NO_2	**N** from +5 to +4
H_2SO_4 (concentrated)	SO_2	**S** from +6 to +4
O_2	OH^-, H_2O	**O** from 0 to −2
Cl_2	Cl^-	**Cl** from 0 to −1
H_2O_2	OH^-, H_2O	**O** from −1 to −2

Reducing Agents (Reduced State)	Usual Oxidized State	Change in Oxidation Number
C	**CO** or CO_2	**C** from 0 to +2 or +4
Metals (**M**)	Metal cations (M^{n+})	**M** from 0 to +n
$SnCl_2$	$SnCl_4$	**Sn** from +2 to −4
H_2SO_3	HSO_4^- or $SO_4^=$	**S** from +4 to +6
H_2S	**S**	**S** from −2 to 0
Fe^{++}	Fe^{+++}	**Fe** from +2 to +3
H_2	H^+ or H_2O	**H** from 0 to +1
HI	I_2	**I** from −1 to 0
H_2O_2	O_2	**O** from −1 to 0

REVIEW

1. Most chemical elements can exist in more than one oxidation state.
2. All elements in their uncombined state have an oxidation number of zero.
3. An atom is reduced when it gains electrons to produce a more numerically negative oxidation state.
4. An atom is oxidized when it loses electrons to produce a more numerically positive oxidation state.
5. Oxygen need not be present in order for oxidation to occur.

6. All oxidation reactions are coupled to reduction reactions so that the electrons produced by one are completely taken up by the other. There are no free, unassociated electrons.
7. The products of redox reactions are often a function of the *p*H of the medium and the concentration of the reactants.
8. Redox equations must be balanced in terms of electrons as well as atoms.
9. Redox equations can be balanced either by assigning oxidation states to individual atoms or by using the "Whiz Bang" method. The results in either case are identical.
10. The products of redox reactions can be expressed by the balanced total ionic equation. Ionically, it is the same for similar concentrations of the same ions in similar mediums. For example, ferrous ion ($\mathbf{Fe^{++}}$) is oxidized to ferric ion ($\mathbf{Fe^{+++}}$) by permanganate ($\mathbf{MnO_4^-}$) irrespective of whether the reactants are ferrous chloride or ferrous nitrate, sodium permanganate or potassium permanganate. The secondary ions do not enter into the reaction.
11. One equivalent weight of the substance being oxidized will react with one equivalent weight of the substance being reduced.
12. The equivalent weight of a substance in a redox reaction is equal to its molecular weight divided by the number of electron changes it undergoes.

REFLEXIVE QUIZ

1. In the reaction between $\mathbf{Sn^{++}}$ and $\mathbf{H_2O_2}$, the $\mathbf{Sn^{++}}$ is ________ to $\mathbf{Sn^{++++}}$.
2. Metallic copper in the uncombined state has an oxidation number of ____.
3. In the reaction $\mathbf{Cl_2 + Mg \rightarrow MgCl_2}$, the uncombined chlorine is ________ to chloride ion.
4. The half equation

$$6\,e^- + \_\_\_\ \mathbf{H^+ + Cr_2O_7^= \rightarrow Cr^{+++}} + \_\_\_\ \mathbf{H_2O}$$

can be balanced by adding ____ $\mathbf{H^+}$ and ____ $\mathbf{H_2O}$.

5. The equivalent weight of hydrogen peroxide in the reaction

$$2\,e^- + 2\,\mathbf{H^+ + H_2O_2 \rightarrow 2\,H_2O}$$

is equal to the molecular weight divided by ____.

6. Hydrogen gas is often used as a reducing agent: $\mathbf{H_2 \rightarrow 2\,H^+}$. In order

to balance this reaction, two electrons must be added to the ____ hand side.

7. In basic solutions, ClO^- is reduced to Cl^-. The equivalent weight of ClO^- is equal to the ionic weight divided by ____.

8.
$$2\,I^- \rightarrow I_2 + 2\,e^-$$
$$H_2 \rightarrow 2\,H^+ + 2\,e^-$$

The sum of these two half reactions does not produce a balanced ionic redox equation because both reactions are __________ reactions.

9. Dilute nitric acid is reduced to **NO** according to the following equation:

$$3\,e^- + 4\,H^+ + NO_3^- \rightarrow NO + 2\,H_2O$$

A one-tenth molar nitric acid solution is therefore _____ normal.

10.
$$Sn^{++} \rightarrow Sn^{++++} + 2\,e^-$$
$$6\,e^- + 6\,H^+ + ClO_3^- \rightarrow Cl^- + 3\,H_2O$$

The tin equation must be multiplied by ____ before the two half equations are added; otherwise the ________ will not cancel.

Answers: (1) Oxidized [94]. (2) 0 [94, 104]. (3) Reduced [94, 104]. (4) 14, 7 [99]. (5) 2 [103, 105]. (6) Right [99–100]. (7) 2 [101–105]. (8) Oxidation [96, 105]. (9) 0.3 [102]. (10) 3, electrons [98].

SUPPLEMENTARY READINGS

1. W. C. Child, Jr., and R. W. Ramette. "The Stoichiometry of an Oxidation-Reduction Reaction." *Journal of Chemical Education* 44 (1967): 109.
2. M. P. Goodstein. "Interpretation of Oxidation-Reduction." *Journal of Chemical Education* 47 (1970): 452.
3. J. A. Bishop. "Redox Reactions and the Acid-Base Properties of Solvents." *Chemistry* 43 (1), (1970): 18.
4. W. L. Jolly. "The Use of Oxidation Potentials in Inorganic Chemistry." *Journal of Chemical Education* 43 (1966): 198.
5. W. M. Latimer. *Oxidation Potentials.* Englewood Cliffs, N.J.: Prentice-Hall, Inc., 1952.

Section 

Electrochemistry

Section 8

Electrochemistry

Electricity is the flow of electrons from one place to another. In a redox reaction the material being oxidized loses electrons. These electrons are transferred to the material being reduced. If the oxidation is separated from the reduction, neither will occur. It is necessary for the two reactions to be coupled in order that the resulting charges can produce a flow of electricity.

Any workable redox couple having the proper physical configuration can be used as a battery. Figure 8-1 illustrates a typical battery. It is constructed of a zinc rod in a zinc sulfate solution and a copper rod in a copper sulfate solution. The solutions are prevented from mixing by

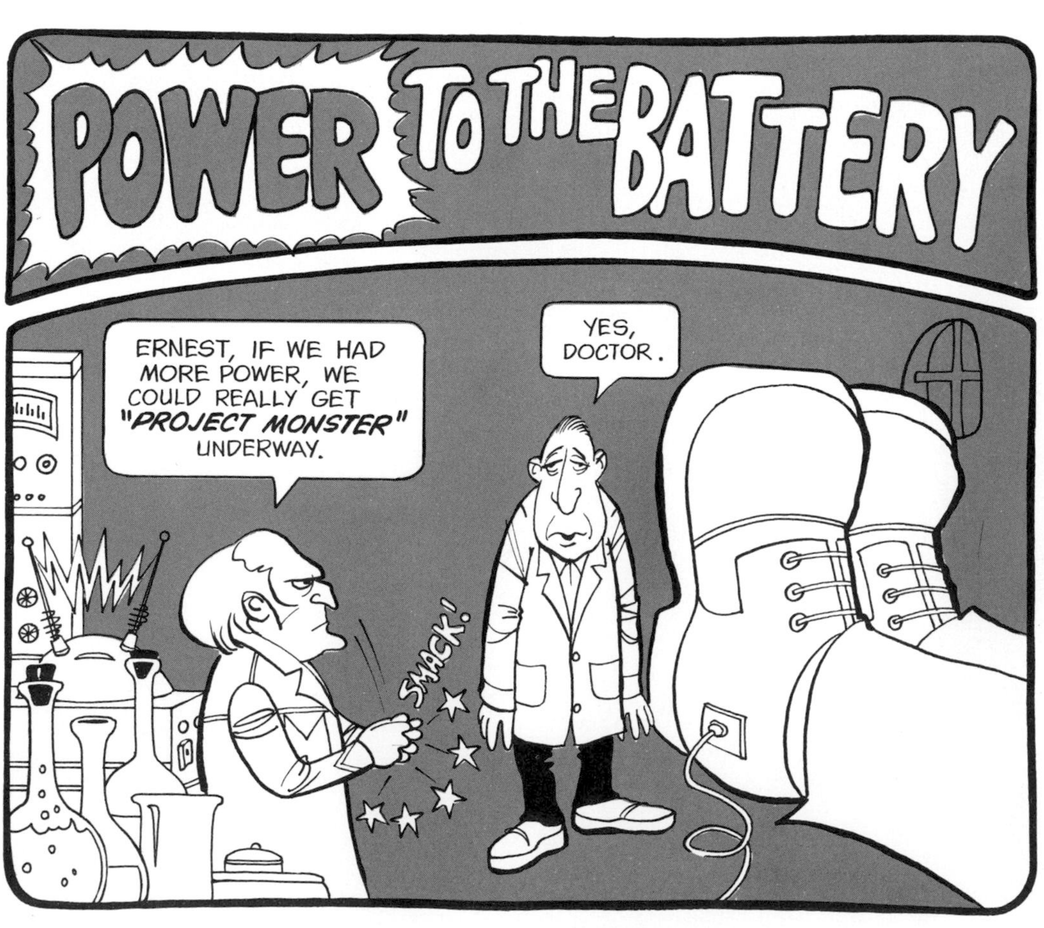

the porous wall, which, in order to maintain electrical neutrality, in both compartments, allows the passage of sulfate ions. As drawn, the battery contains a zinc–zinc sulfate reaction compartment and a copper–copper sulfate reaction compartment. Each of these compartments is capable of undergoing either an oxidation or a reduction reaction.

The zinc reactions in the lefthand compartment could be:

$$(1) \qquad \mathbf{Zn^0} \rightarrow \mathbf{Zn^{++}} + 2\,e^- \quad \text{(Oxidation)}$$

$$(2) \qquad \mathbf{Zn^{++}} + 2\,e^- \rightarrow \mathbf{Zn^0} \quad \text{(Reduction)}$$

Equation 1 states that the zinc rod may be oxidized to zinc ion. This would result in an increase in the concentration of zinc sulfate in solution. Alternatively, equation 2 states that the zinc ion in solution

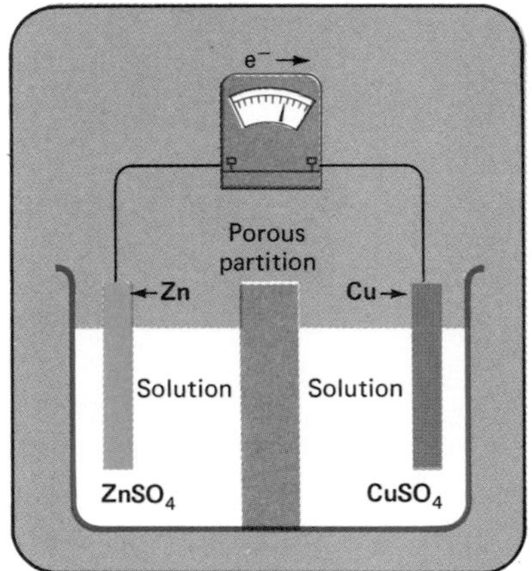

FIGURE 8–1. A Typical Battery

may be reduced to zinc metal. This would increase the amount of metallic zinc in the compartment. The equivalent copper reactions in the righthand compartment could be:

$$(1) \quad \mathbf{Cu^0 \rightarrow Cu^{++}} + 2\,e^- \quad \text{(Oxidation)}$$

or

$$(2) \quad \mathbf{Cu^{++}} + 2\,e^- \rightarrow \mathbf{Cu^0} \quad \text{(Reduction)}$$

As stated by these equations, both compartments may contain either oxidation or reduction reactants.

If a workable battery is to be constructed, however, it is essential that several criteria be met.

1. Only one of the two possible half-reactions may occur in each compartment at any time.
2. When an oxidation reaction occurs in one compartment, a reduction must simultaneously occur in the other.
3. Provision must be made to transfer the electrons from the oxidation compartment, where they are being produced, to the reduction compartment, where they are utilized.
4. Provision must be made for transferring the excess sulfate ions from the reduction compartment to the oxidation compartment,

where they are needed to neutralize the newly formed metallic ions.

Once these criteria are met, the only impediment to producing a battery is the decision as to which compartment should contain the oxidation and which the reduction. Fortunately this may be readily determined through use of a table of standard electrode potentials. (Table 8–1.) This table specifies in volts the ability of each reaction to occur. The more positive the voltage, the greater the tendency for the reaction to occur as written. If the voltages of the two possible reduction reactions are compared, it becomes readily apparent that the reduction of the copper ion is the more probable of the two processes.

$$\mathbf{Cu^{++}} + 2\,e^- \rightarrow \mathbf{Cu^0} \qquad (+.34V)$$

$$\mathbf{Zn^{++}} + 2\,e^- \rightarrow \mathbf{Zn^0} \qquad (-.76V)$$

In a normally operating battery, therefore, reduction will take place in the copper compartment and oxidation will take place in the zinc compartment. It should be noted that all the reactions in Table 8–1 are written as reductions. In order to obtain the voltages of the corresponding oxidation reactions, it is necessary merely to reverse the reaction and change the numerical sign of the voltage.

$$\mathbf{Cu^0} \rightarrow \mathbf{Cu^{++}} + 2\,e^- \qquad (-.34V)$$

$$\mathbf{Zn^0} \rightarrow \mathbf{Zn^{++}} + 2\,e^- \qquad (+.76V)$$

The fact that the voltage of the zinc oxidation reactions is more positive than that of the copper confirms the earlier finding that the reduction will take place in the copper compartment and the oxidation will occur in the zinc compartment.

The voltage of the entire battery can be determined by writing the oxidation and reduction reactions as they would normally occur and adding their respective voltages. These voltages are representative of one molar solutions of reactants at 25°C and are typical of what may be expected for the type of battery.

$$\begin{array}{rlr} & \mathbf{Zn^0} \rightarrow \mathbf{Zn^{++}} + 2\,e^- & (+.76V) \\ \mathbf{Cu^{++}} + 2\,e^- \rightarrow \mathbf{Cu^0} & & (+.34V) \\ \hline \mathbf{Zn^0} + \mathbf{Cu^{++}} \rightarrow \mathbf{Zn^{++}} + \mathbf{Cu^0} & & (+1.10V) \end{array}$$

8–1 Standard Electrode Potentials

TABLE

In Acid Solutions: $[H^+] = 1M$, at 25°C

Half-reaction	Volts
$F_2 + 2e^- \rightleftarrows 2F^-$	2.8
$H_2O_2 + 2H^+ + 2e^- \rightleftarrows 2H_2O$	1.77
$MnO_4^- + 4H^+ + 3e^- \rightleftarrows MnO_2 + 2H_2O$	1.69
$MnO_4^- + 8H^+ + 5e^- \rightleftarrows Mn^{+2} + 4H_2O$	1.51
$PbO_2 + 4H^+ + 2e^- \rightleftarrows Pb^{+2} + 2H_2O$	1.45
$Cl_2 + 2e^- \rightleftarrows 2Cl^-$	1.36
$Cr_2O_7^= + 14H^+ + 6e^- \rightleftarrows 2Cr^{+3} + 7H_2O$	1.33
$MnO_2 + 4H^+ + 2e^- \rightleftarrows Mn^{+2} + 2H_2O$	1.23
$O_2 + 4H^+ + 4e^- \rightleftarrows 2H_2O$	1.23
$Br_2 + 2e^- \rightleftarrows 2Br^-$	1.06
$NO_3^- + 4H^+ + 3e^- \rightleftarrows NO + 2H_2O$	0.96
$2Hg^{+2} + 2e^- \rightleftarrows Hg_2^{+2}$	0.92
$Ag^+ + e^- \rightleftarrows Ag$	0.80
$Hg_2^{+2} + 2e^- \rightleftarrows 2Hg$	0.79
$Fe^{+3} + e^- \rightleftarrows Fe^{+2}$	0.77
$O_2 + 2H^+ + 2e^- \rightleftarrows H_2O_2$	0.68
$MnO_4^- + e^- \rightleftarrows MnO_4^=$	0.56
$I_2 + 2e^- \rightleftarrows 2I^-$	0.54
$Cu^+ + e^- \rightleftarrows Cu$	0.52
$Cu^{+2} + 2e^- \rightleftarrows Cu$	0.34
$AgCl(s) + e^- \rightleftarrows Ag + Cl^-$	0.22
$Cu^{+2} + e^- \rightleftarrows Cu^+$	0.15
$2H^+ + 2e^- \rightleftarrows H_2$	0.00
$Cd^{+2} + 2e^- \rightleftarrows Cd$	−0.40
$Cr^{+3} + e^- \rightleftarrows Cr^{+2}$	−0.41
$Fe^{+2} + 2e^- \rightleftarrows Fe$	−0.44
$Zn^{+2} + 2e^- \rightleftarrows Zn$	−0.76
$Mn^{+2} + 2e^- \rightleftarrows Mn$	−1.18
$Al^{+3} + 3e^- \rightleftarrows Al$	−1.66
$H_2 + 2e^- \rightleftarrows 2H^-$	−2.25
$Mg^{+2} + 2e^- \rightleftarrows Mg$	−2.37
$Na^+ + e^- \rightleftarrows Na$	−2.71
$Ca^{+2} + 2e^- \rightleftarrows Ca$	−2.87
$Ba^{+2} + 2e^- \rightleftarrows Ba$	−2.90
$K^+ + e^- \rightleftarrows K$	−2.93
$Li^+ + e^- \rightleftarrows Li$	−3.05

In Basic Solutions: $[OH^-] = 1M$, at 25°C

Half-reaction	Volts
$MnO_4^= + 2H_2O + 2e^- \rightleftarrows MnO_2 + 4OH^-$	0.60
$O_2 + 2H_2O + 4e^- \rightleftarrows 4OH^-$	0.40
$O_2 + H_2O + 2e^- \rightleftarrows HO_2^- + OH^-$	−0.08
$Cu(NH_3)_2^+ + e^- \rightleftarrows Cu + 2NH_3$	−0.12
$Ag(CN)_2^- + e^- \rightleftarrows Ag + 2CN^-$	−0.31
$Hg(CN)_4^{-2} + 2e^- \rightleftarrows Hg + 4CN^-$	−0.37
$SO_4^= + H_2O + 2e^- \rightleftarrows SO_3^= + 2OH^-$	−0.93
$Zn(NH_3)_4^{+2} + 2e^- \rightleftarrows Zn + 4NH_3$	−1.03
$Zn(OH)_4^{-2} + 2e^- \rightleftarrows Zn + 4OH^-$	−1.22
$Mg(OH)_2 + 2e^- \rightleftarrows Mg + 2OH^-$	−2.69
$Ca(OH)_2 + 2e^- \rightleftarrows Ca + 2OH^-$	−3.03

This battery will produce a voltage of 1.1V. The process theoretically will continue as long as there exists sufficient zinc rod to be oxidized and sufficient copper ion to be reduced. As the battery discharges, the zinc electrode becomes smaller and the copper electrode becomes larger. The concentration of zinc sulfate increases while the copper sulfate decreases. Appreciable decrease in any of the reactants results in a corresponding decrease in battery voltage. When the voltage decreases to the point where it is no longer useful, the battery goes dead. To recharge it, it is necessary to provide an external voltage slightly in excess of 1.10 volts. This voltage reverses the normal battery operation and causes regeneration of the zinc electrode at the expense of the copper.

If the sum of the half-reaction voltages is negative rather than positive, the battery will not work. A negative value indicates that the materials chosen to be reduced oxidize more readily than those chosen to be oxidized, and vice versa. To convert such a battery into a workable one, it is necessary to reverse and interchange both half-reactions.

Example. A battery having lithium metal as the *cathode* (the electrode at which reduction takes place) and silver metal as the *anode* (the electrode at which oxidation takes place) does not work. Using these same two metals, construct a battery that will work. What will the voltage of this battery be?

Cathode	$Li^+ + e^- \rightarrow Li^0$	(−3.05V)	
Anode	$Ag \rightarrow Ag^+ + e^-$	(−0.80V)	Unworkable
	$Li^+ + Ag^0 \rightarrow Ag^+ + Li^0$	(−3.85V)	battery

Anode	$Li \rightarrow Li^+ + e^-$	(+3.05V)	
Cathode	$Ag^+ + e^- \rightarrow Ag^0$	(+0.80V)	Workable
	$Li^0 + Ag^+ \rightarrow Ag^0 + Li^+$	(+3.85V)	battery

Thus, it can be seen that the silver ion is more easily reduced than the lithium ion and that lithium metal is more easily oxidized than silver metal.

If the half-reaction voltages in Table 8–1 are compared, it should become evident that the maximum voltage any single battery can produce will be less than 6. A fluorine-lithium battery, for example, theoretically should produce a voltage of 5.85 [2.8 − (−3.05) = 5.85V]. In point of fact, most commercially available batteries will have voltages much lower than 5.85V. Batteries having voltages approaching 6 volts are usually impractical because of the high reactivity of their components and their associated explosion risks. High voltage batteries found in electronic flash guns and other appliances are manufactured by connecting smaller batteries so that their voltages are additive. This is done by connecting the cathode of one to the anode of another. The practice is familiar to anyone who has ever put two batteries into a flashlight. (Figure 8–2.)

Standard flashlight batteries have a voltage of 1.5 volts; when two are connected end to end so that the cathode of one touches the anode of the other, the electrons are permitted to flow from one to the other, that is, in the same direction. As a result, 3 volts are produced. This voltage is sufficient to cause the bulb to light. If the batteries are placed in the flashlight so that either the anodes or the cathodes are end to end, each battery attempts to make the electrons flow in an opposite direction. (Figure 8–2.) Since each has the same amount of force (1.5V), they cancel each other, no voltage is produced, and the bulb does not light.

The redox half-reactions occurring in a common "dry cell" battery are shown in Figure 8–3.

Anode $$Zn \rightarrow Zn^{++} + 2\,e^-$$

Cathode $$MnO_2 + 2\,H_2O + e^- \rightarrow Mn(OH)_3 + OH^-$$

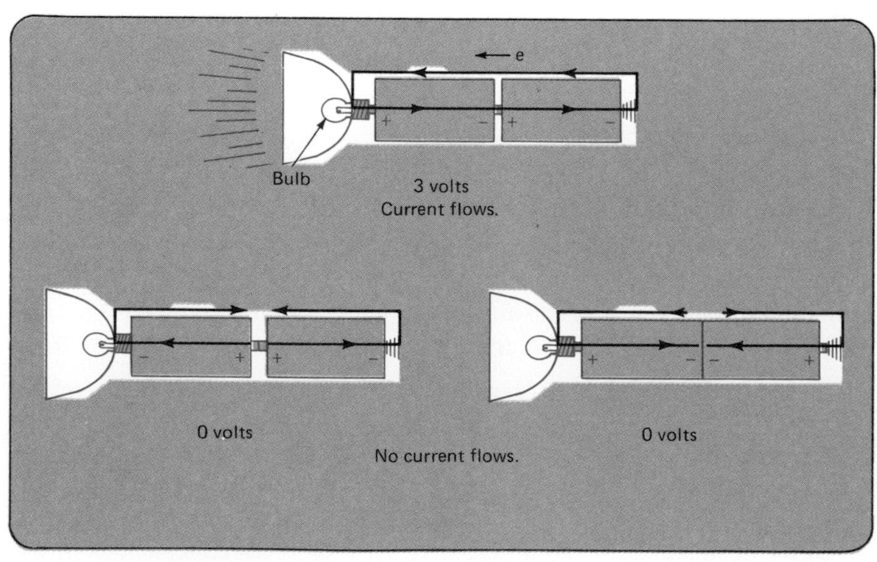

FIGURE

8–2 Electron Flow in a Flashlight

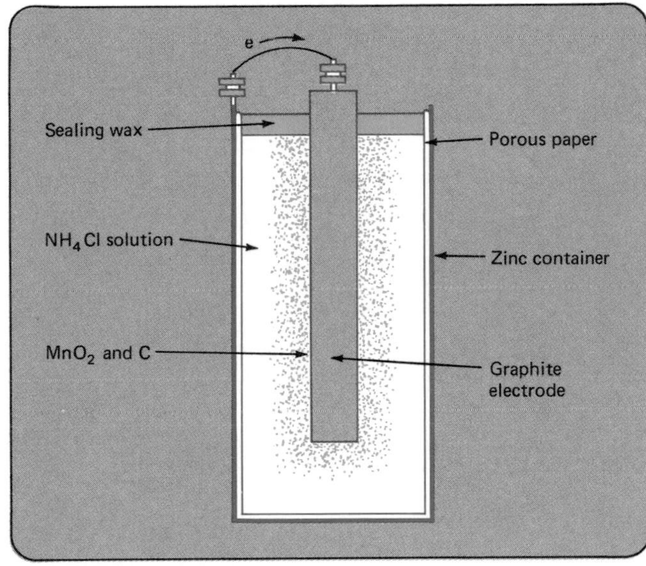

FIGURE 8–3 A Typical "Dry Cell" Battery

The total balanced ionic equation is obtained by multiplying the cathode reaction by 2 and adding.

$$\mathbf{Zn + 2\,MnO_2 + H_2O \rightarrow Zn^{++} + 2\,Mn(OH)_3 + 2\,OH^-}$$

Discharging by the Numbers

Up to this point batteries have been described as mystical, magical marvels capable of utilizing electrons as they are transferred between oxidation and reduction reactions. The mechanics by which they accomplish this feat is described in Figure 8–4 and the accompanying text.

1. Some of the zinc electrode is oxidized.
2. For each zinc atom that is oxidized to a zinc ion ($\mathbf{Zn^{++}}$) and goes into solution two electrons are left on the zinc electrode. This electrode is known as the anode; oxidation takes place here.
3. The mutual repulsion resulting from the excess of electrons at the anode causes them to flow through the wire to the copper electrode in the second chamber.
4. The excess of electrons on the copper electrode causes the copper ions on its surface to be reduced to copper atoms. These atoms are

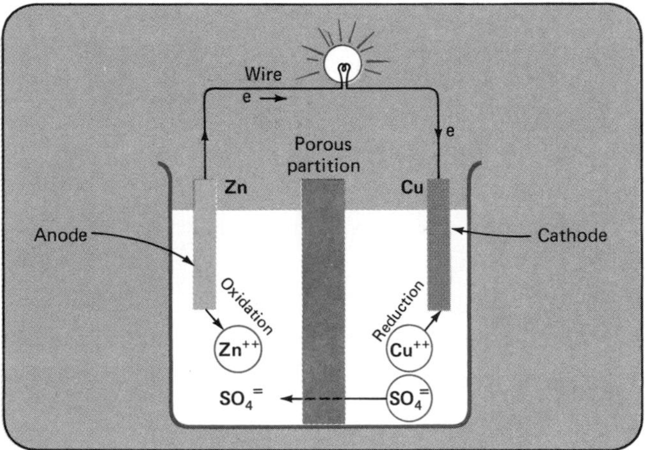

FIGURE

8–4 Electron Transfer in a Battery

deposited upon the electrode, increasing its size. This electrode is referred to as the cathode; reduction takes place at its surface.

5. The reduction of $\mathbf{Cu^{++}}$ to $\mathbf{Cu^{0}}$ has resulted in a concentration of sulfate ions ($\mathbf{SO_4^{=}}$) in the cathode compartment that is in excess of what is necessary to neutralize electrically the remaining cupric ions ($\mathbf{Cu^{++}}$). Similarly, the increase in zinc ion ($\mathbf{Zn^{++}}$) in the anode chamber has destroyed its electrical neutrality. To make both compartments neutral once again it is necessary for the excess sulfate ions in the cathode chamber to diffuse through the porous wall into the anode chamber. The net effect of the flow of sulfate ions coupled with the flow of electrons is to cause a clockwise flow as per Figure 8–4 of negative charge through the battery and its external wire. The battery acts as a pump, pumping charge at a pressure of 1.10V. If the wire is cut and an electrical appliance of proper voltage rating is placed between its ends, useful work can be derived.

Recharging Is Just Like Pedaling Backward.

As constructed, the zinc-copper battery will pump negative charges in a clockwise direction until it is dead. The process is called discharging. Once dead, however, the battery can be recharged. The resurrection is accomplished by reversing the direction of flow of the negative charge. Due to the chemical nature of the electrodes, the reversal of charge flow is counter the normal battery tendencies. These tendencies must, however, be overcome if the battery is to be restored to its original condition. Force must be countered with force and the recharging force must prevail. This is accomplished by making the recharging voltage greater than 1.10V. This is the force exerted upon the electrons by a fresh battery as it drives the electrons in a clockwise direction. To restore the battery to its charged condition the external pumping voltage is applied in a counterclockwise direction.

Consider the situation of Bart, a miller who lives in the desert. Bart drives his mill with an apparatus consisting of a water tank on a tower and a water wheel. His full water tank is equivalent to a fully charged battery; his water wheel is equivalent to an electrical appliance. Whenever the tank spigot on his water tower is opened, water runs out, hits the paddle wheel, and causes it to turn. This continues as long as the tank contains water. When it is empty, the apparatus is analogous to a dead battery. The water has not been lost; it is still present in the pool, but it is no longer capable of turning the wheel. Similarly, the electrical charges in a dead battery have not been lost; they are still present but are no longer capable of powering a device. Just as the water in the tower has a natural tendency to flow down because of gravitational forces, the electrical charges in the battery tend to flow in a definite direction because of electromotive forces (the forces that cause electrons to move). To make the water tower useful once again, it is necessary to expend energy to pump the water from the pool, against the force of gravity, back up to the tank on the tower. Similarly, to recharge a battery it is necessary to pump the negative charges in the reverse direction to return the battery to its original condition. The height to which the water must be pumped in the water tower is equivalent to the voltage that must be overcome to recharge the battery. Figure 8–5 illustrates what happens when a battery recharges.

The recharger transfers electrons to the zinc cathode at a voltage greater than 1.10 volts. This means that the recharger pushes electrons onto the zinc electrode with a force greater than the zinc electrode can push them into the connecting wire. The net effect of all this pushing is to produce an excess of electrons on the zinc electrode. This results in the reduction to metallic zinc of some of the zinc ions in solution. Once reduced, the zinc is deposited on the zinc electrode.

The recharger performs the exact opposite function at the copper anode. Here it pulls off electrons with a force greater than 1.10V. Since

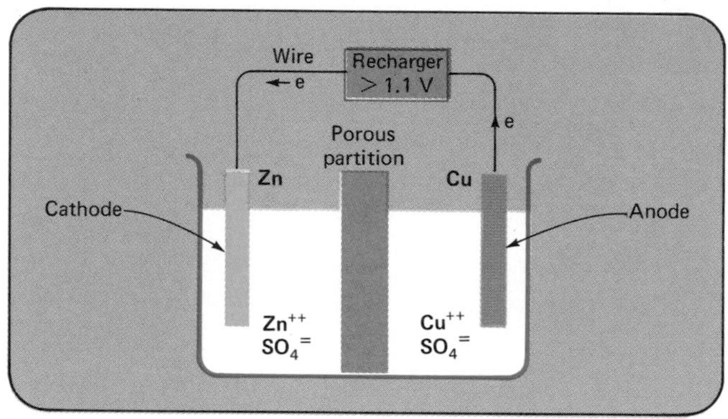

FIGURE 8–5 Recharging a Battery

1.10V is a greater force than the copper metal has to hold on to electrons, some of the copper is oxidized to Cu^{++} ions and goes into solution. While discharging, the zinc electrode functioned as the anode and the copper electrode acted as the cathode. During recharging these functions are reversed. The new electrode functions result in the current reversal necessary for recharging the battery. The counterclockwise flow of charge is made complete by the diffusion of the sulfate ions from the cathode chamber through the porous partition into the anode compartment, where they neutralize the charge of the newly formed Cu^{++} ions.

A recharged battery can once again function as a pump for negative charges. If it is large enough to pump one mole of electrons, sufficient energy will be expended to light a 100 watt bulb for twenty-seven hours.

Forks and Spoons Make Great Cathodes.

A unique and wonderful property of cathodes is that they can reduce metallic ions into bright, shiny metal. When the copper-zinc battery discharged, the metal was copper; when it recharged, the metal was zinc. This property, the ability to create a surface deposition of a layer of metal, has major commercial applications in the process called electroplating.

Gold, silver, and platinum, because of their pleasant appearance, resistance to corrosion, and high electrical conductivity, are metals in great demand. Limited supplies, however, make the costs of these

BARTHOLOMEW, THIS JOB WOULD BE MUCH EASIER IF WE DIDN'T HAVE TO PUMP THE WATER BACK INTO THE TOWER.

precious metals significantly higher than the costs of base metals. If price were no object, perhaps everyone would use a silver or gold or platinum fork. It would impart no undesirable taste to foods and would be appreciably more attractive than a fork made of a base metal. Since the food and the tongue—as well as the eye—come in contact only with the surface of a fork, one uniformly coated with a thin layer of a precious metal will have the same desirable features as a pure metal fork, and at a much lower cost.

Such a coating can be deposited by electroplating an object with the desired metal—say, a fork with silver. The process is simple; the fork is made a cathode in an electroplating cell where the anode is constructed of silver. (Figure 8–6.) A salt solution of silver surrounds both electrodes. The construction of the electroplating cell is similar to the copper-zinc battery, although it does not have a porous wall.

The process is similar to recharging a battery:

1. The current source removes the electrons from the silver anode and pumps them through the wire to the fork.
2. The excess electrons on the fork cause the silver ions in contact with its surface to be deposited as silver metal. The greater the number of electrons pumped onto the fork, the more silver metal will be deposited.
3. At the same time the silver is deposited at the cathode, the removal of electrons at the anode causes an equivalent amount of silver to be oxidized and enter the solution as silver ion. In so doing the newly formed silver ions maintain the chemical concentration and electrical neutrality of the solution.

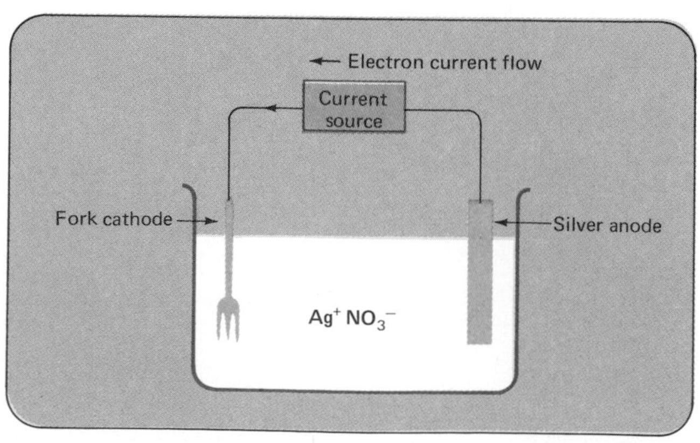

FIGURE 8–6 Cell for Electroplating Silver

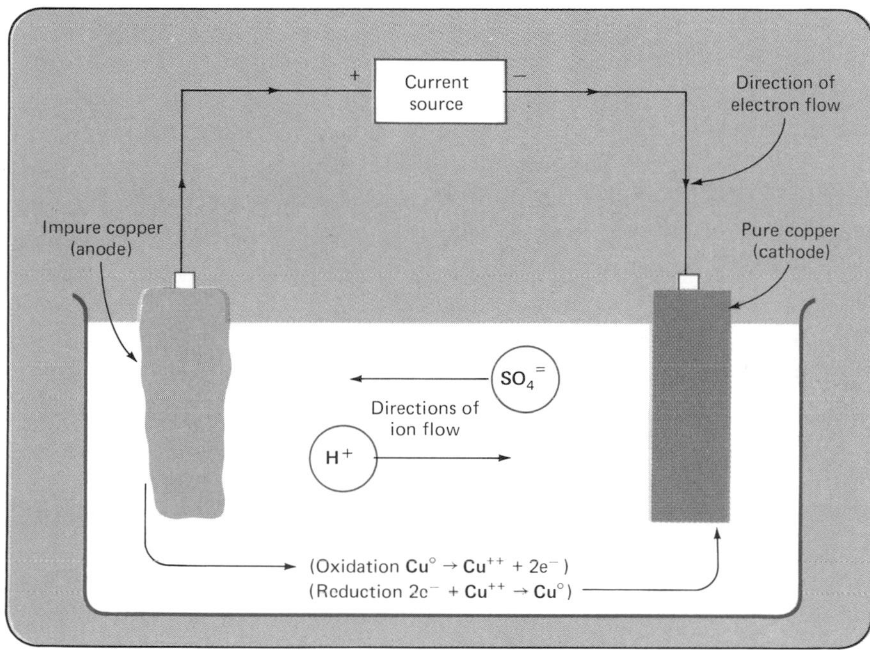

FIGURE 8–7 **Process for Refining Copper**

The only thing that changes in this process is that the anode gets lighter and the cathode gets heavier by the same amount of silver. If the electron flow is kept constant, then the longer the process is allowed to continue, the thicker the coating will be. The cost of silver plating flatware seldom exceeds several cents per utensil. The cost of an equivalent piece in solid silver would be several dollars.

Electroplating is not restricted to ornamental uses. In the burgeoning field of electronics it is often necessary to produce metals in the ultra pure state. Electroplating is a convenient and commercially feasible process by which this can be accomplished. Copper, when refined from its sulfide ore, is nominally 99 percent pure. The presence of trace impurities such as silver, gold, arsenic, antimony, nickel, zinc, and iron, however, often make it unsuitable for particular electronic applications. This copper is further purified by making it the anode in an electrolytic cell containing a copper sulfate–sulfuric acid electrolyte (solution that conducts electrical flow). The cathode is usually a sheet or rod of high purity copper. The process is similar to silver electroplating. The impure copper is oxidized off the anode and is redeposited electrolytically pure at the cathode. Metals such as gold and silver are not so easily oxidized as copper and therefore drop off and fall to the bottom of the solution as the copper is oxidized away.

Other impurities such as zinc and nickel are more easily oxidized than copper. They enter the electrolytic solution when the copper is oxidized, but they are prevented by conditions of voltage and concentration from depositing on the pure copper cathode. (Figure 8–7.)

Perhaps the most famous electrolytic refining process is the one developed by Charles M. Hall for the extraction of aluminum from its ore. Aluminum is the most abundant metal occurring in the earth's crust. In spite of this abundance, however, it did not become a metal of commerce until the early 1900s; until then, extraction of the metal from its ore involved great difficulty and expense. In 1852 the price of aluminum was 545 dollars per pound. By 1941 the price had plummeted to 15 cents per pound. The reason for this drastic price reduction was the invention in 1886 of an economically advantageous extraction process. At the time of his invention, Hall was a twenty-two-year-old student at Oberlin College; his laboratory was a woodshed, and his equipment was either handmade or borrowed.

Hall's process consisted of dissolving aluminum oxide in a molten salt (cryolite) in an electrolytic cell and then reducing it to aluminum metal at the cathode. In solution the aluminum oxide breaks up into Al^{+++} and $O^{=}$ ions. As the current passes through the cell, the aluminum ions migrate toward the cathode, where they are reduced to metallic aluminum.

$$2\,Al^{+++} + 6\,e^{-} \rightarrow 2\,Al$$

The reduction is conducted at a temperature greater than the melting point of aluminum. As a result, when the aluminum metal is formed,

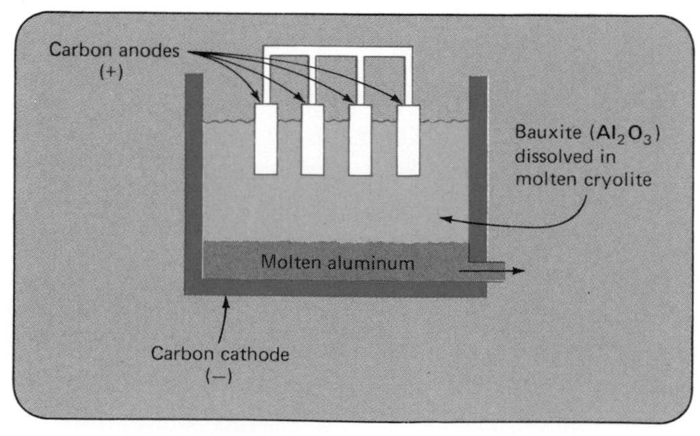

FIGURE

8–8 Electrolytic Cell for Producing Aluminum

it is a liquid. This sinks to the bottom of the cell and is drawn off at appropriate times. (Figure 8–8.)

The second half of the redox reaction is the oxidation of the oxide ion.

$$3\,\mathbf{O}^{=} \rightarrow 3\,\mathbf{O} + 6\,e^{-}$$

The oxidation occurs at the carbon anode and results in its consumption and the ultimate production of $\mathbf{CO_2}$.

$$\mathbf{C} + 2\,\mathbf{O} \rightarrow \mathbf{CO_2}$$

A strange sidelight to this story is that three months after Hall made his find, a second chemist, Paul Heroult, also twenty-two years old, working in France and unaware of Hall's existence, made the same discovery. *C'est la vie.*

REVIEW

1. A battery is a device capable of pumping negative charges from a source to a destination.
2. The electron source is usually an oxidation reaction and its destination a reduction reaction.
3. The driving force for the charge pump is the chemical nature of the electrodes.
4. The designation of reactions as to source and destination is determined by the half-cell voltage of the reaction.
5. If both reactions are written as reductions, the one with the greatest positive half-cell voltage will be the destination.
6. The half-cell voltage of a reduction reaction is determined by reversing the oxidation reaction of interest and changing the numerical sign of the voltage.
7. The voltage of a battery is equal to the sum of the voltage of its two half-cells.
8. In order for a battery to be operative, the sum of its half-cell voltages must be positive.
9. A battery will pump negative charge as long as sufficient reactants exist to maintain a voltage.
10. As the reactants are consumed, the voltage drops and eventually the battery goes dead.
11. A dead battery can be recharged by connecting an external current source to the battery so that it opposes the normal direction of flow of negative charges.

12. In order for recharging to occur, the voltage at which the current is supplied must exceed the normal battery voltage.
13. During the recharging process, the two electrodes interchange their designation. The cathode becomes the anode and vice versa. At all times, however, oxidation takes place at the anode and reduction at the cathode.
14. In a battery the negative ions in solution flow so as to complement the electron flow in the wire. If the electron flow is clockwise, so too will be the flow of negative ions.
15. An object can be electroplated with a particular metal, if it is made the cathode in an electrolytic cell containing an anode and a salt solution of the metal of choice.
16. Impure metals can be refined by constructing an electrolytic cell in which the impure metal is the anode and the pure metal the cathode.

REFLEXIVE QUIZ

1. When a wire is connected across the terminals of a battery the current flows from the ______ to the cathode.
2. The oxidation reaction in a battery takes place in the ______ compartment.
3. If the half cell reaction and voltage of a reduction is known, the voltage of the reverse oxidation can be found by changing the ______ of the number.
4. In order to construct a workable battery, the sum of the oxidation and reduction reaction voltages must be ______.
5.

$$\mathbf{Mg}^{++} + 2\,e^- \rightleftarrows \mathbf{Mg} \quad (-2.37\text{V})$$

$$\mathbf{Ca}^{++} + 2\,e \rightleftarrows \mathbf{Ca} \quad (-2.87\text{V})$$

   a. The cathode of a battery constructed using magnesium and calcium electrodes should be made of ______ metal.
   b. The voltage of this battery will be ______ volts.
6. The purpose of the porous partition in a battery is to allow the negative ions in the cathode chamber to flow into the anode compartment to neutralize the ______ of the newly formed ions.
7. In order to recharge a battery, the recharging voltage must be ______ than the normal battery voltage.
8. In an electroplating cell the more current that flows, the ______ will be the amount of metal deposited.

9. If a metal is to be purified by electrodeposition, it should be made the ________ in the electrolytic cell.

Answers: (1) Anode [108–113]. (2) Anode [108–113]. (3) Sign [108–111, 123]. (4) Positive [111–113]. (5a) Magnesium, (5b) 0.5 [112–113, 123]. (6) Charges [114–118]. (7) Greater [116–118, 124]. (8) Greater [118–123]. (9) Anode [121–124].

SUPPLEMENTARY READINGS

1. E. H. Lyons, Jr. *Introduction to Electrochemistry.* Boston: D. C. Heath, 1967.
2. C. R. Dillard and P. H. Kammeyer. "An Experiment with Galvanic Cells." *Journal of Chemical Education* 40 (1963): 363.
3. A. R. Denaro. *Elementary Electrochemistry.* Washington, D.C.: Butterworth & Co., Ltd., 1965.
4. W. J. Moore. *Physical Chemistry*, 3rd ed. Englewood Cliffs, N.J.: Prentice-Hall, Inc., 1963.
5. D. A. MacInnes. *The Principles of Electrochemistry.* New York: Dover Publications, Inc., 1961.

Nuclear Chemistry

Section 9

Nuclear Chemistry

Tourist's Guide to Nuclear Points of Interest

1. Nuclei are composed of neutrons and protons. Their masses and stability depend on the number of these constituent components. They exist in a wide variety of masses from 1 to more than 260 AMU. In the range from 1 to 83, all masses with the exception of 5 and 8 have been observed. A particular mass does not define a specific element, however, but rather a group of elements called *isobars*. Each isobar has the same total number of neutrons and protons.

Isotopes, on the other hand, are atoms having the same number of protons but a different number of neutrons. All the isotopes of a given element have similar *chemical* characteristics, because they have the same number of protons, and electrons. All elements have more than one isotopic form, only some of which occur in nature; the atomic mass that is listed for an element in the periodic chart is really an average of the masses of its naturally occurring isotopes.

The elements barium, lanthanum, and cerium, for example, all possess isotopes having an atomic mass of 138; these particular isotopes are all isobars of mass 138. The *total* number of neutrons and protons in each atom is the same, 138. The isotope of barium contains 56 protons and 82 neutrons, the isotope of lanthanum contains 57 protons and 81 neutrons, and the isotope of cerium contains 58 protons and 80 neutrons. Because each of these atoms contains a different number of protons, each is an isotope of a distinctly different element. The total number of protons and neutrons in each of these isobars equals 138. Barium atoms contain 56 protons and 82 neutrons; lanthanum atoms contain 57 protons and 81 neutrons; and cerium atoms contain 58 protons and 80 neutrons. This is denoted by writing the isobars in the following manner: ${}^{138}_{56}\mathbf{Ba}$, ${}^{138}_{57}\mathbf{La}$, and ${}^{138}_{58}\mathbf{Ce}$. The upper number indicates the atomic mass; the lower, the charge caused by the number of protons it contains.

2. With the exception of hydrogen, no common nucleus contains fewer neutrons than protons. With the exception of ${}^{1}_{1}\mathbf{H}$ and ${}^{3}_{2}\mathbf{He}$, the mass of every stable nucleus is equal to or greater than twice its charge. As a rule, in most stable common isobars the ratio of neutrons to protons seldom exceeds 1.2:

$$\frac{\text{Number of neutrons}}{\text{Number of protons}} \leqq 1.2$$

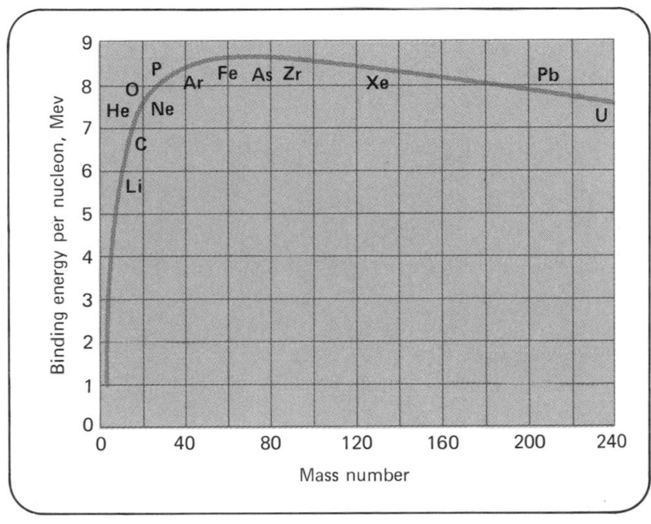

FIGURE 9–1 **Binding Energy per Nucleon as a Function of Atomic Mass**

3. Elements that have an even number of protons in their nuclei generally possess more isobars and are more abundant in nature than those with an odd number. The most stable nuclei usually have an even mass number composed of an even number of protons and an even number of neutrons. Elements having the highest nuclear stability have masses close to that of iron, element 26.

Figure 9–1 illustrates the stabilities of nuclei as a function of their atomic masses. The higher the energy with which each nucleon (proton or neutron) is bound in a nucleus (the binding energy per nucleon), the more stable the nucleus will be. When nuclei are unstable, they decompose radioactively until a stable nucleus is produced. All nuclei having charges greater than bismuth, element 83, and some of lesser atomic weights are unstable. Nuclear instabilities can be correlated with the tendencies of the nuclei to readjust to more favorable neutron-to-proton ratios.

Cures for Unfavorable Neutron-to-Proton Ratios

When afflicted by excessive neutrons, nuclei may relieve their discomfort by either beta (β) or neutron (${}^{0}_{1}n$) emissions.

Beta emission consists essentially of the expulsion from the nucleus of a high speed electron. Since we have just noted that the nucleus

contains only protons and neutrons, this statement may seem strange indeed. The enigma can be explained, however, if the electrons are considered to have arisen from the breakup of a neutron to form a proton and an electron. This simplistic explanation can be depicted by the following reaction:

$$\begin{array}{ccccc} \text{Neutron}^0 & & \text{Proton}^+ & & \text{Electron}^- \\ {}^1_0\text{n} & \rightarrow & {}^1_1\text{p} & + & {}^{\ 0}_{-1}\beta \end{array}$$

Note. In a balanced nuclear reaction, the sum of the subscripts on one side of the equation must equal the sum of the subscripts on the other. The same is true of the superscripts.

The emission of a beta ray from a nucleus has a double-barreled effect on the neutron-to-proton ratio. The ratio decreases because of both the increase in protons and the decrease in neutrons.

Carbon 14 is an example of a nucleus that gains stability by beta emissions.

$$^{14}_{6}\mathbf{C} \rightarrow {}^{14}_{7}\mathbf{N} + {}^{0}_{-1}\beta$$

Note. The superscript masses and the subscript charges on both sides of the equation are equal ($14 = 14 + 0$; $6 = 7 - 1$). The reaction is thus a balanced nuclear equation.

The transformation of a carbon neutron into a proton has converted the carbon to nitrogen. The beta emission has changed the starting material, a black solid, into a colorless, odorless, tasteless gas.

This is the first time that this type of isolationist nuclear reaction has been discussed. Prior to this, the only reactions with which we have dealt have involved the reorganization of groups of atoms or electrons between atoms. Here we are dealing with a change in the number of neutrons and protons within a single atomic nucleus. In the case of $^{14}_{6}\mathbf{C}$, this is a fairly slow process, taking 5,760 years to convert half of the carbon into nitrogen. Because of this property of $^{14}_{6}\mathbf{C}$, this reaction is useful in determining the age of archeological materials. More will be said about this in subsequent sections.

The second way that an ailing nucleus can relieve an abnormally high neutron-to-proton ratio is neutron emission. Such emissions are relatively difficult to detect and are far less common than beta emissions. $^{9}_{3}\mathbf{Li}$, in addition to decomposing by beta emission, can also decompose by neutron emission. Neutron emission converts it to a lighter isotope of lithium whereas beta emission converts it to beryllium.

$$^{9}_{3}\mathbf{Li} \rightarrow {}^{8}_{3}\mathbf{Li} + {}^{1}_{0}\mathrm{n} \quad \text{(Neutron emission)}$$

$$^{9}_{3}\mathbf{Li} \rightarrow {}^{9}_{4}\mathbf{Be} + {}^{0}_{-1}\beta \quad \text{(Beta emission)}$$

Neutron-deficient nuclei can theoretically relieve their empty feelings by (1) positron emission, (2) orbital electron capture, or (3) proton emission.

1. Positrons ($^{0}_{1}\beta$) are high-speed particles having a charge of +1 and a mass equivalent to a beta particle. They may be regarded as arising from the transformation of a proton to a neutron.

$$^{1}_{1}\mathrm{p} \rightarrow {}^{1}_{0}\mathrm{n} + {}^{0}_{1}\beta \quad \text{(Positron emission)}$$

The emission of a positron results in the lowering by 1 of the nuclear charge and the atomic number. Because of its negligible mass, however, the emission of a positron has no effect on the mass number of the nucleus. Positron emission has been observed only from man-made radioisotopes.

$$^{19}_{10}\mathbf{Ne} \rightarrow {}^{19}_{9}\mathbf{F} + {}^{0}_{1}\beta$$

It should be noted at this point that the nuclear conversion reactions used to explain both $_{-1}^{0}\beta$ and $_{1}^{0}\beta$ emission are merely convenient pedagogical tools. They should not be interpreted as the conversion of neutrons and protons existing individually within the nucleus.

2. Instead of emitting a positron, a nucleus may accomplish exactly the same end by capturing one of the orbital electrons. In so doing, it would effectively convert a proton into a neutron. The process is more common in artificially produced isotopes than in natural ones. It is called "*K* capture" because the electrons that are captured originally were in the first (*K*) quantum shell.

$$^{55}_{26}\mathbf{Fe} + e^{-} \rightarrow {}^{55}_{25}\mathbf{Mn}$$

Here again, just as in the case of positron emissions, the mass number remains unchanged but the atom is converted into a new element.

3. Although a low neutron-to-proton ratio could be cured by proton emissions, such a process is highly unlikely and can for all practical purposes be disregarded.

The Alpha Reducing Plan for Overweight Nuclei

As we noted earlier, no isotope having a mass greater than $^{209}_{83}\mathbf{Bi}$ is stable. All isotopes with greater masses decompose to produce lighter nuclei. The most popular way to unload this excess mass is through *alpha emission.* An alpha (α) particle consists of two protons and two neutrons. It has a charge of +2 and a mass of 4 AMU. It is equivalent to a helium atom that has been stripped of its electrons. When a nucleus undergoes alpha emission, the atom is converted to a new element having an atomic number two less and an atomic mass four less than the original isotope.

$$^{226}_{88}\mathbf{Ra} \rightarrow {}^{222}_{86}\mathbf{Rn} + {}^{4}_{+2}\alpha$$

Alpha emissions are common only from the heaviest nuclei. This is because few smaller nuclei will possess sufficient energy to expel a particle of this magnitude.

When You're Hot, You're Hot.

When a nucleus contains energy in excess of that necessary to provide optimum stability, it dissipates this energy by emitting gamma (γ) rays. Gamma rays are electromagnetic radiation similar to light but of a higher energy and penetrating power. They are usually emitted in conjunction with either alpha or beta radiation and most probably are the result of a nuclear attempt to readjust to the loss of these particles. Since they have neither mass nor charge their loss does not result in an isotopic change.

$$^{54}_{25}\text{Mn}^* \rightarrow {}^{54}_{25}\text{Mn} + \gamma$$

Note. The asterisk (*) indicates an excited nucleus.

Given Their Half-Lives to Live Over . . .

A shipwrecked sailor was stranded with only one canteen of fresh water. Having no idea when he might be found, he decided to ration his water supply carefully. After giving the problem much thought, he devised a system that allowed him to drink some water each day and still be assured that no matter how long it took until he was rescued, he would still have some left. The system was quite simple. Each day he would drink half his remaining water. The first day, he drank half a container. The second day, he drank half the remaining half, and so on. By the end of the sixth day, when he was rescued, he had $\frac{1}{64}$ ($\frac{1}{2} \times \frac{1}{2} \times \frac{1}{2} \times \frac{1}{2} \times \frac{1}{2} \times \frac{1}{2} = \frac{1}{64}$) of the original amount of water in his canteen. No matter how long it might have been before he was rescued, the sailor would have never run out of water. One half of something (the previous day's water supply), no matter how small, is still something.

Radioactive isotopes disappear in much the same way as the sailor's water. Rather than decomposing all at once or at a fixed rate, they decompose fractionally for each interval of passing time. The amount of time it takes for half of any isotope to decompose is called its half-life ($T_{\frac{1}{2}}$). This means that whether the starting amount was ten ounces or ten tons, when one half-life had elapsed, half the amount would be left; five ounces in the first case, five tons in the second. Half-lives of radioisotopes may vary from extremes of 10^{-15} seconds to 10^{15} years. (See Appendix I for exponential notation.) The shortness of the former and the eternity of the latter are both sufficiently great to

boggle the mind. Most half-lives, however, are not so extreme. They usually are between several seconds and several years.

If the product of a decomposition is itself not stable, it too will undergo decomposition in an attempt to achieve stability. In any series of decomposition, the starting material is referred to as the parent, and the resulting products, the daughters. The decomposition of bismuth 210 to a stable isotope of lead can be used to illustrate such a series.

$$^{210}_{83}\mathbf{Bi} \xrightarrow[{}^{0}_{-1}\beta]{T_{\frac{1}{2}} = 5 \text{ days}} {}^{210}_{84}\mathbf{Po} \xrightarrow[{}^{4}_{2}\alpha]{T_{\frac{1}{2}} = 138 \text{ days}} {}^{206}_{82}\mathbf{Pb}$$

Bismuth 210 decomposes by beta emission to polonium 210. Since the mass of a beta particle is equal to that of an electron and negligible by comparison to either a neutron or proton, the mass of the daughter remains essentially the same as the mass of the parent. The loss of a negative charge from the nucleus, however, increases the number of protons and the atomic number by 1. The new daughter, as a consequence of this, is a polonium atom rather than a bismuth atom.

The half-life of this process is five days. If one started with pure $^{210}_{83}\mathbf{Bi}$, after five days he would have a mixture of one-half $^{210}_{83}\mathbf{Bi}$ and one-half $^{210}_{84}\mathbf{Po}$. At the end of ten days, the mixture would be one-quarter $^{210}_{83}\mathbf{Bi}$ and three-quarters $^{210}_{84}\mathbf{Po}$. After fifty days, less than 1/1,000 of the original amount of $^{210}_{83}\mathbf{Bi}$ would be left.

It should not be forgotten, however, that polonium 210 is itself unstable and decomposes with a half-life of 183 days to lead 206. As a result of this, the amount of polonium present is always less than the amount of bismuth that disappeared. Because the polonium 210 has a longer half-life than bismuth 210, it will decompose at a much slower rate. When it does decompose, however, the amount of lead 206 produced will exactly equal the amount of polonium that disappeared. This is because lead 206 is a nonradioactive stable isotope.

As the daughter of an alpha decomposition, lead 206 has an atomic number 2, and an atomic weight 4, less than the parent polonium 210. With the passage of time, the amount of bismuth decreases, the amount of polonium first increases, and then decreases; and the amount of lead constantly increases. This sequence of events can be visualized in terms of Figure 9–2, a graph of the decomposition of two radioisotopes where the parent half-life is half as long as the daughter's. A_0 is the initial amount of parent. Curves A, B, and C could correspond respectively to the amounts of bismuth, polonium, and lead in this discussion.

The effect of time upon the storage of a radioisotope must always be taken into account. It is of particular importance when the isotope is short-lived. A bottle containing a salt of silver 111, for instance, if allowed to stand for any length of time, will be converted to a salt of cadmium 111.

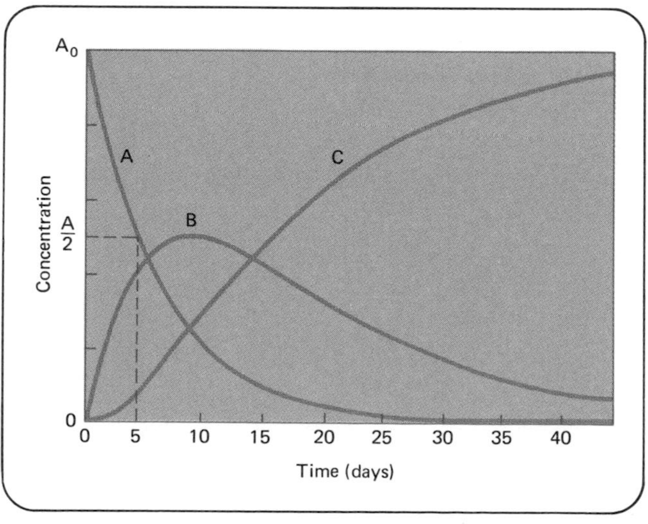

FIGURE 9–2 Concentration-Time Curves for Substances A, B, and C in a Series of Radioactive Decomposition Reactions

Example. $^{111}_{47}\mathbf{Ag}$ decomposes by beta emission to $^{111}_{48}\mathbf{Cd}$. Write a balanced equation for this decomposition. If the half-life of $^{111}_{47}\mathbf{Ag}$ is 7.6 days, what fraction of the original amount of silver will be left after 38 days?

$$^{111}_{47}\mathbf{Ag} \xrightarrow{T_{\frac{1}{2}} = 7.6 \text{ days}} {}^{111}_{48}\mathbf{Cd} + {}^{0}_{-1}\beta$$

To determine the amount of silver remaining after 38 days, it is necessary to establish how many half-lives have elapsed:

$$\frac{38 \text{ days}}{7.6 \text{ days}} = 5 \text{ half-lives}$$

The amount of silver remaining after five half-lives have elapsed will be

$$\begin{matrix} (1) & (2) & (3) & (4) & (5) \\ \tfrac{1}{2} \times & \tfrac{1}{2} \times & \tfrac{1}{2} \times & \tfrac{1}{2} \times & \tfrac{1}{2} = \tfrac{1}{32} \end{matrix}$$

of the starting amount. The amount of cadmium produced will be 31/32 of the maximum amount. As time progresses, this amount will increase. It should be realized, however, that the largest amount was produced during the first half-life. Each successive half-life produced only half as much. For any finite time interval, the maximum amount will never be attained.

Tut, Tut, He's Not That Old.

Radioactive carbon 14 ($^{14}_{6}\mathbf{C}$) has been produced in the atmosphere by the action of cosmic ray neutrons on nitrogen 14 ($^{14}_{7}\mathbf{N}$) since the beginning of the solar system.

$$^{14}_{7}\mathbf{N} + {}^{1}_{0}\mathrm{n} \rightarrow {}^{14}_{6}\mathbf{C} + {}^{1}_{1}\mathbf{H}$$

The carbon produced by this transmutation (change of one element to another) combines with atmospheric oxygen to form radioactive carbon dioxide.

$$^{14}_{6}\mathbf{C} + \mathbf{O}_2 \rightarrow {}^{14}_{6}\mathbf{CO}_2$$

This carbon dioxide mixes with existing nonradioactive carbon dioxide ($^{12}_{6}\mathbf{CO}_2$) to become part of the pool from which all living plants, and ultimately animals, draw for their food supply.

Radiocarbon dating is dependent upon two premises:

1. The amount of radioactive carbon dioxide existing presently in the atmosphere does not differ appreciably from what existed 35,000 years ago.
2. The ratio of radioactive to nonradioactive carbon contained in all living things is equal to the ratio found in the atmosphere.

$$\frac{{}^{14}_{6}\mathbf{C}\text{ (living things)}}{{}^{12}_{6}\mathbf{C}\text{ (living things)}} = \frac{{}^{14}_{6}\mathbf{CO}_2\text{ (atmosphere)}}{{}^{12}_{6}\mathbf{CO}_2\text{ (atmosphere)}}$$

These are not unreasonable assumptions. Plants utilize atmospheric carbon dioxide to manufacture carbohydrates by the process of photosynthesis. Animals acquire this ratio by eating the plants. With regard to the production of atmospheric carbon 14, there is no reason, aside from the current atomic age, to believe that any cataclysmic event occurred within the past 35,000 years that could have altered the rate of production of carbon 14.

Carbon 14 is a beta emitter with a half-life of 5,760 years.

$$ {}^{14}_{6}\mathbf{C} \xrightarrow{T_{\frac{1}{2}} = 5{,}760\text{ yrs.}} {}^{14}_{7}\mathbf{N} + {}^{0}_{-1}\beta $$

Disintegrations of the ${}^{14}_{6}\mathbf{C}$ in the atmospheric carbon dioxide pool will produce 15 beta emissions per gram of carbon per minute. All things that are currently alive have this same level of radioactivity. When organisms die, however, their intake of food and associated carbon 14 ceases. As a result of this, the maximum amount of carbon 14 an organism, either animal or vegetable, can ever contain will be the amount present at the time it dies. After that time, whatever carbon 14 remains in an animal skeleton or in the structural fabric of vegetation will decompose to nitrogen 14. If a piece of wood or an animal skin is exhumed and found to emit 15 disintegrations in 2 minutes ($7\frac{1}{2}$ disintegrations per minute), it can be assumed that the source from which it was derived has been dead one carbon 14 half-life (5,760 years). On the basis of this type of measurement, the cypress beams used to construct the tomb of Egyptian Pharaoh Sneferu have been estimated to be 4,597 years old, and the age of the Dead Sea Scrolls is estimated at 1,917 years, give or take a couple of hundred.

Radiocarbon dating is useful only for objects whose maximum age does not exceeed 35,000 years. This time interval corresponds to approximately six half-lives. A sample of organic matter this old should display an activity of

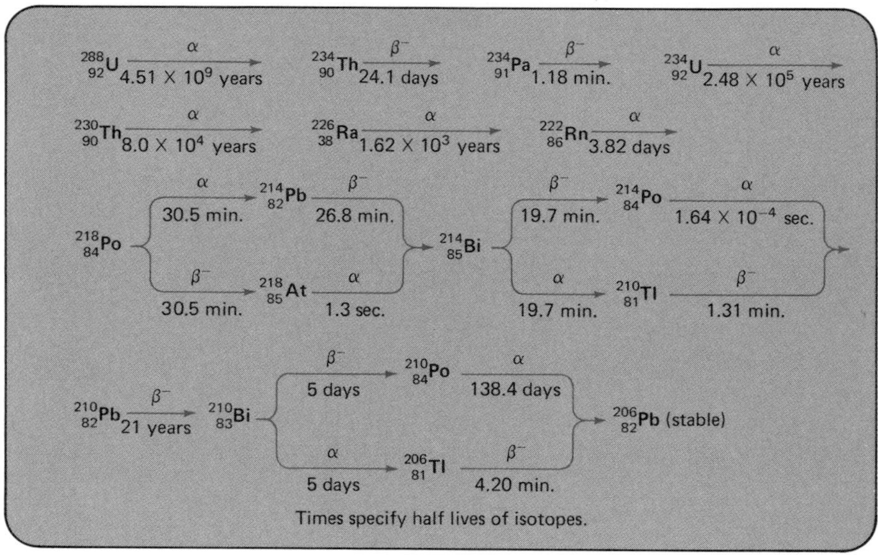

FIGURE 9–3 Disintegration Series of $^{238}_{92}U$

$$\frac{15 \text{ Disintegrations/minute}}{2 \times 2 \times 2 \times 2 \times 2 \times 2} \approx \tfrac{1}{4} \text{ Disintegration/minute}$$

approximately one disintegration every four minutes per gram of carbon. Since a great number of samples used for dating purposes contain less than a thousandth of a gram or less of carbon, the precision involved in determining the radioactivity of a sample that may decompose only once every four thousand minutes makes this task less than feasible.

Radiocarbon dating cannot be used for determining the age of most rocks. This is because most rocks are older than 35,000 years and few contain appreciable amounts of carbon. The dating of rocks is normally performed by measuring the concentration of their component long-lived radioisotopes. $^{238}_{92}U$, $^{235}_{92}U$, $^{232}_{92}U$, and $^{40}_{19}K$ are particularly useful for this purpose. The half-lives of these isotopes are all 10^9 years or more.

The disintegration of uranium is a chemical classic example to illustrate how this dating procedure operates. When a uranium atom disintegrates, it goes through a series of decays until it produces a stable isotope of lead. (Figure 9–3.) Eight alpha particles are liberated as a result of this process. They react with the surrounding rock and produce helium atoms.

$$^{4}_{2}\alpha + 2\,e^{-} \rightarrow {}^{4}_{2}\mathbf{He}$$

In favorable cases, when the rock is very fine-grained and the helium pressure is low, the helium remains trapped within the rock and serves as an indicator of the amount of uranium that has decomposed. Using this system, the ages of rocks as old as 3×10^{9} years have been estimated. By a similar procedure, the age of the earth has been estimated at 5×10^{9} years.

The Taming of the Tiger

Working with radiation is like having a tiger by the tail. Extreme caution must be exercised at all times and one is seldom given the opportunity of making a serious mistake twice. The work, however, is not without its rewards. In recent years significant benefits have accrued in the fields of medicine, agriculture, and chemistry that could have been made possible only through the use of radiation and radioisotopes.

Tiger Medicine

In medicine, radioisotopes are useful as both diagnostic and curative tools. For diagnostic purposes, technetium 99m, an excited form of technetium; iodine 131; and tritium currently have wide application.

Technetium is an artificially produced element. All of its known isotopes are radioactive. One excited form $^{99m}_{43}\mathbf{Tc}$ is a gamma emitter, decomposing with a half-life of 6.1 hours.

$$^{99m}_{43}\mathbf{Tc} \xrightarrow{T_{\frac{1}{2}} = 6.1 \text{ hours}} {}^{99}_{43}\mathbf{Tc} + \gamma$$

In the form of sodium pertechnetate ($\mathbf{Na^{99m}TcO_4}$) it can be used as a diagnostic probe for determining the health of cells in the brain and other organs. When used to diagnose brain anomalies, it is injected intravenously as a component of a physiologically compatible salt solution. Once in the body, it is incorporated into the cells in varying amounts. The amount absorbed by a tumorous cell will differ from that absorbed by a healthy one. By scanning the area of the brain for differences in gamma emission, the presence, size, and location of tumors and other pathological conditions can be determined.

Iodine 131 is a beta and gamma emitter decaying to xenon 131 with a half-life of eight days. Its two main diagnostic applications are the evaluation of conditions of the heart and the thyroid gland. By using a scintillation counter (an instrument that determines level of radioactivity) placed on the patient's chest and intravenously injecting serum albumin containing iodine 131, the rate of flow of blood through the heart can be monitored. Since the cause of most "heart attacks" is deterioration of the heart muscle as a result of an insufficient blood supply, knowledge of the rate of flow of blood through the heart is useful in predicting the possibility of future "heart attacks." The data collected by monitoring iodine 131 are combined with other information concerning the patient's normal blood volume, heart volume, and flow rate to arrive at a prognostic index. When this index is greater than 7, the chances are better than even that the patient will have a "heart attack" within the following six months. If the index is less than 7, the chances are only one in ten.

Because the thyroid gland is the only significant user of iodine in the body, iodine 131 can be employed both as a diagnostic aid and as a treatment for specific clinical thyroid conditions. The patient is given a drink of water containing a small quantity of $Na^{131}I$. If after several hours it is determined that the thyroid has taken up significantly more or less than 12 percent of the iodine administered, a malfunctioning thyroid is indicated. Should the diagnosis indicate a malignancy, larger amounts of iodine 131 can be administered to cause the selective destruction of malignant tissue. This treatment is useful even after the cancerous cells have broken away from the thyroid and have distributed themselves in other parts of the body.

Tritium ($^{3}_{1}H$), a beta emitter, is a useful tool in medical research. Water containing tritium instead of normal hydrogen ($^{1}_{1}H$) can be used to monitor the fluid functions. Its relatively long half-life (12.4 years), however, restricts its applications in human medicine.

Besides those mentioned, a number of other radioisotopes have been employed in medicine to destroy cancerous growths. Notable among these are radium 226 and cobalt 60. It must never be forgotten, however, that radiation capable of destroying unwanted tissue is also capable of destroying healthy tissue. Even a tame tiger is dangerous.

Tiger Crops

In India, the castor bean has long been raised as a commercial crop, valuable for its high oil content. After the beans are harvested, their hulls are removed, the seeds are pressed, and oil is extracted. The oil prepared from the first pressing is normally used for medicinal purposes. The oil from subsequent pressings has a wide variety of uses in both the paint and the plastics industries and as a lubricant. The residue after pressing is used for fertilizer and, if detoxified, can be used as cattle feed. The only fault the Indians have found with the castor plant is that it takes 270 days to mature. Consequently, the rainy season would often arrive before the plants matured, and much of the crop would be lost. If the plant could be made to mature in less time, the chance of losing a crop would be lessened. By exposing the castor plant to cobalt-60 radiation, a mutant strain that matures in 120 days was produced. This strain not only provided insurance against rain damage but also allowed the same land to be planted with other crops during the time that was saved. In a similar manner, improved strains of cereal crops that mature earlier and produce greater yields have been produced by bombardment with mutation-inducing neutrons and gamma rays.

Tiger Chemistry

Classical methods of chemical analysis seldom allow the detection of chemical constituents whose concentrations are less than one part per million. The use of the spectroscope, an instrument that detects substances by their interaction with light, may extend this limit by a factor of a hundred, but no procedure other than activation analysis is capable of detecting elements in quantities as small as 10^{-12} grams. Activation analysis consists essentially of subjecting the substance that is to be analyzed to neutron bombardment in a nuclear reactor; radiation emitted by the sample as a result of this bombardment is then correlated with characteristic elemental radiations. Sixty-nine of the elements can be detected by this method. (Figure 9–4.)

The effectiveness of activation analysis was dramatically demonstrated when it was used to determine the cause of death of Napoleon Bonaparte. For many years, the cause of Napoleon's death was a matter of conjecture. In exile on St. Helena, he suffered great stomach distress that he attributed to poison fed him by his jailers. Many years after his death in 1821, the advent of activation analysis made it possible to determine whether his suspicions were justified. Because arsenic when ingested is concentrated in the hair, several of Napoleon's hairs, which were preserved by his valets Marchand and Noverraz, were subjected to heavy bombardment by neutrons. The analysis of the resultant radiation proved conclusively that he was indeed poisoned during his exile.*

Another chemical use of nuclear reactions is the production of new and exotic elements. Prior to 1940, the heaviest known element was uranium, element 92. In 1940, McMullen and Abelson prepared the

*S. Forshufvud: *Who Killed Napoleon?* London: Hutchinson and Co. (Publishers) Ltd., 1962.

FIGURE 9–4 Elements Subject to Activation Analysis Detection

																	F
β																	—
γ																	1
	Na	**Mg**											**Al**	**Si**	**P**	**S**	**Cl**
β	5×10^{-3}	5×10^{-1}											1×10^{-1}	5×10^{-2}	5×10^{-1}	5	1×10^{-2}
γ	5×10^{-3}	5×10^{-1}											1×10^{-2}	500	—	200	1×10^{-1}
	K	**Ca**	**Sc**	**Ti**	**V**	**Cr**	**Mn**	**Fe**	**Co**	**Ni**	**Cu**	**Zn**	**Ga**	**Ge**	**As**	**Se**	**Br**
β	5×10^{-2}	1.0	1×10^{-2}	5×10^{-1}	5×10^{-3}	—	5×10^{-5}	50	5×10^{-3}	5×10^{-2}	1×10^{-3}	1×10^{-1}	5×10^{-3}	5×10^{-3}	1×10^{-3}	—	5×10^{-3}
γ	5×10^{-2}	5	5×10^{-2}	5×10^{-2}	1×10^{-3}	1	5×10^{-5}	200	1×10^{-1}	5×10^{-1}	1×10^{-3}	1×10^{-1}	5×10^{-3}	5×10^{-2}	5×10^{-3}	5	5×10^{-3}
	Rb	**Sr**		**Zr**	**Nb**	**Mo**		**Ru**	**Rh**	**Pd**	**Ag**	**Cd**	**In**	**Sn**	**Sb**	**Te**	**I**
β	5×10^{-2}	5×10^{-3}		1	5×10^{-3}	5×10^{-1}		1×10^{-2}	1×10^{-3}	5×10^{-4}	5×10^{-3}	5×10^{-2}	5×10^{-5}	5×10^{-1}	5×10^{-3}	5×10^{-2}	5×10^{-3}
γ	5	5×10^{-3}		1	1	1×10^{-1}		5×10^{-2}	5×10^{-4}	5	5×10^{-3}	5×10^{-1}	1×10^{-4}	5×10^{-1}	1×10^{-2}	5×10^{-2}	1×10^{-2}
	Cs	**Ba**	**La**	**Hf**	**Ta**	**W**	**Re**	**Os**	**Ir**	**Pt**	**Au**	**Hg**		**Pb**	**Bi**		
β	5×10^{-1}	5×10^{-2}	1×10^{-3}		5×10^{-2}	1×10^{-3}	5×10^{-4}	5×10^{-2}	1×10^{-4}	5×10^{-2}	5×10^{-4}	—		10	5×10^{-1}		
γ	5×10^{-1}	1×10^{-1}	5×10^{-3}	1	5×10^{-1}	5×10^{-3}	1×10^{-3}	—	1×10^{-3}	1×10^{-1}	5×10^{-4}	1×10^{-2}		—	—		

	Ce	**Pr**	**Nd**		**Sm**	**Eu**	**Gd**	**Tb**	**Dy**	**Ho**	**Er**	**Tm**	**Yb**	**Lu**
β	1×10^{-1}	5×10^{-4}	1×10^{-1}		5×10^{-4}	5×10^{-6}	1×10^{-2}	5×10^{-2}	1×10^{-6}	1×10^{-4}	1×10^{-3}	1×10^{-2}	1×10^{-3}	5×10^{-5}
γ	1×10^{-1}	5×10^{-2}	1×10^{-1}		5×10^{-3}	5×10^{-4}	5×10^{-2}	1×10^{-1}	5×10^{-6}	1×10^{-4}	1×10^{-3}	1×10^{-1}	1×10^{-3}	5×10^{-5}
	Th		**U**											
β	5×10^{-2}		5×10^{-3}											
γ	5×10^{-2}		5×10^{-3}											

TABLE

9–1 Transuranium Elements Produced by Nuclear Reactions

Element	Atomic Number	Reaction
Neptunium, **Np**	93	$^{238}_{92}U + ^{1}_{0}n \rightarrow ^{239}_{93}Np + ^{0}_{-1}e$
Plutonium, **Pu**	94	$^{238}_{92}U + ^{2}_{1}H \rightarrow ^{238}_{93}Np + 2\,^{1}_{0}n$
		$^{238}_{93}Np \rightarrow ^{238}_{94}Pu + ^{0}_{-1}e$
Americium, **Am**	95	$^{239}_{94}Pu + ^{1}_{0}n \rightarrow ^{240}_{95}Am + ^{0}_{-1}e$
Curium, **Cm**	96	$^{239}_{94}Pu + ^{4}_{2}He \rightarrow ^{242}_{96}Cm + ^{1}_{0}n$
Berkelium, **Bk**	97	$^{241}_{95}Am + ^{4}_{2}He \rightarrow ^{243}_{97}Bk + 2\,^{1}_{0}n$
Californium, **Cf**	98	$^{242}_{96}Cm + ^{4}_{2}He \rightarrow ^{245}_{98}Cf + ^{1}_{0}n$
Einsteinium, **Es**	99	$^{238}_{92}U + 15\,^{1}_{0}n \rightarrow ^{253}_{99}Es + 7\,^{0}_{-1}e$
Fermium, **Fm**	100	$^{238}_{92}U + 17\,^{1}_{0}n \rightarrow ^{255}_{100}Fm + 8\,^{0}_{-1}e$
Mendelevium, **Mv**	101	$^{253}_{99}Es + ^{4}_{2}He \rightarrow ^{256}_{101}Mv + ^{1}_{0}n$
Nobelium, **No**	102	$^{246}_{96}Cm + ^{12}_{6}C \rightarrow ^{254}_{102}No + 4\,^{1}_{0}n$
Lawrencium, **Lr**	103	$^{252}_{98}Cf + ^{10}_{5}B \rightarrow ^{257}_{103}Lr + 5\,^{1}_{0}n$
Kurchatovium, **Ku**	104	$^{242}_{94}Pu + ^{22}_{10}Ne \rightarrow ^{260}_{104}Ku + 4\,^{1}_{0}n$
Hahnium, **Ha**	105	$^{249}_{98}Cf + ^{15}_{7}N \rightarrow ^{260}_{105}Ha + 4\,n$

first man-made element, neptunium ($^{239}_{93}Np$). The process involved bombarding a target of $^{238}_{92}U$ with high-speed deuterons ($^{2}_{1}H$). The initial reaction involved the conversion of $^{238}_{92}U$ to $^{239}_{92}U$:

$$^{238}_{92}U + ^{2}_{1}H \rightarrow ^{239}_{92}U + ^{1}_{1}H$$

Uranium 239 has a half-life of 23.5 minutes. It decomposes by beta emissions to yield neptunium.

$$^{239}_{92}U \rightarrow ^{239}_{93}Np + ^{0}_{-1}\beta$$

Subsequent to this initial breakthrough, a dozen new elements came in fairly rapid succession. A listing of all of them and their methods of production appears in Table 9–1.

Decomposing Is Not All the Same Old Rot.

An energetically unstable nucleus having an unfavorable neutron-to-proton ratio does not always decompose to give the same products. Aluminum 28, which is artificially produced by bombarding aluminum

27 with high-energy neutrons, is an example of a nucleus having more than one mode of decomposition.

$$^{27}_{13}\mathbf{Al} + ^{1}_{0}\mathrm{n} \rightarrow ^{28}_{13}\mathbf{Al}^{*} \rightarrow \begin{cases} ^{26}_{13}\mathbf{Al} + 2\,(^{1}_{0}\mathrm{n}) \\ ^{28}_{13}\mathbf{Al} + \gamma \\ ^{27}_{12}\mathbf{Mg} + ^{1}_{1}\mathrm{p} \\ ^{24}_{11}\mathbf{Na} + ^{4}_{2}\alpha \end{cases}$$

Each daughter of the energetically unstable aluminum 28 is itself radioactive and further adjusts its neutron-to-proton ratio by either positron or beta emission to give a stable nonradioactive nucleus.

$$^{26}_{13}\mathbf{Al} \rightarrow ^{26}_{12}\mathbf{Mg} + ^{0}_{1}\beta \quad \text{(positron)}$$

$$^{28}_{13}\mathbf{Al} \rightarrow ^{28}_{14}\mathbf{Si} + ^{0}_{-1}\beta \quad \text{(beta)}$$

$$^{27}_{12}\mathbf{Mg} \rightarrow ^{27}_{13}\mathbf{Al} + ^{0}_{-1}\beta \quad \text{(beta)}$$

$$^{24}_{11}\mathbf{Na} \rightarrow ^{24}_{12}\mathbf{Mg} + ^{0}_{-1}\beta \quad \text{(beta)}$$

In each of these disintegrations, an isotopic nucleus has gone to a more stable state through the release of radiation and its associated energy. This energy, when controlled, can be utilized to provide commercially acceptable sources of power.

Fission

A representative reaction from which nuclear power is currently obtained is the decomposition of uranium 235. Uranium 235 is a natural isotope that occurs to the extent of 0.7 percent mixed with the more abundant uranium 238. Under normal conditions, uranium 235 is an alpha emitter with an extremely long half-life (10^9 years). If it is bombarded with neutrons, however, the resulting product splits into two large fragments and either two or three neutrons. An equation representative of this reaction is:

$$^{235}_{92}\mathbf{U} + ^{1}_{0}\mathrm{n} \rightarrow ^{90}_{38}\mathbf{Sr} + ^{143}_{54}\mathbf{Xe} + 3\,^{1}_{0}\mathrm{n}$$

This process is known as fission. The major resulting fragments approximate half the size of the original atom. The binding energy of the protons and neutrons in these fragments is less than the energy they possessed as components of the uranium-235 nucleus. This difference is the energy released during a nuclear reaction. It is over a

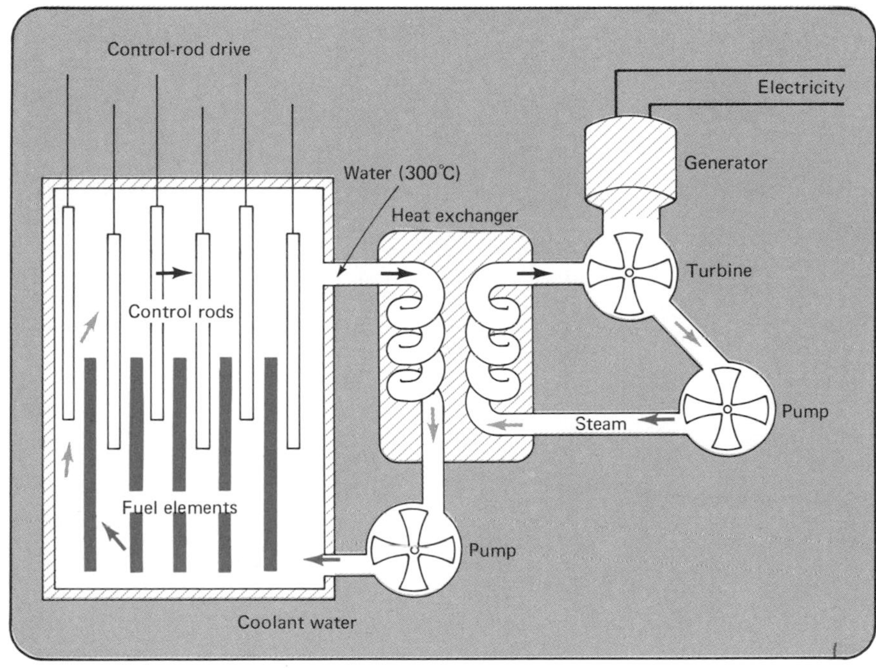

FIGURE 9–5 **Electricity Produced by Atomic Energy**

million times greater than the energy that can be obtained by burning an equivalent weight of coal. This energy is tapped by circulating a coolant through the nuclear reactor. The heat of the reaction raises the temperature of the coolant which through the use of a heat exchanger can be used to drive a conventional electrical generator or some similar device. (Figure 9–5.)

To convert nuclear fission to practical sources of energy, the reaction must be made self-sustaining. This can be done only if each nuclei after capturing a neutron decomposes to give more than one neutron. Once these new neutrons are emitted, they themselves must also be captured, and so forth. This is accomplished by constructing a reactor of proper material and geometry sufficiently rich in uranium 235 to ensure optimum capture of emitted neutrons. Fast neutrons, as they are produced in fission reactions, are not easily captured by other uranium 235 atoms. Nuclear reactors, because of this, incorporate moderators such as graphite or water in their construction. The purpose of these moderators is to slow the neutrons sufficiently so that they can be more readily caught by the uranium-235 nuclei. Figure 9–6 is a cross-sectional schematic of a reactor that uses graphite as a moderator.

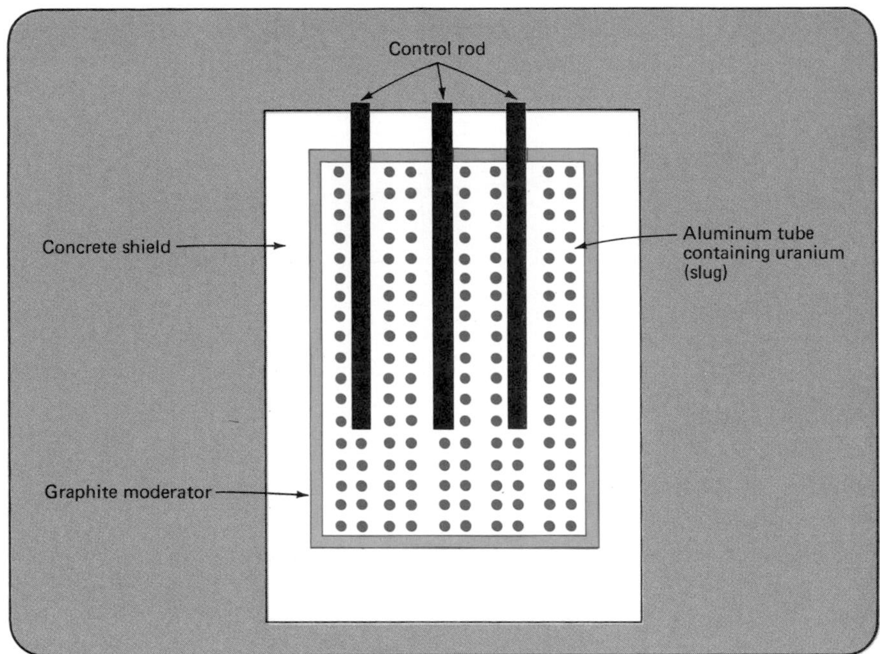

FIGURE

9–6 Schematic Diagram of a Nuclear Reactor

When more than the desired number of neutrons are produced, cadmium or boron steel rods are used to control the reaction. Their function is to capture excess neutrons. By inserting or removing these rods, the ratio of neutron production to absorption can be maintained at a value close to 1. If the ratio falls below this value, the reaction will eventually stop. If it increases much above 1, the reactor may explode. The neutron-controlling reactions by which these rods operate are:

$$^{113}_{48}\mathbf{Cd} + ^{1}_{0}\mathrm{n} \rightarrow ^{114}_{48}\mathbf{Cd} + \gamma$$

$$^{10}_{5}\mathbf{B} + ^{1}_{0}\mathrm{n} \rightarrow ^{11}_{5}\mathbf{B} + \gamma$$

Neither the cadmium nor the boron reactants or products are radioactive. The neutrons that are captured go into the formation of a stable non-neutron-producing isotope. In so doing, they lower the ratio of neutrons absorbed to neutrons produced by the uranium atoms.

Too Many Neutrons Can Mean Big Trouble.

The standard mail-order Beginner's Atomic-Bomb Kit consists of two pieces of uranium each shaped like half a large grapefruit and weighing about 45 pounds, a pad of legal-size paper, and a ballpoint pen.

Using the ballpoint pen and the pad of paper, the proud new kit-owner first makes out his will. Once this is completed, he brings the two pieces of uranium together very fast and everything self-destructs. The destructive capability of this type of bomb is equivalent to approximately 20,000 tons of TNT.

Although this macabre description may sound ridiculous, the concept in principle is correct. The active components in a nuclear bomb need not be much larger than a grapefruit nor weigh much more than 90 pounds. The reaction that gives the bomb its destructive force is essentially the same reaction produced in a nuclear energy plant. The sole difference is that in a power plant the energy is produced over a longer time interval and therefore can more readily be tapped for useful purposes.

The speed of a bomb's reaction is the result of the rapid increase in the ratio of neutrons absorbed to neutrons produced. Instead of a value of 1, which would provide just enough neutrons to sustain the reaction, this ratio may approach 3. Such a high value generates an inordinately large amount of fission in an extremely short time. This sudden release of energy is what gives the bomb its explosive force. If each fission produced only one neutron, then after a series of ten neutron capture-fission reactions there would still be only one neutron to induce the next fission reaction. If, however, as in the case of uranium 235, each fission produces three neutrons, then after a series of ten fissions, the number of neutrons available to produce further fission would be in excess of fifty-nine thousand. All these neutrons would be produced in a time much smaller than a millionth of a second.

Figure 9–7 illustrates how one stray neutron in three fission steps may result in six neutrons, each of which is capable of inducing a fission. It should be noted from this diagram, that not all the neutrons produced result in the fission of a uranium-235 atom. Some escape without interacting with any uranium, and others interact with uranium 238 to produce plutonium. This occurs in three steps:

$$(1) \qquad {}^{238}_{92}\mathbf{U} + {}^{1}_{0}\mathrm{n} \rightarrow {}^{239}_{92}\mathbf{U} + \gamma$$

${}^{239}_{92}\mathbf{U}$ is unstable and decays by two steps to ${}^{237}_{94}\mathbf{Pu}$

$$(2) \qquad {}^{239}_{92}\mathbf{U} \xrightarrow{T_{\frac{1}{2}} = 24 \text{ minutes}} {}^{239}_{93}\mathbf{Np} + {}^{0}_{-1}\beta$$

$$(3) \qquad {}^{239}_{93}\mathbf{Np} \xrightarrow{T_{\frac{1}{2}} = 2.3 \text{ days}} {}^{239}_{94}\mathbf{Pu} + {}^{0}_{-1}\beta + \gamma$$

Plutonium 239, like uranium 235, undergoes fission after capturing a neutron. The plutonium reaction series has produced a fissionable isotope from nonfissionable uranium 238. Reactors constructed to

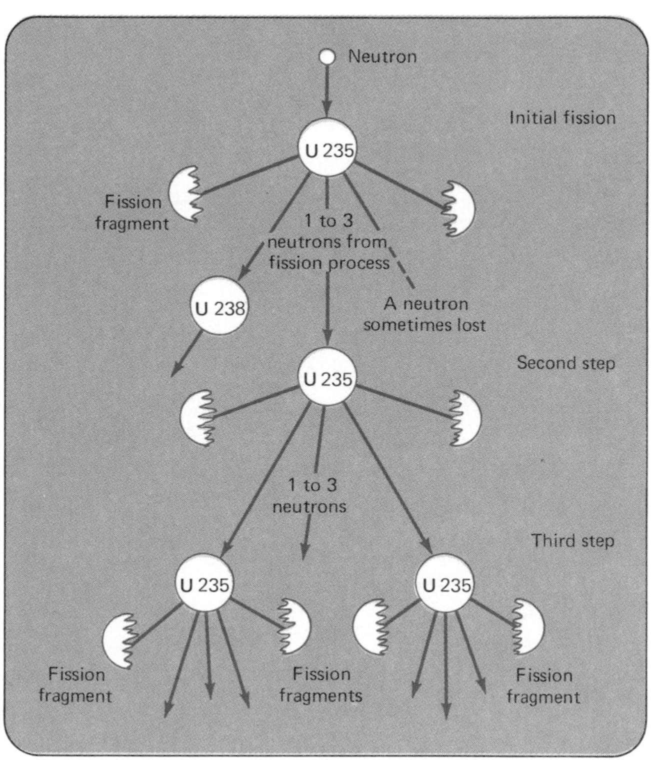

FIGURE 9–7 Six Neutrons from One Due to Three Fissions

take advantage of this series are called breeder reactors. Breeding reactions are of commercial importance because they provide an additional source of nuclear fuel to complement the natural sources of uranium 235.

Build a Bigger Bomb. Make a Bigger Bang.

In 1961, the USSR exploded a 50-megaton (equivalent to 50 million tons of TNT) thermonuclear device. It was the third such device ever tested and at the time the most powerful man-made force ever created. Thermonuclear devices, unlike fission devices, derive their energy from the building of atoms rather than the breaking of them—fusion rather than fission. This is the process by which our sun generates its energy.

That energy can be derived from the fusion of nuclei should not come as a surprise: the binding energy curve (Figure 9–1) indicates that nuclei with both very low and high atomic weights have lower binding energy per nucleon than do nuclei of intermediate weights. Because this difference is more marked at the lower end of the atomic weight scale, the curve suggests that greater energy should be liberated when light nuclei combine than when heavy nuclei are split. This has, in fact, been found to be true. The release in energy that occurs during the fusion of two hydrogen nuclei to form a helium nucleus is about four times greater per given mass of starting material than the energy liberated during the fission of uranium.

Depending upon the form of the hydrogen in a thermonuclear device, a number of possible descriptive reactions can be written. All involve the conversion of hydrogen to helium. A typical one is

$$^{2}_{1}\mathrm{H} + ^{3}_{1}\mathrm{H} \rightarrow ^{4}_{2}\mathrm{He} + ^{1}_{0}\mathrm{n}$$

In order for this reaction to occur, it is necessary to raise the temperature of the reactants to minimum of 50,000,000°C. At present, the only method of producing such temperatures is through the detonation of an associated fission device. Temperatures in excess of 400,000,000°C can be produced in this manner. On the sun and other stars similar processes occur. The following three reactions have been proposed as the sun's source of energy:

$$^{1}_{1}\mathrm{H} + ^{1}_{1}\mathrm{H} \rightarrow ^{2}_{1}\mathrm{H} + ^{0}_{1}\beta$$

$$^{2}_{1}\mathrm{H} + ^{1}_{1}\mathrm{H} \rightarrow ^{3}_{2}\mathrm{He} + \gamma$$

$$^{3}_{2}\mathrm{He} + ^{3}_{2}\mathrm{He} \rightarrow ^{4}_{2}\mathrm{He} + 2\,^{1}_{1}\mathrm{H}$$

REVIEW

1. The stability of atomic nuclei is dependent upon the number of neutrons and protons they contain.
2. Except hydrogen $^{1}_{1}\mathrm{H}$, all common nuclei contain at least as many neutrons as protons.
3. Nuclei with a disproportionately high neutron-to-proton ratio tend to be beta emitters.
4. Daughter isotopes of beta emitters have the same mass as their parent but an atomic number one greater than their parent.
5. Daughters of alpha emitters have a mass four less and an atomic number two less than their parent.

6. Isotopes that are deficient in neutrons may remedy the situation by either positron emission or *K* electron capture. Both processes increase the atomic number by 1 and leave the isotopic mass unchanged.
7. Nuclei can release energy through gamma-ray emission; this process leaves the mass and charge unchanged.
8. When balancing nuclear equations, atoms, subscripts, and superscripts must balance.
9. The end product of a radioactive decomposition chain is a stable non-radioactive isotope having a high nuclear binding force.
10. All isotopes heavier than bismuth 83 are radioactive.
11. The time necessary for an isotope to decay to half its original amount is known as its half-life.
12. Radioisotopes can be used to estimate the age of a substance in which they are contained by determining the number of half-lives that have elapsed since they were created.
13. Radioisotopes can be used in medicine for both diagnostic and therapeutic purposes.
14. Radiation can be used to alter the genetic makeup of plants to produce mutant strains that are commercially more desirable.
15. Radiation can be used in chemistry to produce new elements and to detect ultra small quantities of a large number of natural elements.
16. Radioisotopes can decay in more than one way to produce more stable nuclei.
17. Fission occurs when a nucleus breaks into two new isotopes each having roughly half the mass of the parent. Large amounts of energy are released by this process.
18. Fusion occurs when two nuclei combine to produce a nucleus having a mass greater than either parent. The energy generated by this process is several times greater than that produced by fission.

REFLEXIVE QUIZ

1. The stability of an isotopic nucleus is dependent upon its mass and its neutron-to-________ ratio.
2. Beta rays are emitted by nuclei having more ________ than are necessary for optimum stability.
3. The emission of a beta ray increases the atomic number of an atom by ______ but leaves its atomic mass unchanged.

4. Energetically unstable nuclei can dispose of their excess energy without a change in mass or atomic number by emitting ________ rays.
5. ________ emission and K electron capture are both methods of increasing the neutron-to-proton ratio.
6. The most expedient method of reducing the mass of heavy nuclei is ________ particle emission.
7. The half-life of bismuth 210 is five days. In fifteen days, it will decompose to ________ its original amount.
8. The emission of an alpha particle changes the weight of a nucleus by ________ and its atomic number by two.
9. In an atomic reactor, isotopes that are bombarded with ________ undergo fission-releasing energy.
10. To make a nuclear reactor self-sustaining, the ratio of neutrons produced to neutrons absorbed must equal approximately ____.

Answers: (1) Proton [128–129, 149]. (2) Neutrons [129–132]. (3) One [129–132, 149]. (4) Gamma [133, 150]. (5) Positron [132, 150]. (6) Alpha [132]. (7) $\frac{1}{8}$ [133–136, 150]. (8) Four [132, 149]. (9) Neutrons [144–148]. (10) 1 [144–146].

SUPPLEMENTARY READINGS

1. G. T. Seaborg. "Some Recollections of Early Nuclear Age Chemistry." *Journal of Chemical Education* 45 (1968): 278.
2. H. R. Lukens. "Neutron Activation Analysis." *Journal of Chemical Education* 44 (1967): 668.
3. W. H. Wahl and H. H. Kramer. "Neutron Activation Analysis." *Scientific American* (April 1967), p. 68.
4. W. F. Libby. "Radiocarbon Dating." *Chemistry in Britain* 5 (1969): 548.
5. C. Holden. "Nuclear Waste." *Science* 172 (1971): 249.
6. O. Hahn. "Discovery of Fission." *Scientific American* (February 1958), p. 76.
7. G. T. Seaborg and J. L. Bloom. "The Synthetic Elements." *Scientific American* (April 1969), p. 56.

Section 

The Gas Molecules of the Universe vs. the Scientific Community

Section 10

The Gas Molecules of the Universe vs. the Scientific Community

Exhibit I. Memoirs of an Imprisoned Gas Molecule

"The recreational facilities in a sealed container are very limited. The problem is further compounded when the container must be shared with billions of other molecules. In an attempt to escape the pressures imposed by such environment the sanest of us is often driven to wild extremes. It is not uncommon to see distraught molecules fly the full length of the container, hurl themselves against the wall in despair, then turn around and proceed to repeat the process upon an opposing wall. Given the same number of molecules, the smaller the container, the greater the pressure. At times pressures are so great that containers have been known to rupture. The pressures become most unbearable when the temperature goes up. With an increase in temperature some among us go into an absolute frenzy. Boyle, Charles, Dalton, Graham, and Gay-Lussac have written of our plight, but the public is apathetic."

Exhibit II. Statement of the Scientific Community

The laws governing the confinement of gases are rooted in common sense. They are as valid today as they were fifty years ago. With the exceptions of extremes in temperature and pressure they provide a high degree of predictability and are in need of minimal reform.

Exhibit III. A Day in the Life of Vic Vapor, All-American Gas Molecule

Vic lives in a cubic box one millimeter on a side. He shares this box with almost 25,000,000,000,000,000 other molecules. The large assemblage does not disturb him, however, because he himself is quite small by comparison to the dimension of the box. What really annoys him is the lack of something meaningful to do. No matter what the time of day, all he does is fly the length of the box and collide with the walls. Infrequently he collides with another gas molecule, but the meetings are brief and seldom rewarding.

A day in the Life of VIC VAPOR

These collisions are the cause of all the pressure. Each time a gas molecule hits the wall it pushes against it with a small but definite force. When a large number of molecules push against a wall at the same time a significant pressure is produced. The more molecules hitting a wall at a given time, the greater the pressure on that wall. In small boxes the distance the molecules have to fly before they hit a wall is reduced. As a result of this, a given molecule can hit a larger number of walls or the same number of walls more frequently and in this manner produce a greater pressure.

Molecules fly faster when the temperature rises. This increased speed allows them to fly between walls in less time. The faster they fly, the greater the number of impacts and the greater the resulting pressure.

Exhibit IV. Volume

The volume of a rectangular container can be determined by multiplying its length by its width by its height. If the internal dimension of a container is 100 centimeters × 10 centimeters × 1 centimeter, its total volume is 1,000 cubic centimeters or one liter. (A cubic container having dimensions 10 cm. × 10 cm. × 10 cm. also has a volume of one liter.) A liter is slightly more than a quart.

Exhibit V. Pressure

Pressure is the amount of force exerted on a given area of surface. It is the secret of successful thumbtacks.

If you attempt to press a steel rod as wide as a thumbtack into a plank of wood, the chances are your thumb will give before the wood. This is because the pressure on the wood is the same as on your thumb.

Since the wood is more resilient, your thumb buckles first. If, however, you used a thumbtack instead of a metal rod, the chances are more likely that the wood would be pierced. Although the force on the top of the tack and the rod are identical, the pressure at the bottom of each is vastly different. This is because the area of the tack point is several hundred times smaller than the area at its top and the same force is being applied to both.

Pressure is defined as force per given area. When the area is reduced and the force is maintained, the pressure rises. The same force applied to half the area produces double the pressure. If the force is applied to an area several hundred times smaller, then the resultant pressure will be several hundred times greater. The same effect occurs in a closed container of gas. If one dimension of a rectangular container is halved and the other two left the same, the wall surface and the volume will be halved and the pressure will be doubled. The total force exerted by the gas is the same in both cases, but because the area is only one half as great, the pressure—the force per unit area—is doubled.

Pressure is commonly measured in two units: atmospheres and millimeters of mercury. One atmosphere is equal to the pressure exerted at sea level of the blanket of air that surrounds the earth. This pressure is great enough to support a column of mercury 760 millimeters high (1 atm. = 760 mm. **Hg**).

Note. You can prove the existence of this pressure to yourself by taking a tumbler, filling it with water to the brim, covering it with a piece of cardboard, and while holding the cardboard in place gingerly inverting it. When you remove your hand, you will find much to your surprise that the water will not run out but will be held in place by

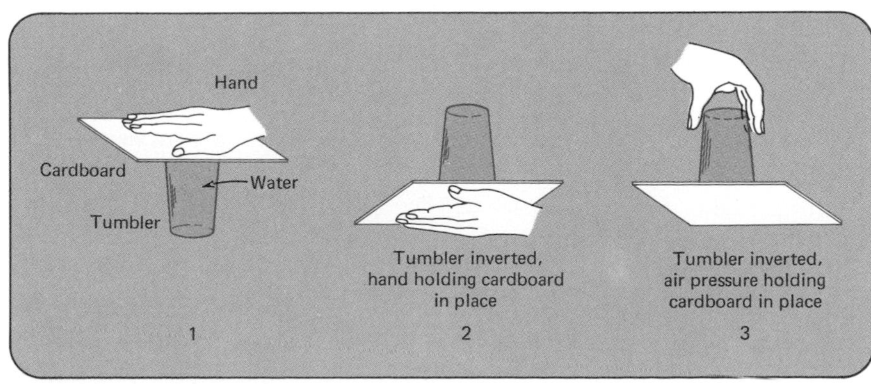

FIGURE 10–1 An Experiment with Air Pressure

the atmospheric pressure. What is even more surprising is that the atmospheric pressure will support a column of water 32 feet high. If you want to make this test with a 32-foot glass, do it *very* carefully.

Exhibit VI. Temperature

Temperature is a queer beast. Nearly everyone has some idea of what it is. More often than not, most are wrong. On a cold winter's morning you jump out of bed, hurry to the bathroom, step barefoot on a cold tile floor, then quickly seek the comfort of a bath mat. Most people will insist that the mat is warmer than the floor. Not so. If a thermometer is placed first on the floor and then on the mat, it will soon be determined that the readings are identical. What most people mistakenly confuse with temperature is the rate at which the body either gives off or takes up heat.

Heat is an extensive property, whereas temperature is an intensive property. There is more heat in a large pot of boiling water than there is in a burning match; the temperature or heat intensity of the match, however, is much higher than the water. Heat may be thought analogous to force and temperature to pressure. The match flame is the thumbtack of the thermal world. The human body can sense only the transfer of heat not temperature.

When you step on a tile floor the feeling of cold is produced by heat leaving the body and going into the floor. Because heat flows more slowly into a bath mat than into tile, a bath mat provides insulation against the heat loss and therefore feels warmer.

For scientific purposes, the most common units of temperature are degrees Celsius (centigrade) or degrees Kelvin. The former is related to Fahrenheit in the following manner:

$$\text{Celsius temperature} = \frac{5}{9}\,(\text{Fahrenheit temperature} - 32)$$

This relationship was derived by comparing the temperatures at which water boils and freezes on both scales.

Water freezes at 0° Celsius or 32° Fahrenheit. It boils at 100° Celsius or 212° Fahrenheit. Because the difference in temperature between these two points is the same, no matter whether it is measured with a Fahrenheit or a Celsius thermometer, 180 Fahrenheit degrees can be equated to 100 Celsius degrees. One Celsius degree, therefore, equals $\frac{180}{100}$ or $\frac{9}{5}$ of a Fahrenheit degree. When this factor is combined with

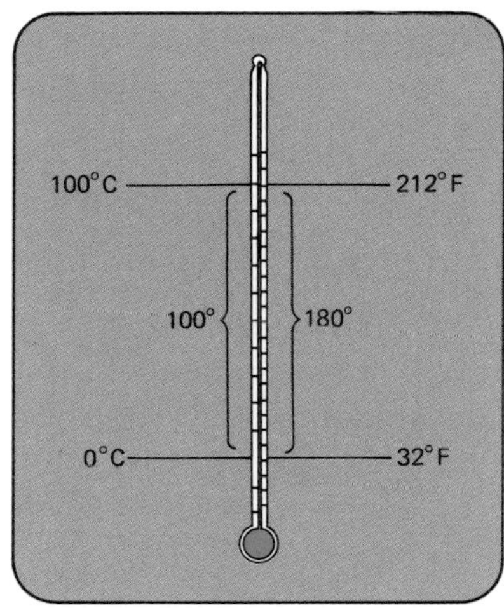

FIGURE

10–2 Comparison of Celsius and Fahrenheit Scales

the fact that the freezing point of water is 32° higher on the Fahrenheit scale the following conversion equation may be written:

$$F^\circ = \frac{9}{5}C^\circ + 32$$

This equation is just another form of the first, rearranged to allow the conversion of Celsius into Fahrenheit degrees.

Both Fahrenheit and Celsius scales suffer from negative temperatures. When the temperature gets sufficiently cold the thermometer will read below zero. To avoid this difficulty the Kelvin scale was devised. The Kelvin scale is unique in that nothing is or can get colder than zero degrees. Zero degrees Kelvin is absolute zero. All temperatures are greater than this value and therefore there are no negative temperatures. 0°K is equal to −273°C. Thus, when converting from Celsius to Kelvin, it is necessary to add 273° to the Celsius temperature. The freezing and boiling points of water are, therefore, 273°K and 373°K, respectively.

Exhibit VII. Avogadro's Winning Number

In 1811, Avogadro proposed that equal volumes of all gases at the same temperature and pressure have the same number of molecules. One gram molecular weight of any gas at 0°C and one atmosphere of pressure occupies 22.4 liters and contains 6×10^{23} molecules. This is Avogadro's number (N) of molecules.

Exhibit VIII. Dalton's Law of Partial Pressures

John Dalton observed, around 1810, that when more than one type of gas molecule is enclosed in any container, each kind behaves as if it were there all alone. This outright snobbery is the general practice among gas molecules. Not only does it occur between molecules of different types, but it is equally prevalent among molecules of the same type. The result of this lack of association is that the total pressure of any mixture of gases can be determined by adding together the pressures of the individual components.

Exhibit IX. Graham's Law of Diffusion

Thomas Graham (1805–1869) observed that if a mixture of gas is allowed to escape through a hole in its container, it is found that the lighter molecules escape faster than the heavier ones. This is because all gas molecules at the same temperature have the same kinetic energy (energy of motion). If that energy is used to push a light molecule, then that molecule will travel much faster than a heavier molecule given the same push. All things considered, a man can run much faster normally dressed than with a heavy weight on his back. His energy is the same. Only the mass that he is moving has changed. When gas molecules move, their speed is inversely proportional to the square root of their masses.

$$\frac{V_a}{V_b} = \sqrt{\frac{M_b}{M_a}}$$

The equation states that, at a given temperature, if the mass (M) of two molecules and the velocity (V) of one is known, then the velocity of the other can be determined. The same equation can be restated in terms of the time it takes to move a mass a given distance:

$$\frac{t_b}{t_a} = \sqrt{\frac{M_b}{M_a}}$$

t_a, V_a, and M_a indicate the time of travel, velocity, and mass of the a-type particles. The notations with the b subscript indicate the same quantities for molecular type b.

Example. A cylinder containing equal quantities of methane, CH_4 (mol. wt. 16), and helium, **He** (mol. wt. 4), springs a leak. One half of the helium leaks out in five minutes. How long will it take one half of the methane to leak out?

If helium is considered gas *a* and methane gas *b*, then substituting into the time equation yields the following relationships:

$$\frac{t_b}{t_a} = \sqrt{\frac{M_b}{M_a}}$$

$$\frac{t_b}{5} = \sqrt{\frac{16}{4}} = \sqrt{4} = 2$$

$$t_b = 2 \times 5 = 10 \text{ minutes}$$

Exhibit X. Eyewitness Accounts

These observations are the sworn testimony of all parties concerned with the gas laws. Neither the scientific community nor the gas molecules contest their validity.

1. *Boyle's Law.* Robert Boyle (1627–1691) observed that the volume of a gas is inversely proportional to the pressure applied. If the pressure on an enclosed gas is halved, its volume will double. If the pressure is doubled, the volume will halve. At any temperature the product of the pressure and the volume for a given number of gas molecules is a constant:

$$P_1 \times V_1 = P_2 \times V_2$$

The subscripts 1 and 2 denote the first and second set of conditions. To facilitate solving for P_2 or V_2 the relationship can be rewritten in terms of these variables.

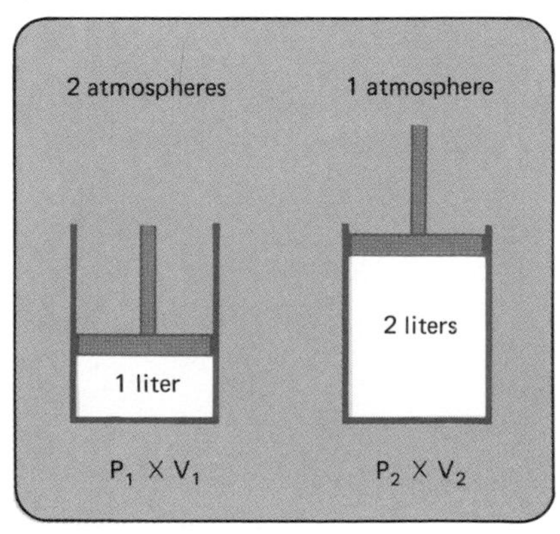

FIGURE 10–3 Boyle's Law

$$P_2 = P_1 \times \frac{V_1}{V_2} \qquad V_2 = V_1 \times \frac{P_1}{P_2}$$

When volume changes, pressure can be determined by multiplying the initial pressure by the ratio of the old volume to the new one. In a similar manner, the volume resulting from a pressure change can be determined by multiplying the initial volume by the ratio of the old pressure to the new one.

Example. If 6 liters of a gas at 2 atmospheres pressure are compressed by increasing the pressure to 3 atmospheres, what will be the volume of the gas at the new pressure?

In this problem the unknown quantity is V_2. The known quantities are: $V_1 = 6$ liters; $P_1 = 2$ atmospheres; $P_2 = 3$ atmospheres. Substituting these values in the equation allows the determination of V_2.

$$V_2 = V_1 \times \frac{P_1}{P_2}$$

$$V_2 = 6 \text{ liters} \times \frac{2 \text{ atm.}}{3 \text{ atm.}} = 4 \text{ liters}$$

2. *Charles's Law.* M. Jacques A. C. Charles noticed, in 1787, that as the temperature of a gas increased so too did the volume.

3. *Gay-Lussac's Law.* In 1809, M. J. L. Gay-Lussac by diligent measurement determined that a gas increased $\frac{1}{273}$ its volume at 0°C for

every degree that the temperature rose. Since 0°C equals 273°K, it can be stated that the volume of a gas is proportioned to the absolute temperature. The volume, therefore, at any temperature can be determined by multiplying the initial volume by the ratio of new to the initial temperature. Both temperatures must, however, be expressed in degrees Kelvin.

$$V_2 = V_1 \times \frac{T_2}{T_1}$$

As the temperature decreases, so does the gas volume. It may seem from this equation that the volume of a gas at zero degrees Kelvin should be zero. This, however, is not the case. Matter does not disappear at absolute zero. What actually happens is that most gases are converted to liquids or solids long before the temperature approaches 0°K.

Example. If the temperature of a gas is raised from 10°C to 100°C and the pressure and number of gas molecules remain constant, how much larger will its new volume be?

Before any computations can be made, both temperatures must be converted to their Kelvin equivalents: 10°C = 283°K; 100°C = 373°K. The volume on heating this gas will be

$$V_2 = V_1 \times \frac{T_2}{T_1} = V_1 \times \frac{373}{283} \cong 1.32\ V_1$$

approximately 1.32 times greater than starting volume. It should be noted that although the Celsius temperature increased tenfold, the absolute temperature increased only by a factor of 1.32. If the initial volume had been four liters, then after heating the final volume would have been 1.32 × 4 liters, or 5.2 liters.

4. *Avogadro's Law.* Count Avogadro observed that equal volumes of any gas at the same temperature and pressure contain the same number of molecules. By extension, twice the number of molecules at the same temperature and pressure occupies twice the volume.

The temperature, pressure, and volume relationships just discussed are applicable to gases where there is no change in the number of molecules. When there is a change, the effect can be represented by the following equation

$$V_2 = V_1 \times \frac{N_2}{N_1}$$

N_1 and N_2 represent the initial and final number of molecules or moles of gas. If N_2 is twice N_1, then V_2 will be twice V_1. The final volume is equal to the initial volume multiplied by the ratio of the final number of moles of gas to the initial number.

Example. If 9 moles of gas occupy 100 liters, how much volume would 27 moles of gas occupy under the same conditions of temperature and pressure?

$$V_2 = V_1 \times \frac{N_2}{N_1}$$

$$\left.\begin{array}{l} N_1 = 9 \text{ moles} \\ V_1 = 100 \text{ liters} \end{array}\right\} \text{initial conditions}$$

$N_2 = 27$ moles

$$V_2 = 100 \times \frac{27}{9} = 300 \text{ liters}$$

Summation of Arguments of the Scientific Community

The observations of the four scientists, Mr. Boyle in England, Messrs. Charles and Gay-Lussac in France, and Count Avogadro in Italy should not be dismissed lightly. Not only are they significant in their own right, but when combined they provide a unified description of the complete activities of gas molecules. The process of combining these observations consists merely of performing in sequence the operations of interest and observing their cumulative effect on the volume of gas.

If a gas is subjected to a pressure change followed by a temperature change, both Boyle's and Gay-Lussac's equations will have to be used to evaluate the final volume. This is done by taking the final volume V_2 after the pressure change and making it the initial volume in the temperature equation.

$$V_{P_2} = V_{P_1}\left(\frac{P_1}{P_2}\right)$$

$$V_{T_2} = V_{T_1}\left(\frac{T_2}{T_1}\right) = V_{P_2}\left(\frac{T_2}{T_1}\right) = V_{P_1}\left(\frac{P_1}{P_2}\right)\left(\frac{T_2}{T_1}\right)$$

Once the two equations are combined the subscripts indicating the origin of the volume terms can be eliminated.

$$V_2 = V_1\left(\frac{P_1}{P_2}\right)\left(\frac{T_2}{T_1}\right)$$

This equation is a statement of the "Combined Gas Law" for a constant number of molecules of any gas. In order to be able to determine the volume resulting from a change in the number of molecules as well as a change in the pressure and temperature, Avogadro's relationship must be incorporated into an "Ideal Gas Law" equation. This is done in the same manner as the pressure and temperature equations were combined. The final volume of the pressure-temperature equation is made the initial volume in Avogadro's equation:

$$V_{N_2} = V_{N_1}\left(\frac{N_2}{N_1}\right) \qquad V_{(PT)_2} = V_{(PT)_1}\left(\frac{P_1}{P_2}\right)\left(\frac{T_2}{T_1}\right)$$

$$V_{N_2} = V_{(PT)_1}\left(\frac{N_2}{N_1}\right)\left(\frac{P_1}{P_2}\right)\left(\frac{T_2}{T_1}\right)$$

With the excessive subscripts eliminated, the equation becomes

$$V_2 = V_1\left(\frac{N_2}{N_1}\right)\left(\frac{P_1}{P_2}\right)\left(\frac{T_2}{T_1}\right)$$

This complete "Ideal Gas Law" equation incorporates all the witnesses' observations. It is capable of predicting any change in pressure, temperature, or volume of any number of molecules of any gas. Although mathematically it was constructed in one particular sequence, this sequence has no physical significance. The final gas volume is the same irrespective of the order in which the operations occur. (That is, the final gas volume will be the same whether the pressure is first

HI! I'M YOUR GALLOPING CHEMIST. WATCH ME AS I PRESENT THE EXCITING SAGA...

Changing Volumes

IN THIS THRILLING TALE YOU WILL SEE VOLUME CHANGE BEFORE YOUR EYES AS IT IS SUBJECTED TO PRESSURE! THEN WATCH IT AS IT IS EXPOSED TO HEAT. CHEER THE HEART-RENDING RESULTS WHEN VOLUME IS EXPOSED TO BOTH HEAT AND PRESSURE! WATCH THIS...

UUH!... I'M GOING TO REDUCE THE VOLUME OF THIS BOX IF IT KILLS ME!
V_{P_1}
WHEW! WHAT A JOB BUT I'VE GOT IT DOWN TO V_{P_2}
PROVING THAT $V_{P_1}\left(\frac{P_1}{P_2}\right) = V_{P_2}$ IS AN EFFECTIVE FORMULA.
V_{P_2}

NOW FOR THE FEAT SUPREME. I WILL HEAT THIS BOX LABELED V_{T_1}
V_{T_1}

... AND INCREASE ITS SIZE TO V_{T_2} THAT SHOWS YOU THAT THE FORMULA... $V_{T_1}\left(\frac{T_2}{T_1}\right) = V_{T_2}$ WORKS!
V_{T_2}

doubled and then the number of molecules is tripled or the number of molecules is first tripled followed by a doubling of the pressure.)

With rearranged terms, this equation can be used to find unknown pressures, temperatures, and numbers of gas molecules, as well as unknown volumes.

$$P_2 = P_1\left(\frac{V_1}{V_2}\right)\left(\frac{N_2}{N_1}\right)\left(\frac{T_2}{T_1}\right)$$

$$T_2 = T_1\left(\frac{V_2}{V_1}\right)\left(\frac{N_1}{N_2}\right)\left(\frac{P_2}{P_1}\right)$$

$$N_2 = N_1\left(\frac{V_2}{V_1}\right)\left(\frac{T_1}{T_2}\right)\left(\frac{P_2}{P_1}\right)$$

The difficulty with these four equations is that it is necessary to know seven different quantities in order to solve for one unknown. This situation can be simplified by rearranging any of these equations in terms of initial and final conditions, and then substituting standard values for the initial conditions. Rearranging yields

$$\frac{P_1V_1}{N_1T_1} = \frac{P_2V_2}{N_2T_2}$$

With standard conditions substituted for initial conditions, the left side of the equation becomes a constant, and the equation is greatly simplified. This is a justifiable operation since the equation is valid for any number of moles of any gas at any temperature and pressure. The standard condition most usually chosen is that one mole of gas at 273°K and one atmosphere of pressure occupies 22.4 liters. This is the gram molecular volume of all gases under these conditions. When these values are substituted for the initial conditions, we have

$$\frac{(1 \text{ atm.})(22.4 \text{ l.})}{(1 \text{ mole})(273°)} = \frac{P_2V_2}{N_2T_2} = 0.082 \text{ atm.l./mole°}$$

Since the value 0.082 atm.l./mole° is a constant the equation can be written as:

$$PV = 0.082\ nT$$

In this form the Ideal Gas Law equation contains only four unknown and no subscripts. The constant 0.082 is the result of using atmospheres as units of pressure, liters as units of volume, degrees Kelvin as the unit of temperature, and moles (n) to describe the number of molecules. If units other than these are employed, this constant will have a different value.

Example. What is the pressure of 2 moles of gas in a volume of 41 liters at 300°K?

$$\frac{.082(n)(T)}{V} = \frac{.082(2)(300)}{41} = 1.2 \text{ atm.}$$

Decision

It is the decision of the court in the case of the Gas Molecules of the Universe vs. the Scientific Community that the "Ideal Gas Law" and all its associated laws are valid as presented. Further, the court would like to thank Messrs. Boyle, Gay-Lussac, Avogadro, Graham, and Dalton for their assistance.

A word in passing concerning Mr. Dalton's law. Implicit in the decision that the Ideal Gas Law is valid for any gas is the understanding that it is applicable to gas mixtures as well as pure gases. When dealing with a mixture the number of moles to be considered is the sum of the component moles.

Example. If 5 moles of nitrogen are mixed with 3 moles of oxygen in a 32.8 liter container at 200°K, what is the resultant pressure?

$$P = \frac{n \times .082 \times T}{V}$$

Substituting 8 as the total number of moles yields

$$P = \frac{8 \times .082 \times 200}{32.8} = 4 \text{ atmospheres}$$

The court recognizes the claims of the scientific community.

REVIEW

1. There are four variables that can be used to describe a confined gas. They are temperature, pressure, volume, and number of gas molecules.
2. Temperature is a measure of heat intensity.
3. Pressure is the force per area exerted by gas molecules colliding with the wall of their container.
4. The volume of a gas is equal to the size of the container that encloses it. This is because all gases expand to fill their containers.
5. The number of gas molecules or moles is self-explanatory. It is the intrinsic amount of gas.
6. Boyle's law states that the volume of a gas is inversely proportional to its pressure. When a gas is compressed, its volume goes down and its pressure goes up.
7. Gay-Lussac's law states that the volume of a gas is proportional to the absolute temperature. As the temperature goes up, so too does the volume.
8. Avogadro's law states that the volume of a gas is proportional to the number of molecules. As the number of moles increases, so too does the volume.
9. $PV = 0.082\ nT$ is the "Ideal Gas Law." It applies for all gases and their mixtures in any amount providing the units of pressure, volume, amount, and temperature are respectively atmospheres, liters, moles, and degrees Kelvin.
10. Graham's law of diffusion states that the velocity of a gas at any temperature is inversely proportional to the square root of its mass. A gas with a molecular weight four times greater than another will travel at one half the velocity of the other.
11. To convert degrees Fahrenheit to degrees Celsius, it is necessary to subtract 32° and take $\frac{5}{9}$ of the remaining number.
12. To convert degrees Celsius to degrees Kelvin, it is necessary to add 273°.

REFLEXIVE QUIZ

1. The four variables that describe a gas are *P*, *V*, ____, and *T*.
2. The pressure of a gas in a container will ________ as the temperature increases.
3. If the number of molecules in a container is doubled, the pressure of the gas will ________.
4. The pressure resulting from the addition to a container of nitrogen at 2 atmospheres of an equal volume of oxygen at 4 atmospheres is ____ atmospheres.
5. The temperature of a bowl of soup is 50° Celsius. This temperature is equal to _____° Kelvin.
6. To maintain 2 atmospheres of air pressure in auto tires both summer and winter, it is necessary to _____ some air when winter comes.
7. Two atmospheres pressure is equivalent to _____ mm. of mercury.
8. A man running a temperature of 104°F has a temperature of _____°C.
9. Oxygen has a molecular weight of 32 as compared to 2 for hydrogen. Because of this weight difference, hydrogen molecules will travel ____ times faster than oxygen at the same temperature.
10. The gages on a sealed container having a volume of 112 liters read 1 atmosphere of pressure and 273° Kelvin. With the Ideal Gas Law, the number of molecules in this container can be computed as ____ $\times 10^{24}$.

Answers: (1) *N* [156–159, 169]. (2) Increase [162, 169]. (3) Double [159, 169]. (4) 6 [160]. (5) 323°K [158–159, 169]. (6) Add [162–163, 169]. (7) 1520 mm. [156–157]. (8) 40°C [158–159]. (9) 4 [160–161, 169]. (10) 3 [168–169].

SUPPLEMENTARY READINGS

1. M. B. Hall. "Robert Boyle." *Scientific American* (August 1967), p. 97.
2. E. A. Mason and B. Knonstadt. "Graham's Laws of Diffusion and Effusion." *Journal of Chemical Education* 44 (1967): 740.
3. E. H. Brown. "Some Early Thermometers." *Journal of Chemical Education* 11 (1934): 448.
4. N. Feifer. "The Relationship between Avogadro's Principle and the Law of Gay-Lussac." *Journal of Chemical Education* 43 (1966): 411.
5. R. G. Neville. "The Discovery of Boyle's Law, 1661–62." *Journal of Chemical Education* 39 (1962): 356.
6. R. N. Pease. "The Kinetic Theory of Gases." *Journal of Chemical Education* 16 (1939): 242.

Section 11

Inorganic Chemistry

Section 11

Inorganic Chemistry

Using only the periodic chart, a chemist would be hard pressed to describe all the chemical reactions encountered in everyday science and industry. This is because most reactions occur between compounds rather than elements. With the exception of noble gases, noble metals such as gold and platinum, and some forms of nonmetals such as carbon and sulfur, nearly all elements when found naturally are combined with other elements.

Natural compounds, for the most part, can be arbitrarily divided into two major categories: (1) organic and (2) inorganic.

1. *Organic compounds* comprise the larger of the two categories; they are composed mainly of carbon and hydrogen and may contain lesser amounts of oxygen, nitrogen, sulfur, and halogen atoms. The reason they are more numerous than inorganic compounds can be attributed to the ability of a carbon atom to bond to other carbon atoms to form chains and networks which are skeletons from which large molecules can be produced.
2. *Inorganic compounds* are all the molecules that do not contain carbon skeletons.

Chemical reactions can occur either between molecules of the same category or between molecules of both categories. That is, the reactants may be all organic molecules or all inorganic molecules, or some may be organic and others inorganic. If the reactants are of both categories, then the reaction may result in the conversion of organic molecules to inorganic molecules or vice versa.

Because inorganic compounds as a category are less complex than organic compounds, we shall study them first. Students should realize that this order of study is merely to ease the learning process.

Note. The terms *organic* and *inorganic* are universally accepted, and I have therefore used them in this book. Nevertheless, I think they are unfortunate. All living things contain both organic and inorganic components. What may be inorganic in this part of the universe might be organic in another. A better pair of terms to describe the major categories might be *carbon skeletal* and *noncarbon skeletal* compounds.

Synthesizing Gold from Lead Is No Easy Task.

The inorganic compounds that have the greatest impact on our lives are, for the most part, constructed from elements of the first four periods. They are usually combinations of metallic or hydrogen atoms and the nonmetallic atoms oxygen, nitrogen, sulfur, and phosphorus or the halogens either singly or in groups. Those compounds in which hydrogen acts as the metallic portion of the molecule are acids. Those containing atoms other than hydrogen are salts. Particular metallic salts can be produced by a reaction between the appropriate acid and the particular base.

Example. Magnesium chloride, a salt, can be produced from hydrochloric acid by reactions with the base magnesium hydroxide.

$$2\,HCl = Mg(OH)_2 \rightarrow MgCl_2 + 2\,H_2O$$

This is in fact what occurs when a person takes milk of magnesia to relieve acid indigestion. Milk of magnesia is merely an aqueous suspension of magnesium hydroxide. It reacts with the excess hydrochloric acid in the stomach and, in so doing, relieves the discomfort.

In theory, the significance of the acid-salt conversion capability of inorganic compounds is that, once a particular acid is produced, any desired salt of this acid can be synthesized merely by combining it with the appropriate base. For many compounds the reverse conversion is also possible. (That is, given a particular salt, the parent acid can be regenerated by reaction with a strong acid.)

When solutions of salts ionize, they display characteristics of the individual component ions rather than radically new compound properties. If two ions normally undergo a particular reaction, the history of which ions they were associated with prior to the reaction is of little importance. In the production of barium sulfate, for example, it makes no difference whether the sulfate ion comes from zinc sulfate, potassium sulfate, or sulfuric acid. Any of these and many similar compounds will react with barium hydroxide, barium chloride, barium nitrate, or similar compounds to produce insoluble barium sulfate. The reaction takes place between the barium and the sulfate ions; other ions in the solution are essentially bystanders.

From this simple illustration it can be seen that although the compounds in the two groups can be paired in any of nine combinations, insoluble barium sulfate is always the end product. For this reason, knowledge of the properties of component ions is extremely useful to the understanding of inorganic reactions. When the component ions are derived from metals, it is safe to predict that other ions derived from metals in the same periodic group will have similar properties. This degree of predictability is attainable because metals mostly form simple ions. This is not, however, the case with nonmetals. Nonmetals, besides combining with metals directly, can also combine among themselves to produce negative anionic groups. Five of these groups are of considerable interest:

1. The oxygen-peroxide group.
2. The nitrogen-oxygen-hydrogen group.
3. The sulfur-oxygen-hydrogen group.
4. The phosphorous-oxygen-hydrogen group.
5. The halogen-oxygen-hydrogen group.

Since this text is an introductory primer, this list has of necessity been abbreviated. Students wishing a more comprehensive treatment can find it in any advanced inorganic text.

The Oxygen-Peroxide Group

When an oxygen atom combines with another atom, any of four different types of products may result. Depending on the particular element and the amount of oxygen, most of the resulting oxide may be classified as (1) basic, (2) acidic, (3) amphoteric, or (4) peroxides.

1. Basic oxides are compounds that react with acids to form salts. They are the result of the combination of oxygen with the most active metals. When dissolved in water, they increase the hydroxide ion concentration.

$$Na_2O + H_2O \rightarrow 2\,NaOH = 2\,Na^+ + 2\,OH^-$$

2. Acidic oxides are compounds that react with bases to form salts. They are the result of the combination of oxygen with active non-metals. When dissolved in water, they increase the hydrogen ion concentration.

$$SO_3 + H_2O \rightarrow SO_4^{=} + 2\,H^+$$

3. Amphoteric oxides are compounds that react with either acids or bases to form salts. They are the result of the combination of oxygen with elements whose nature is neither strongly electronegative (tending to form negative ions) nor strongly electropositive (tending to form positive ions) and whose bonds are neither markedly ionic nor markedly covalent. When mixed with water, they do not appreciably alter either the hydrogen or the hydroxide ion concentration.

$$\underset{\text{Base}}{Al_2O_3} + \underset{\text{Acid}}{6\,HCl} \rightarrow 2\,AlCl_3 + 3\,H_2O$$

$$\underset{\text{Acid}}{Al_2O_3} + \underset{\text{Base}}{2\,NaOH} \rightarrow 2\,NaAlO_2\ \text{(hydrated)} + H_2O$$

4. Peroxides are compounds that contain more oxygen than would be expected in an oxide compound. This is possible because some of the oxygen, rather than combining with the other atomic type, combines with itself to form an **O—O** bond. No generalized statement can be made concerning the effects of peroxide compounds on the *p*H of a solution. Depending on the particular peroxide, the *p*H of its solution may be acidic, basic, or close to neutral.

A listing of some oxides typical of these four groups appears in Table 11–1.

Since the acid-base properties of oxides have already been discussed, in Section 6, we shall turn now to examine a new type of oxygen compound, the peroxides. Although peroxides of a number of metals and nonmetals exist, by far the most familiar to the man in the street is hydrogen peroxide. The designation *peroxide* signifies that the compound contains more than the expected proportion of oxygen atoms. Hydrogen peroxide, H_2O_2, contains more oxygen than water, H_2O, which is hydrogen oxide.

Concentrated hydrogen peroxide, of purity 90 percent or higher, was used to propel the German V-2 robot bombs of World War II. Dilute hydrogen peroxide having a concentration of 5 percent or less was used to produce the blond bombshells of the same era.

Pure hydrogen peroxide is a liquid with a freezing point close to that of water, −0.89°C, and a boiling point of approximately 151°C. The value is approximate because at atmospheric pressure it explodes before reaching this temperature. Its electronic formula can be represented by

$$\mathrm{H:\overset{\cdot\cdot}{\underset{\cdot\cdot}{O}}:\overset{\cdot\cdot}{\underset{\cdot\cdot}{O}}:H}$$

Because of its electronic structure, one might easily assume that the hydrogen peroxide molecule is linear. X-ray studies, however, indicate that this is not so. The overall geometric configuration reresembles an arthritic snake preparing to strike. If the two oxygen and

TABLE 11–1 Some Typical Oxides

Basic	Acidic	Amphoteric	Peroxides
K_2O	B_2O_3	BeO	Na_2O_2
MgO	SO_3	SnO	K_2O_2
CaO	Cl_2O_7	TiO_2	SO_4
La_2O_3	CO_2	Cr_2O_3	ZnO_2

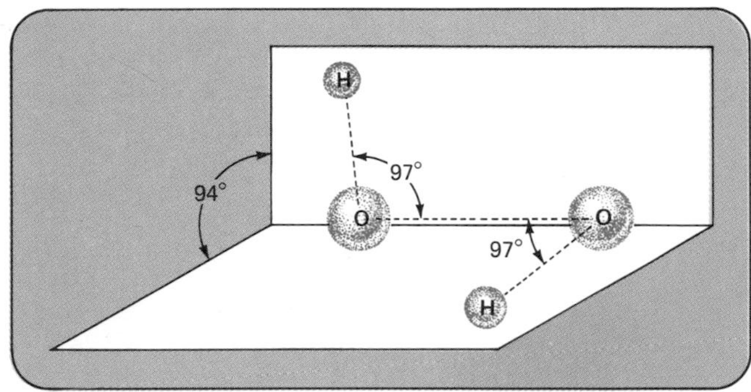

FIGURE 11-1 Geometry of Hydrogen Peroxide

one hydrogen are kept within a single plane, then the second hydrogen rises above the plane to appear as the striking head.

The current commercial method of producing hydrogen peroxide is through the electrolysis of a cold 50 percent solution of sulfuric acid. It may, however, be produced in the laboratory by the action of sulfuric acid upon barium peroxide.

$$\mathbf{Ba}^{++} + \mathbf{O_2}^{=} + 2\,\mathbf{H}^{+} + \mathbf{SO_4}^{=} \rightarrow \mathbf{H_2O_2} + \mathbf{BaSO_4}\downarrow$$

The reaction is a substitution where *anions* (negative ions) and *cations* (positive ions) are interchanged. The driving force once again is the formation of insoluble barium sulfate. The precipitation of barium sulfate from solution leaves only the slightly acid hydrogen peroxide.

Chemically, hydrogen peroxide is very unstable. It decomposes to water and oxygen with the evolution of a considerable amount of energy:

$$2\,\mathbf{H_2O_2} \rightarrow 2\,\mathbf{H_2O} + \mathbf{O_2} + 92{,}400 \text{ calories}$$

Note. A *calorie* is a unit of heat energy. One calorie provides sufficient heat to raise the temperature of one gram of water 1°C; 92,400 calories is sufficient energy to raise the temperature of a liter of water 92.4°C.

Since the decomposition of hydrogen peroxide can be catalyzed by dust, transition elements, heavy metal ions, and organic matter, it is essential if hydrogen peroxide is to be stored for any length of time, that the container be scrupulously free of these substances. To prevent concentrated hydrogen peroxide from exploding, it is stored in specially constructed aluminum tanks where its temperature is constantly monitored. If the temperature in the tank should rise above that of the surroundings, quick-release dump valves at the bottom of

the tank open to prevent explosions. A number of substances such as phosphates, acetanilide, and barbituric acid are known to inhibit the decomposition of hydrogen peroxide and are therefore often added to commercial preparations.

Hydrogen peroxide is a powerful oxidizing agent. When it is reacted, however, with stronger oxidizing agents, such as permanganate, it itself is oxidized, and the permanganate is reduced.

Example. Acting as an oxidizing agent, hydrogen peroxide oxidizes iodide ion to iodine and is itself reduced to water.

$$2\,\mathbf{I}^- \rightarrow \mathbf{I}_2 + 2\,e^- \qquad \text{(Oxidation)}$$

$$2\,e^- + 2\,\mathbf{H}^+ + \mathbf{H_2O_2} \rightarrow 2\,\mathbf{H_2O} \qquad \text{(Reduction)}$$

$$2\,\mathbf{H}^+ + \mathbf{H_2O_2} + 2\,\mathbf{I}^- \rightarrow \mathbf{I}_2 + 2\,\mathbf{H_2O}$$

Acting as a reducing agent, hydrogen peroxide reduces permanganate to manganous ions.

$$2\,(5\,e^- + 8\,\mathbf{H}^+ + \mathbf{MnO_4}^- \rightarrow \mathbf{Mn}^{++} + 4\,\mathbf{H_2O}) \qquad \text{(Reduction)}$$

$$5\,(\mathbf{H_2O_2} \rightarrow \mathbf{O_2} + 2\,\mathbf{H}^+ + 2\,e^-) \qquad \text{(Oxidation)}$$

$$2\,\mathbf{MnO_4} + 16\,\mathbf{H}^+ + 5\,\mathbf{H_2O_2} \rightarrow 2\,\mathbf{Mn}^{++} + 8\,\mathbf{H_2O} + 5\,\mathbf{O_2} + 10\,\mathbf{H}^+$$

Simplifying, we have

$$2\,\mathbf{MnO_4} + 6\,\mathbf{H}^+ + 5\,\mathbf{H_2O_2} \rightarrow 2\,\mathbf{Mn}^{++} + 8\,\mathbf{H_2O} + 5\,\mathbf{O_2}$$

It is interesting to note that when hydrogen peroxide decomposes, it is both oxidized and reduced to form water and oxygen.

$$H_2O_2 \rightarrow O_2 + 2\,H^+ + 2\,e^- \qquad \text{(Oxidation)}$$

$$2\,e^- + 2\,H^+ + H_2O_2 \rightarrow 2\,H_2O \qquad \text{(Reduction)}$$

$$2\,H_2O_2 \rightarrow 2\,H_2O + O_2$$

The Nitrogen-Oxygen-Hydrogen Group

There are four million billion tons of nitrogen in the atmosphere surrounding our planet earth. This is approximately 24 million tons over each square mile. No matter how you slice it, that's a lot of gas. Nitrogen can combine with oxygen, the other major component of air, to form eight different oxides: N_2O, NO, N_2O_3, NO_2, N_2O_4, N_2O_5, NO_3, and N_2O_6. Of these eight, the last two, NO_3 and N_2O_6, are laboratory curiosities and of little importance. The remaining six compounds contain nitrogen in all the possible oxidation states from +1 to +5. (Table 11–2.) These numbers are based on the assignment of a −2 oxidation state to each oxygen atom. The sum of the oxidation states in any compound must total zero.

Note. Two systems are in common usage in the naming of compounds composed of more than one atom of a given type. The first and simpler is to utilize Greek prefixes to distinguish the number of each type of atoms. If no prefix is indicated, then it is assumed that the number is 1. N_2O and N_2O_3, for instance, are respectively named dinitrogen oxide

TABLE

11–2 Some Oxides of Nitrogen

Name	Chemical Formula	Oxidation Number
Nitrous oxide or Dinitrogen oxide	N_2O	+1
Nitric oxide	NO	+2
Dinitrogen trioxide	N_2O_3	+3
Nitrogen dioxide	NO_2	+4
Dinitrogen tetroxide	N_2O_4	+4
Dinitrogen pentoxide	N_2O_5	+5

All the oxides with the exception of dinitrogen pentoxide are gases at room temperature.

and dinitrogen trioxide. The prefixes *mono, di, tri, tetra,* and *penta* indicate that the compound contains 1, 2, 3, 4, and 5 atoms of a given type.

The second system, which is dependent upon the oxidation number of the more metallic of the two elements in a binary compound, is the one more frequently used by nonacademically oriented chemists. The name of the compound with the higher oxidation state is assigned an *ic* suffix. The name of the compound in which the more metallic element is in the lower oxidation state ends in *ous*. Thus, **NO**, in which the nitrogen has an oxidation number of +2, is called nitric oxide, and N_2O, in which the nitrogen is in a +1 oxidation state, is called nitrous oxide. The *ic* and *ous* endings, quite unlike the Greek prefixes, merely indicate relative oxidation states without specifying exact numbers of atoms.

In a given series of compounds, usually the two most common are assigned the *ic* and *ous* suffixes. If the compounds are acids, their respective salts are assigned the suffixes *ite* and *ate*. (HN_2 = nitrous acid; HNO_3 = nitric acid; then $NaNO_2$ = sodium nitr*ite*; $NaNO_3$ = sodium nitr*ate*. If there exist other compounds in the series whose degree of oxidation is less than that found in the *ous* or more than that found in the *ic* molecules, the prefixes *hypo* and *per* are used. This nomenclature will become more understandable when the oxy-compounds of the halogens are discussed. Until that point let it suffice to say that we have already encountered the *per* prefix in the section on peroxides. In a peroxide, you may recall, the oxygen associated with the atom occurs in greater than the anticipated amount, producing an atom with a higher degree of oxidation.

Laughing gas is not all that funny.

Nitrous oxide (N_2O), the binary compound of nitrogen and oxygen in which nitrogen possesses its lowest oxidation state, was discovered in 1772 by Joseph Priestly. Priestly found that a candle burned with a greater flame in what he called "diminished nitrous air" than it did in the atmosphere. To test the biological value of this gas, which has twice as much oxygen as does air, he enlisted the services of a youthful mouse assistant. The mouse, when placed in a container of pure N_2O was unable to extract the necessary oxygen from the compound and promptly died. This short-lived scientific collaboration did not, however, deter its use on people. In 1799 Sir Humphrey Davy noticed that N_2O could be used to anesthetize animals, providing they were given oxygen shortly thereafter. As a result of these findings

Horace Wells, an American dentist, employed it for dental surgery. His attempts in 1845 to demonstrate the efficacy of N_2O as an anesthetic to the Boston medical community, however, proved to be a failure. The criticism that arose as a result forced him to give up his practice. Three years later, a broken man, he committed suicide. A most tragic ending for a man who had pioneered the use of "laughing gas."

Three-way action

Nitric oxide (NO) is colorless as a gas but, in the liquid or solid state, it is blue. It is capable of reacting in any of three ways to produce a diverse assortment of compounds.

1. *Electron sharing.* The sharing of electrons produces volatile, colored, covalent nitrosyl compounds. Nitrosyl bromide ($NOBr$) can be synthesized by passing NO into bromine at −15°C.
2. *Electron gain.* This results in the production of the NO^- ion. Sodium nitrosyl ($NaNO$) is an example of a compound containing such an anion. It is produced by the action of nitric oxide on sodium in liquid ammonia ($Na + NO \rightarrow NaNO$).
3. *Electron loss.* The loss of an electron produces the nitrosonium ion (NO^+). Nitrosonium compounds such as $NO^+ClO_4^-$, $NO^+HSO_4^-$, and $NO^+BF_4^-$ can be prepared in nonhydroxide containing solvents. Nitrosonium compounds react with hydroxide containing solvents to produce nitrous acid (HNO_2).

Nitrates cured my cornbeef and it wasn't even sick.

Dinitrogen trioxide (N_2O_3) exists only in the solid state. It is the anhydride of nitrous acid (HNO_2), which also is an unstable molecule. The stable salts that can be produced from this acid are comparatively few. They are limited to the group I and II metals, zinc, cadmium, and the single oxidation states of silver and mercury. The +3 oxidation state of nitrogen in nitrous acid and its nitrite salts allows these compounds to act as either oxidizing or reducing agents.

Note. The +3 oxidation state of nitrogen is derived by assigning a −2 to each oxygen and a +1 to the hydrogen atom. Although a weak acid, nitrous acid is a fairly strong oxidizing agent, having a standard half-cell potential of 1.00 volts.

$$HNO_2 + H^+ + e^- \rightarrow NO + H_2O \qquad E = 1.00\ V$$

In the presence of strong oxidizing agents such as chlorine, nitrous acid is oxidized to nitrate ions (NO_3^-).

$$2\,e^- + Cl_2 \rightarrow 2\,Cl^-$$

$$\frac{H_2O + NO_2^- \rightarrow NO_3^- + 2\,H^+ + 2\,e^-}{Cl_2 + NO_2^- + H_2O \rightarrow 2\,Cl^- + NO_3^- + 2\,H^+}$$

Note. The salts of nitrous acid (oxidation state +3) are *nitrites*. The salts of nitric acid (oxidation state +5) are *nitrates*.

Sodium and potassium nitrites are quite stable and show little tendency to decompose, even when they are fused. They are of commercial importance in the manufacture of organic chemicals and dyes. Sodium nitrate is used in the curing of meat and medicinally as a vasodilator. When used to dilate blood vessels, its effects are similar to nitroglycerine, only longer lasting.

Pair up and pale that nasty color.

Dinitrogen tetroxide (N_2O_4) is a colorless dimer (double molecule) of nitrogen dioxide (NO_2). It can be prepared pure only when it is a solid. As a gas it is invariably contaminated with the colored monomer. The yellowish brown color of the monomer is the result of unpaired electrons. When a compound contains either an odd number of electrons or electrons so loosely bound that they are virtually unpaired, the interaction of the electromagnetic properties of light with these electrons produces color. Since NO_2 contains 23 electrons, 8 from each oxygen and 7 from the nitrogen, it is only possible to produce 11 pairs and one odd electron. It is the interaction of this electron with light that gives NO_2 its characteristic color. When two NO_2 molecules combine to form a dimer, the resultant molecule contains 46 electrons. Since this is an even number, all the electrons can be paired. For this reason N_2O_4 is a colorless rather than colored molecule. The color change due to the reaction

$$\underset{\text{Colorless}}{N_2O_4} \leftrightarrows \underset{\text{Colored}}{2\,NO_2}$$

is often utilized to demonstrate visually the effect that temperature can have on a chemical equilibrium.

Aqua fortis is strong stuff.

Dinitrogen pentoxide (N_2O_5) is the anhydride of nitric acid, the most important oxy-nitrogen acid.

$$2\,HNO_3 \xrightarrow{\text{dehydration}} N_2O_5 + H_2O$$

Pure nitric acid is a colorless fuming liquid with a sharp odor. Its chemical nature is broad because it is a strong oxidizing agent as well as a strong acid. Although solutions of nitric acid will react with bases to form nitrate salts, hydrogen is almost never evolved when it reacts with metals. This is because nitric acid can oxidize hydrogen to water. Although aqua fortis ("strong water" in Latin), which is pure nitric acid, will not dissolve either gold or platinum, a mixture of one part nitric acid to three parts of hydrochloric acid called aqua regia ("royal water" in Latin) will.

Nitric acid and its associated nitrate salts are the major sources for the production of nitrogen fertilizers for agriculture and explosives for construction and less desirable uses. All chemical explosives depend upon nitrates or nitro groups incorporated in organic molecules for their destructive ability. The action of nitric acid upon glycerine produces glyceral trinitrate, commonly referred to as nitroglycerine. The detonation of glycerol trinitrate produces elementary nitrogen plus more than enough oxygen to burn all the carbon and hydrogen.

$$4\,C_3H_5(NO_3)_3 \rightarrow 10\,H_2O + 12\,CO + 6\,N_2 + 7\,O_2 + \text{(BOOM!)}$$

Dynamite is produced by adding either wood, flour, or cellulose nitrate to glycerol trinitrate. This addition serves two purposes. The dry powder absorbs the liquid glycerol trinitrate, making it easier to handle, and, at the same time, provides additional fuel for the excess oxygen to consume. Dynamite was first manufactured by Alfred Nobel and is one of the inventions upon which he built his fortune. It is from this fortune that monies for the Nobel Prizes are provided.

The negative side of the nitrogen picture

In compounds of nitrogen and more electronegative elements, nitrogen assumes a positive oxidation number. This was the case for the oxy nitrogen compounds. (Electronegativity of oxygen = 3.5; that of nitrogen = 3.0.) When nitrogen is associated with elements having lower electronegativities, however, it assumes a decidedly negative character. Such is the case for compounds of nitrogen and hydrogen. (Electronegativity of hydrogen = 2.2.) In these compounds the oxidation number of nitrogen may be either −1 for hydroxylamine (NH_2OH), −2 for hydrazine (N_2H_4), or −3 for ammonia (NH_3).

Hydroxylamine (NH_2OH) is a colorless solid that is very soluble in water and has slightly basic properties. It has both oxidizing and reducing capabilities. This is not too surprising since the molecule can be considered the result of combining half a hydrogen peroxide molecule with half a hydrazine molecule. It should not be forgotten, however, that although a mule displays many of the characteristics of both a horse and a donkey, it is an entity unto itself. The same is true of the hydroxylamine molecule.

Hydrazine (N_2H_4) is the nitrogen analog of hydrogen peroxide. Its chemistry as a result is in many ways closely related to the chemistry of hydrogen peroxide. It can react as either an oxidizing or a reducing agent. In water it forms a weakly basic solution capable of reducing free oxygen to water.

$$N_2H_4 + O_2 \rightarrow N_2 + 2\,H_2O$$

Ammonia (NH_3) is the most common of the nitrogen-hydrogen binary compounds. Its characteristic odor is usually quite evident when nitrogenous plant or animal matter decomposes. During the time of the alchemists it was prepared by the destructive distillation of deer horns. The medical term *spirits of hartshorn* is derived from this synthetic procedure.

In 1912, the German chemists Haber and Bosch developed a procedure for the direct synthesis of ammonia from nitrogen and hydrogen. This process is accountable for most of the ammonia produced in the United States today. It can be summarized by the following equation:

$$N_2 + 3\,H_2 \xrightleftharpoons[\text{Fe}]{\text{200 atm. 455°C}} 2\,NH_3$$

The data surrounding the arrow indicate that the nitrogen and the hydrogen are combined at 200 atmospheres pressure and 475 degrees Celsius over an iron catalyst. Since the reaction is reversible, these conditions are employed in order that the optimum amount of product per given amount of starting material can be produced. In the United States alone, millions of tons of ammonia are produced by this process each year.

It is not surprising to find that the ammonium salts of a given anion bear close resemblances to their lithium, sodium, and potassium counterparts. This is because the charge on the ammonium ion (NH_4^+) is the same, and the size is quite similar to that of the alkali earth ions. Liquid ammonia, like water, is an ionizing solvent and a good medium for ionic reactions.

$$2\,H_2O \rightleftharpoons H_3O^+ + OH^-$$

$$2\,NH_3 \rightleftharpoons NH_4^+ + NH_2^-$$

The ammonium ion (NH_4^+) in liquid ammonia is equivalent to the hydronium ion (H_3O^+) in water. In a similar manner the amide ion (NH_2^-) in liquid ammonia is equivalent to the hydroxyl ion (OH^-) in water. Any compound that ionizes in liquid ammonia to produce amide ions would therefore be considered a base in that system. Similarly, an acid in the liquid ammonia system will ionize to produce ammonium ions. The products of neutralization reactions in this medium are salts and ammonia. This is exactly analogous to aqueous neutralizations where the products are salts and water.

The following two equations allow the comparison of neutralization reactions in both mediums. Ammonium chloride (NH_4Cl), an acid, is reacted with sodium amide ($NaNH_2$), a base, in liquid ammonia. Hydrochloric acid (HCl), an acid, is reacted with sodium hydroxide ($NaOH$), a base, in water.

$$NH_4^+ + Cl^- + Na^+ + NH_2^- \rightarrow 2\,NH_3 + Na^+ + Cl^- \qquad \begin{pmatrix}\text{Neutralization}\\ \text{in ammonia}\end{pmatrix}$$

$$H_3O^+ + Cl^- + Na^+ + OH^- \rightarrow 2\,H_2O + Na^+ + Cl^- \qquad \begin{pmatrix}\text{Neutralization}\\ \text{in water}\end{pmatrix}$$

Because of the ease of preparation and low cost, most industrial nitrogen compounds are synthesized from ammonia. As a case in point, nitric acid is manufactured by burning ammonia over a platinum catalyst to produce nitric oxide.*

$$4\,NH_3 + 5\,O_2 \xrightarrow{Pt} 4\,NO + 6\,H_2O$$

The nitric oxide (NO) is then reacted with air to produce nitrogen dioxide (NO_2).

$$2\,NO + O_2 \rightarrow 2\,NO_2$$

Nitrogen dioxide when reacted with water produces a mixture of nitric acid and nitric oxide.

$$3\,NO_2 + H_2O \rightarrow 2\,HNO_3 + NO$$

The biproduct nitric oxide is recycled and reoxidized. The nitric acid can be harvested as such or allowed to react with ammonia, the starting material, to produce ammonium nitrate (NH_4NO_3), an extremely fine fertilizer.

$$NH_3 + HNO_3 \rightarrow NH_4NO_3$$

*A *catalyst* is a substance that shortens the time it takes for a chemical reaction to come to equilibrium. Unlike reactants and products, the catalyst is unchanged by the reaction.

TABLE 11–3 The Oxidation States of Nitrogen

Oxidation Number	Compounds
+5	N_2O_5, HNO_3
+4	NO_2, N_2O_4
+3	HNO_2
+2	NO
+1	N_2O
0	N_2
−1	NH_2OH
−2	N_2H_4
−3	NH_3

Although primarily prepared and sold as a high nitrogen fertilizer, ammonium nitrate, because of its nitrate content, is a material deserving respect in handling. It can decompose with explosive violence. In Texas City in 1947, two 10,000-ton freighters were being loaded with ammonium nitrate for shipment to France when a fire broke out aboard one of the ships. As a consequence, the ammonium nitrate detonated, leveling Texas City and resulting in the loss of 600 lives. This unfortunate accident was only possible because the amount of ammonium nitrate being loaded exceeded its critical mass (the minimum mass necessary for an explosion). The mass of ammonium nitrate as packed for use as a lawn and garden product, however, is much less than this quantity and can be safely used without fear of detonation.

So far, nine different oxidation states of nitrogen have been discussed. Table 11–3 lists these states and compounds representative of each.

The Sulfur-Oxygen-Hydrogen Group

Sulfur has been known since ancient times. Records as early as 1000 B.C. describe its use as both a medicine and a fumigant. Its effectiveness as a fumigant for entire cities is cited in the Bible: "The Lord rained upon Sodom and Gomorrah brimstone and fire" (Genesis 19:24).

Brimstone gets its English name from the German *Brennstein*, which means "stone that burns." Its ability to burn fascinated the

ancients. They believed, according to Paracelsus, that sulfur was one of the three constituents of the human body, along with salt and mercury. The alchemists believed it to be the active principle in fire. With these two millstones around its neck, it is no wonder that chemistry progressed very slowly until relatively recent times.

Anyone who has ever been subjected to the fumes emanating from the combustion of sulfur can have little doubt as to why the process is associated with hell. It is a hell of a choking stench.

Sulfur is widely distributed in nature. In the volcanic districts of Iceland, Italy, Mexico, and Sicily it is found in the free state. Although the exact mechanism of the production of native sulfur around volcanic sites is not completely understood, there is general agreement that it is the result of the reaction of hydrogen sulfide ($\mathbf{H_2S}$) with sulfur dioxide ($\mathbf{SO_2}$). Before 1900, about 90 percent of the world's sulfur was furnished by Sicily, which has numerous fumaroles and hot springs that bring the sulfur from its deeply buried deposits to the surface. Once the sulfur has surfaced, it is easily collected by scraping the surrounding rocks or extracting it from its earth mixture by heating.

Mining the large sulfur deposits along the Gulf Coast of the United States presented a bit more of a challenge. There the sulfur is deposited in calcite formations, 1000 feet or more beneath a surface of clay, rock, and quicksand. The challenge was met by Herman Frasch, who created his own Louisiana and Texas fumaroles. These were made by boring through the overlying material to the sulfur beds and sinking three concentric pipes. Superheated water (at 170°C) was forced down the outermost pipe and, when it melted the sulfur, hot compressed air was forced down the innermost. The resulting hot-air–liquid-sulfur mixture, being lighter than water, was forced through the central pipe to the surface, where it was collected. (Figure 11–2.) Sulfur obtained in this manner may have a purity as great as 99.9 percent. The Frasch process was the leading method of obtaining sulfur in the United States until 1970. Since then, most of the sulfur produced in this country has been a biproduct of the purification of natural gas.

Get your red-hot sulfur before it turns black.

Pure sulfur is not a simple substance. It is *allotropic.* This means that sulfur can exist in more than one form per state, for example, three different solid forms. From experience we expect that a liquid, when heated, will become less viscous: north of the equator molasses moves more readily in August than in December. The behavior of liquid sulfur, however, is most curious and quite the opposite. Just above its melting point, it is a very fluid straw-colored liquid. On further heating it becomes quite viscous and changes its color to deep red. If heated still further so that its temperature approaches its boiling point, the viscosity decreases again and the color changes to black. It should be stressed that throughout such heatings and color changes the sulfur remains 100 percent chemically pure. Only its physical characteristics change.

If sulfur consisted of perfectly spherical elastic molecules, one would be hard put to explain its change with temperature. Just above the melting point, however, sulfur exists almost entirely as $\mathbf{S}_8$ molecules. (Figure 11–3.) At this temperature, it is a straw-colored liquid of low viscosity. It is so lightly colored because all the valence electrons are tightly bound in covalent bonds with adjoining sulfurs; interaction with light is therefore minimal.

As the temperature is raised, molecular vibrations and collisions become more and more violent until finally some of the $\mathbf{S}_8$ rings are broken open. Because these broken rings have an odd electron at their ends, they are capable of interacting with visible light. This causes the darkening of the liquid. The viscosity increases because the free ends of these broken rings are capable of joining the ends of other broken rings to produce longer chains, which become entangled more easily. When the temperature is raised above 160°C, thermal agita-

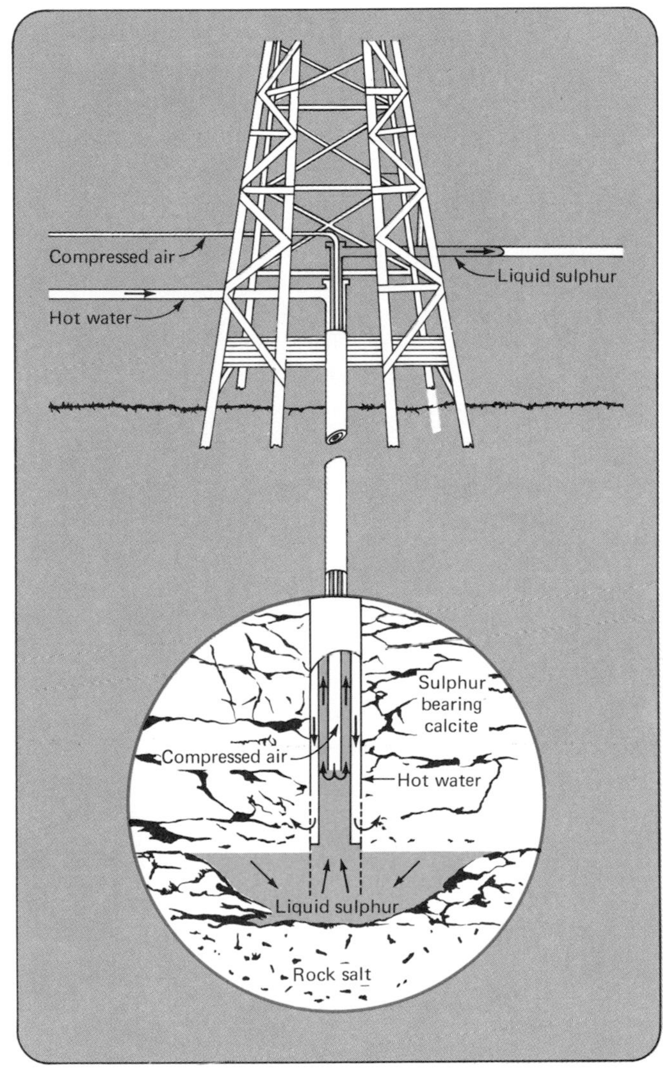

FIGURE

11–2 The Frasch Process of Mining Sulfur

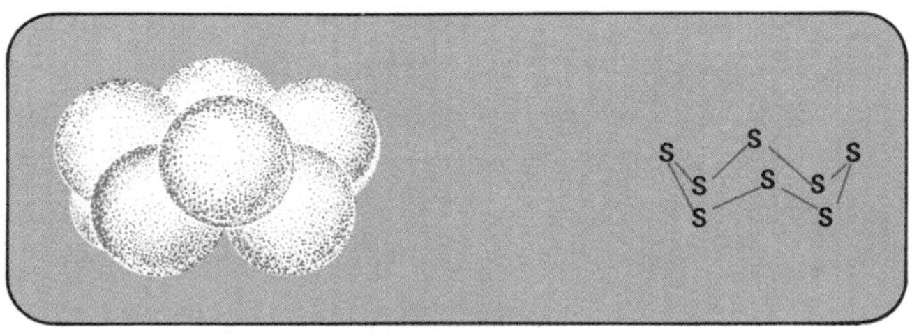

FIGURE

11–3 Structure of an S_8 Molecule

TABLE 11–4 The Oxidation States of Sulfur

Oxidation Number	Compounds
+6	SO_3, H_2SO_4, SF_6
+5	—
+4	SO_2, H_2SO_3, SF_4
+3	—
+2	SF_2
+1	S_2F_2 (contains S–S bonds)
0	S_8
−1	Na_2S_2 (contains S–S bonds)
−2	H_2S

tion causes a breakup of the entangled chains, and the viscosity starts to decrease. Because the broken chains have more unpaired electrons than the entangled chains, they absorb more light, turning the liquid black.

At its boiling point, sulfur again becomes very fluid and evaporates as a mixture of S_2 and more complex molecules. If at any time during heating the sulfur is allowed to cool for an extended period, it will be converted back to its original pale yellow solid form, unchanged for its heating experience.

Sulfur and its congeners, selenium and tellurium, are the heavier elements of group VII, the group headed by oxygen. The range of the oxidation states of sulfur is +6 to −2. (Table 11–4.) Although it is as great as that for nitrogen, unless sulfur-to-sulfur bonds exist within the molecules only the even-numbered states need be considered.

Of these seven oxidation states the one of greatest commercial importance is the highest. This importance is due to sulfuric acid. Sulfuric acid enters into the manufacture of almost everything produced in an industrial nation, from soluble phosphate fertilizers to natural and synthetic textiles, petroleum, and batteries. It is safe to say that a country's annual sulfuric acid consumption is one of the best indicators of its economic status. More than 25 million tons are produced and used in the United States per year. (Table 11–5.)

About 1740, when sulfuric acid was first manufactured commercially in England, its price was about a shilling (approximately 20 cents) per pound. In the years from 1793 to 1860 advances in processing techniques reduced the price to approximately one cent per pound, where it has stayed for more than a century. The various processes for producing sulfuric acid (H_2SO_4) all depend on the hydration of sulfur trioxide (SO_3). This is normally accomplished by passing the sulfur trioxide into sulfuric acid and then hydrating the resultant pyrosulfuric acid ($H_2S_2O_7$).

$$SO_3 + H_2SO_4 \rightarrow H_2S_2O_7$$

$$H_2S_2O_7 + H_2O \rightarrow 2\,H_2SO_4$$

The sulfur trioxide is synthesized by passing oxygen and sulfur dioxide (SO_2), produced from the burning of sulfur in air, over vanadium pentoxide or platinum catalysts.

$$2\,SO_2 + O_2 \xrightarrow{\text{catalyst}} SO_3$$

Sulfuric acid is a colorless, oily liquid 1.8 times denser than water. It freezes at 10.4°C and begins to boil with decomposition at 290°C. Because of its oily appearance it is sometimes referred to as "oil of vitriol." The molecular structure of sulfuric acid, both electronic and spatial, is depicted in Figure 11–4. Each sulfur atom is surrounded tetrahedrally by four oxygens. These tetrahedra are joined to each

TABLE 11–5 Annual Consumption of Sulfuric Acid in the United States

Industry	Tons (Pure)	Use
Fertilizer	11,000,000	Manufacture of superphosphate and ammonium sulfate
Chemicals	6,100,000	Manufacture of inorganic acids (nitric, hydrochloric and hydrofluoric), organic compounds, explosives
Paints and pigments	2,600,000	
Metallurgy	2,000,000	Surface cleaning and refining of metals
Petroleum	1,800,000	Decolorization and purification of petroleum products
Synthetic fibers and films	1,600,000	
Miscellaneous	600,000	
Total	25,700,000	

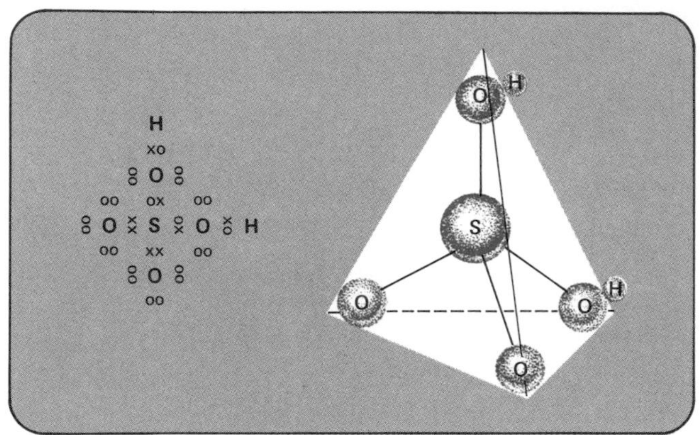

FIGURE

11–4 Molecular Structure of Sulfuric Acid

other through hydrogen bonds with their associated hydrogen atoms. It is this hydrogen bond formation that causes sulfuric acid's high boiling point.

A diprotic acid (one containing two protons), sulfuric acid ionizes in two steps. In dilute solutions the first ionization is almost complete.

$$(1)\quad \mathbf{H_2SO_4 \rightarrow H^+ + HSO_4^-}$$

$$(2)\quad \mathbf{HSO_4^- \rightleftarrows H^+ + SO_4^=}$$

Dilute sulfuric acid displays typical acid properties, reacting with bases to form bisulfate (HSO_4^-) and sulfate ($SO_4^=$) salts,

$$\mathbf{NaOH + (excess)\ H_2SO_4 \rightarrow NaHSO_4 + H_2O}$$

$$\mathbf{2\,NaOH + H_2SO_4 \rightarrow Na_2SO_4 + 2\,H_2O}$$

and with metals above hydrogen in the electromotive series to liberate hydrogen gas.

$$\mathbf{Fe + H_2SO_4 \rightarrow FeSO_4 + H_2}$$

In this reaction the iron is oxidized to the ferrous state (Fe^{++}) and the hydrogen is reduced to free hydrogen. The oxidation number of the sulfur, however, remains the same.

At high concentrations, sulfuric acid can react in three additional modes. It can be used (1) to produce more volatile acids from their salts, (2) to act as an oxidizing agent, or (3) to act as a dehydrating agent. The first of these is merely a substitution reaction.

$$\mathbf{2\,NaCl + H_2SO_4 \rightarrow Na_2SO_4 + 2\,HCl\uparrow}$$

Since the boiling point of hydrochloric acid (**HCl**) is lower than that of sulfuric, its vapors can be extracted from the reaction mixture by heating.

When concentrated sulfuric acid is hot, it is a formidable oxidizing agent capable of oxidizing metals both above and below hydrogen in the electromotive series. Zinc, for example, is above:

$$\mathbf{Zn} + 2\,\mathbf{H_2SO_4} \rightarrow \mathbf{ZnSO_4} + \mathbf{SO_2} + 2\,\mathbf{H_2O}$$

And silver is below:

$$2\,\mathbf{Ag} + 2\,\mathbf{H_2SO_4} \rightarrow \mathbf{Ag_2SO_4} + \mathbf{SO_2} + 2\,\mathbf{H_2O}$$

Note, however, that these two oxidizing reactions are independent of sulfuric acid's acid properties. When sulfuric acid reacts as an acid, it liberates hydrogen without a change in the oxidation state of the sulfur. In the reactions with zinc and silver, no hydrogen is liberated and there is a marked change in the oxidation state of some of the sulfur atoms. In going from $\mathbf{SO_4^{=}}$ to $\mathbf{SO_2}$, the oxidation state of the sulfur is reduced from +6 to +4. To balance this reduction the metals are oxidized from zero to their respective ions, $\mathbf{Zn^{++}}$ and $\mathbf{Ag^{+}}$.

Besides oxidizing metals, hot concentrated sulfuric acid is also capable of oxidizing nonmetals such as carbon and sulfur.

$$\mathbf{C} + 2\,\mathbf{H_2SO_4} \rightarrow \mathbf{CO_2} + 2\,\mathbf{SO_2} + 2\,\mathbf{H_2O}$$

$$2\,\mathbf{S} + 4\,\mathbf{H_2SO_4} \rightarrow 6\,\mathbf{SO_2} + 4\,\mathbf{H_2O}$$

In the second equation, sulfur is both reduced and oxidized. The sulfur in the sulfate ion is reduced from +6 to +4 and the free sulfur is oxidized from 0 to +4.

Concentrated sulfuric acid has a great affinity for water. Not only is it capable of removing water vapor from the atmosphere, but it is also capable of extracting water from entire molecules. The latter property can be demonstrated by adding concentrated sulfuric acid to a concentrated sugar solution. The clear, colorless sugar solution in a very short time is converted to a mass of charcoal and dilute acid. The reaction can be represented by

$$\mathbf{C_{12}H_{22}O_{11}} \xrightarrow[\text{conc.}]{\mathbf{H_2SO_4}} 12\,\mathbf{C} + \mathbf{H_2O}$$

The suffixes *ic* for the acid and *ous* for the salts indicate that the oxidation state is the higher of the two most common oxy-sulfur acids. Sulfurous acid ($\mathbf{H_2SO_3}$) and its sulfite ($\mathbf{SO_3^{=}}$) salts comprise the lower state. Sulfurous acid is produced by the hydration of its anhydride sulfur dioxide ($\mathbf{SO_2}$). The oxidation state of sulfur in these compounds is +4. Although pure sulfurous acid has never been isolated, its bisulfite (HSO_3^-) and sulfite ($SO_3^=$) salts are quite stable. The two types of salts arise because sulfurous, like sulfuric, is a diprotic acid. The synthesis and ionizations of this weak acid can be represented by

$$SO_2 + H_2O \rightleftarrows H^+ + HSO_3^-$$

$$HSO_3^- \rightleftarrows H^+ + SO_3^=$$

Sulfite ions have a fairly strong attraction for protons and therefore produce slightly basic solutions. Bisulfite ions, on the other hand, have a greater tendency to release protons than to take them up. Their solutions are slightly acid. Sulfites and bisulfites of alkali metals can be prepared by reacting their hydroxide solutions with sulfur dioxide. If an excess of sulfur dioxide is used, bisulfites are formed.

$$KOH + (\text{excess})\ SO_2 \rightarrow KHSO_3$$

If, however, only one mole of SO_2 is used for each two moles of hydroxide, then sulfites are formed.

$$2\,KOH + SO_2 \rightarrow K_2SO_3 + H_2O$$

The identical sulfite product is also attainable by treatment of the bisulfite product of the first reaction with an equivalent quantity of hydroxide.

$$KHSO_3 + KOH \rightarrow K_2SO_3 + H_2O$$

Sulfites, sulfurous acid, and sulfur dioxide are used industrially as bleaching agents and food preservatives. Sulfur dioxide is used to kill resident fungi and bacteria on dried fruits because it is far more toxic to plants than it is to man. Calcium bisulfite is used extensively in the manufacture of paper pulp both as a bleach and to separate the cellulose fibers.

The other two even oxidation states of sulfur are +2 and −2. In the +2 state, sulfur can combine with the more electrophilic (having the desire to acquire electrons) halogens to form highly reactive halides such as SF_2 and SCl_2. The fluoride is so reactive it will attack both glass and quartz and, as a consequence, must be stored in platinum containers. It has never been prepared in its pure state. By contrast, the −2 oxidation state of sulfur, the one that might be predicted on the basis of its location in the periodic chart, is relatively stable. Hydrogen sulfide (H_2S), a weak acid, and its congener acids, hydrogen selenide (H_2Se) and hydrogen telluride (H_2Te) are all diprotic, colorless, foul smelling, highly toxic compounds. Hydrogen sulfide is the odor most commonly associated with rotten eggs. Its toxicity is comparable to hydrogen cyanide (**HCN**). Because the nose can detect it, however, when it is present in extremely low concentrations, it is considered less of a hazard. With the exception of the ammonium and alkali metal sulfides, nearly all the common metal sulfides decompose or are insoluble in water. The insolubility of silver sulfide (Ag_2S) tarnish can be attested to by anyone who ever had to polish silver flatware.

The odd ones

Sulfur can exist in compounds in which its oxidation number is either +1 or −1. These odd oxidation numbers are the result of structures containing sulfur-to-sulfur bonds. Sulfur monochloride (S_2Cl_2), which is of commercial importance in the manufacture of rubber, contains sulfur in the +1 state. It is an excellent solvent for sulfur, iodine, certain metal halides, and many organic compounds. Its structure is depicted in Figure 11–5.

Sulfur in the −1 oxidation state is truly a queer duck. It is usually produced by dissolving sulfur in aqueous solutions of alkali metal sulfides. If such a solution is added to cold concentrated hydrochloric acid, a water-insoluble, yellow oil is produced. This oil, when fractionally distilled, can be separated into the component compounds H_2S_2 and H_2S_3, and smaller amounts of H_2S_4, H_2S_5, and H_2S_6. By analogy to hydrogen peroxide, the compound H_2S_2 is called hydrogen persulfide. It is comparable to hydrogen peroxide in structure, but it is much less ionic and therefore a poorer solvent for ionic compounds. It is, however, an excellent solvent for covalent compounds and sulfur.

The polysulfides (those compounds having from three to nine sulfurs

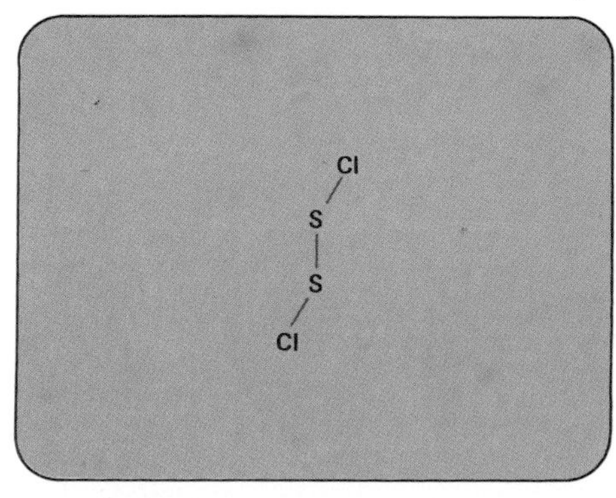

FIGURE 11–5 Geometry of Sulfur Monochloride

associated with each two hydrogens) possess a structure similar to the persulfide structure but contain as many as nine sulfur atoms sandwiched between the two hydrogens. These compounds of apparent fractional oxidation number all revert to hydrogen sulfide and sulfur in the presence of bases or at elevated temperatures.

The Phosphorus-Oxygen-Hydrogen group

Jean Louis Rodolphe Agassiz, Swiss naturalist, doctor, recipient of the Legion of Honor, and holder of the chair of Natural History at Harvard, promulgated the belief that fishermen, as a consequence of eating fish containing phosphorus, were smarter than farmers. Because phosphorus is an essential constituent of brain and nerve tissue, Agassiz and many others during the 1860s believed that the ingestion of fish (which actually contains its highest phosphate content in its bone structure) would increase the intelligence of the eater.

A dull fellow wishing to sharpen his wits approached a fishmonger with his problem. On the fish seller's suggestion he purchased, cooked, and ate a pound of fish heads. While inquiring of his neighbor as to whether he could discern any beneficial effects of the treatment, he was perturbed to discover that his neighbor had been charged only a dollar a pound for fish, whereas he had been charged five dollars a pound for the heads. Enraged at being both tricked and cheated, he returned to complain to the fishmonger. The merchant, on seeing his anger, smiled and pointed out to him that it was truly remarkable that only one pound of fish heads could produce such a marked improvement in the dullard's intelligence.

Phosphorus was discovered, by accident, in 1669 by a Hamburg alchemist, Hennig Brand. Hoping to produce the "philosopher's stone," which would allow him to transmute silver to gold, Brand heated in a retort a mixture of urine residue and white sand. (Note: We know today this was a poor choice.) The vapor that was produced condensed into a white, translucent solid having the marvelous properties of glowing in the dark and igniting spontaneously. (The name phosphorus is derived from the Greek word meaning light bearer.) The discovery of this remarkable material attracted a great deal of attention. It was considered one of the wonders of nature and displayed before the crowned heads of Europe. Because of the difficulty in preparing it, however, its price at the time of its discovery was almost as much as that for gold.

The urine constituent from which Brand most likely derived his phosphorus was mirocosmic salt ($NaNH_4HPO_4$). It is the final product excreted by humans after ingesting foods containing phosphorus. The heat and sand probably converted this salt to tetraphosphorous decaoxide (P_4O_{10}), which, on reaction with the organic material in urine, was reduced to the native phosphorus. The complexity of the composition of urine makes it impossible to write balanced equations for these reactions.

The chemistry of phosphorus is very like that of nitrogen except that it is more electropositive. As a result of this, its hydrides are less stable and its oxides and halides are more stable than the corresponding nitrogen compounds. The component bonds in most phosphorus compounds are covalent; however, when phosphorus combines with extremely electropositive elements such as potassium or cesium, it is capable of forming ionic phosphides such as potassium phosphide (K_3P). Deposited mainly as the minerals calcium phosphate ($Ca_3(PO_4)_2$) and apatite ($Ca_5(PO_4)_3F$), it makes up about 0.12 percent of the earth's crust. It is obtained from these minerals by heating with sand (SiO_2) and coke (C) to a very high temperature in an electric furnace.

$$2\,Ca_3(PO_4)_2 + 6\,SiO_2 \rightarrow 6\,CaSiO_3 + P_4O_{10}$$

$$P_4O_{10} + 10\,C \rightarrow P + 10\,CO$$

The extraction technique is basically the same as that used by Brand, but the yields are far superior.

Like sulfur, phosphorus and two of its group V companions, arsenic and antimony, each have several allotropic forms. Phosphorus itself occurs in white, red, and black forms. The white form is composed of P_4 molecules in which each phosphorus atom has an unshared pair of electrons and completes its octet by forming three single covalent bonds with the other three phosphorus atoms of the molecule. The molecular geometry resulting from this type of bonding is tetrahedral, making the molecule nonpolar and therefore soluble in nonpolar solvents. (Figure 11–6.) Red phosphorus can be prepared by heating white phosphorus to 250°C in the absence of air. (Structurally, red phosphorus can be considered a polymer composed of P_4 molecules.) It is less soluble and less reactive than white phosphorus. Black phosphorus is the least soluble and least reactive of the phosphorus allotropes. It can be prepared by subjecting phosphorus to high pressures. These pressures produce crystals of a layered, flaky material resembling graphite and, like graphite, capable of conducting electricity. (Figure 11–7.)

Although phosphorus is in the same periodic group as nitrogen, its oxidation states are not so extensive. (Table 11–6.) Arsenic, antimony, and bismuth—other group members—have even smaller numbers of oxidation states.

Note. Orthophosphoric acid (H_3PO_4) results from the complete hydration of tetraphosphorous decaoxide (P_4O_{10}).

$$P_4O_{10} + 6\,H_2O \rightarrow 4\,H_3PO_4$$

Many texts refer to P_4O_{10} as phosphorus pentoxide. Although such a designation might adequately describe the empirical formula, the actual molecule contains twice the number of atoms. Its structure is depicted in Figure 11–8. Orthophosphoric acid is a colorless solid and, unlike nitric acid, is a weak acid with minimal oxidizing ability.

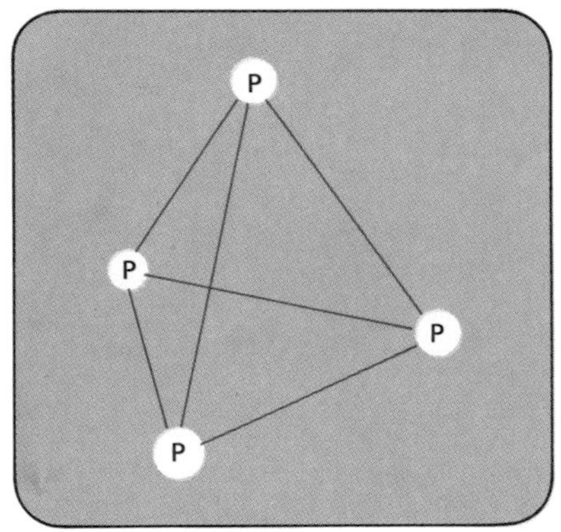

FIGURE 11–6 Structure of a P_4 Molecule

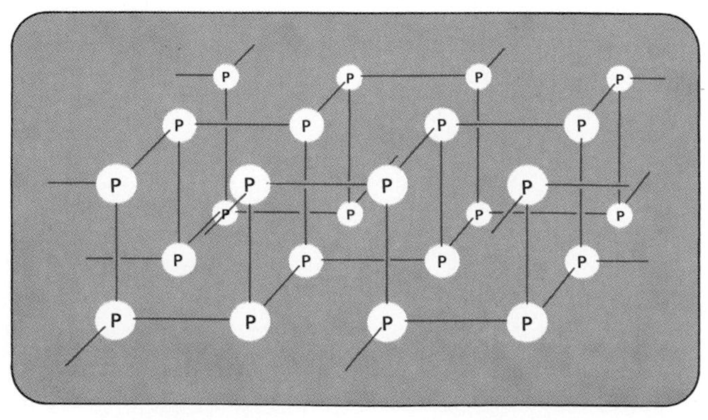

FIGURE 11–7 Black Phosphorus Crystal Structure

TABLE

11–6 The Oxidation States of Phosphorus

Oxidation Number	Compounds
+5	P_4O_{10}, H_3PO_4, PCl_5
+4	—
+3	P_4O_6, H_3PO_3, PCl_3
+2	—
+1	H_3PO_2
0	P_4
−1	—
−2	P_2H_4
−3	PH_3

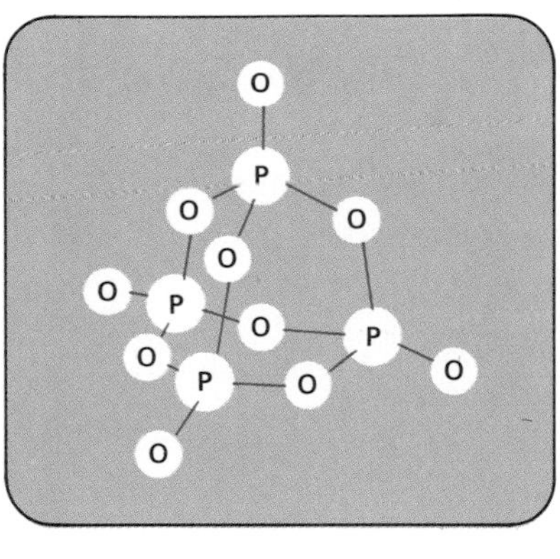

FIGURE

11–8 P_4O_{10} Molecular Structure

Normally it is sold as an 85 percent aqueous solution. Its claim to fame is that it is the only common triprotic inorganic acid. Its three-step dissociation process can be depicted by the following equations:

(1) $H_3PO_4 \rightleftarrows H^+ + H_2PO_4^-$

(2) $H_2PO_4^- \rightleftarrows H^+ + HPO_4^=$

(3) $HPO_4^= \rightleftarrows H^+ + PO_4^{\equiv}$

This triple ionization process allows for the production of three different types of salts when phosphoric acid is neutralized by a base:

1. Sodium dihydrogen phosphate (**NaH_2PO_4**)
2. Sodium monohydrogen phosphate (**Na_2HPO_4**)
3. Sodium phosphate (**Na_3PO_4**)

In the synthesis of phosphoric acid, we have noted that orthophosphoric acid results from the complete hydration of P_4O_{10}. When this hydration is less than complete, condensed phosphoric acids, having more than one phosphorous atom per molecule, are formed. Two large groups of condensed phosphoric acids are the metaphosphoric acids and the polyphosphoric acids. The metaphosphoric acids conform to the general formula $H_nP_nO_{3n}$. Examples are:

$H_3P_3O_9$, trimetaphosphoric acid

$H_4P_4O_{12}$, tetrametaphosphoric acid

The polyphosphoric acids conform to the general formula $H_{n+2}P_nO_{3n+1}$. Examples are:

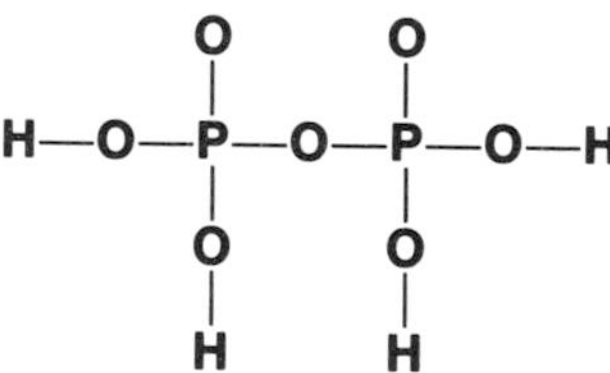

$H_4P_2O_7$, diphosphoric acid

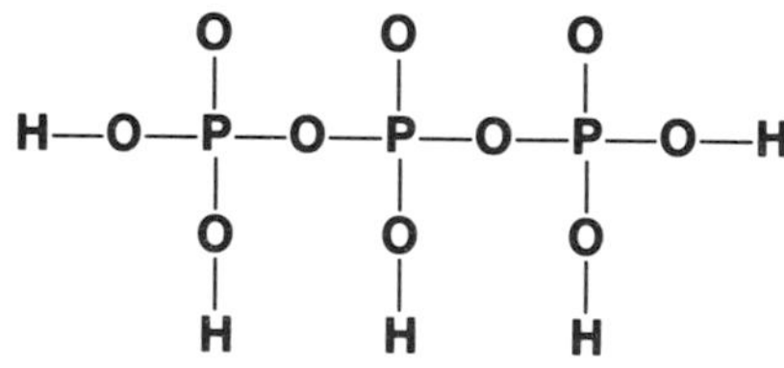

$H_5P_3O_{10}$, triphosphoric acid

It is difficult to believe that these marked structural differences are merely the result of differences in the number of water molecules associated with each molecule of acid. Yet if one mole of water is added to a mole of diphosphoric acid ($H_4P_2O_7$) or three moles of water are added to a mole of trimeta phosphoric acid ($H_3P_3O_9$), then the equivalent of two and three moles of orthophosphoric acid will be produced.

$$H_4P_2O_7 + H_2O \rightarrow 2\,H_3PO_4$$

$$H_3P_3O_9 + 3\,H_2O \rightarrow 3\,H_3PO_4$$

In each of these structures the molecule is based on tetrahedral units where each phosphorous atom is bonded to four oxygen atoms. All are phosphoric acids.

Phosphorous acid, like its nitrogen and sulfur counterparts, has one fewer oxygen and a central atom oxidation state two less than the *ic* acid. It can be prepared by adding water to its anhydride P_4O_6.

$$P_4O_6 + 6\,H_2O \rightarrow 4\,H_3PO_3$$

Condensed phosphorous acids, like condensed phosphoric acids, can be produced by limiting the water used to hydrate the anhydride.

The formula of phosphorous acid is often written as H_3PO_3. Nevertheless, it is a weak diprotic acid rather than a triprotic acid. This is because the three hydrogens are not equivalent. Two hydrogens are attached to oxygen and are thus capable of ionizing, but the third is attached to phosphorus and is not.

```
       O
       |
HO—P—OH
       |
       H
```

Phosphorous acid ionizes according to the following equations:

$$H_2(HPO_3) \rightleftarrows H^+ + H(HPO_3)^-$$

$$H(HPO_3) \rightleftarrows H^+ + (HPO_3)^=$$

On neutralization with sodium hydroxide ($NaOH$) both sodium dihydrogen phosphite (NaH_2PO_3) and sodium hydrogen phosphite (Na_2HPO_3) can be prepared. It is impossible, however, to prepare sodium phosphite because of the difference in bonding of the third hydrogen.

There exists yet a third oxy-phosphorous acid (H_3PO_2) in which the phosphorous oxidation number is +1. Since this oxidation number is less than +3, which is the value of phosphorous acid, it is called hypophosphorous ("less than phosphorous" in Greek) acid. This acid, although it contains three hydrogen in its formula, is only monobasic. The reason is that two hydrogen are attached directly to the phosphorous and only the hydrogen attached to the oxygen is ionizable.

```
          H
          |
H—O—P—O
          |
          H
```

Hypophosphorous acid ionizes in the following manner:

$$H(H_2PO_2) \rightleftarrows H^+ + H_2PO_2^-$$

It reacts with bases to produce only one hypophosphite salt.

$$H(H_2PO_2) + NaOH \rightarrow NaH_2PO_2 + H_2O$$

Phosphine smells like a salami gone bad.

The formulas of the hydrides of phosphorus bear a marked resemblance to those of nitrogen. P_2H_4, which contains phosphorus in the −2 oxidation state, corresponds in formula although not behavior to hydrazine (N_2H_4). Phosphine (PH_3), the −3 state, corresponds to ammonia (NH_3).

P_2H_4 is a colorless liquid having a density approximately equal to water. It can be formed by either the hydrolysis of Ca_3P_2 or the catalytic action of HCl on PH_3. Highly unstable, it spontaneously ignites in air to produce P_4O_{10} and water.

Phosphine, unlike ammonia, cannot be made by the direct union of the constituent elements. It is prepared by heating white phosphorus in a concentrated solution of sodium hydroxide.

$$3\,NaOH + P_4 + 3\,H_2O \rightarrow 3\,NaH_2PO_2 + PH_3\uparrow$$

PH_3, a colorless, highly poisonous gas with an extremely putrid odor, readily decomposes on heating into its elemental constituents.

$$4\,PH_3 \xrightarrow{\text{heat}} P_4 + 6\,H_2$$

In a manner similar to ammonia, phosphine unites with hydrogen halides (see following discussion) to produce phosphonium compounds.

$$PH_3 + HCl \rightarrow PH_4Cl$$

Unlike ammonia, however, these compounds do not ionize in solution to give PH_4^+ ions. Instead they decompose, releasing phosphene gas, which, because it is a much weaker base than ammonia, is only slightly soluble in water.

The Halogen-Oxygen-Hydrogen Group

The group VII elements—fluorine, chlorine, bromine, iodine, and astatine—are collectively known as the halogens ("salt former" in Greek). They are quite an active group. They form halides with metals, a multiplicity of oxides and acids with oxygen and hydrogen, and if things should get dull they combine with each other. In the native state they all exist as diatomic molecules (F_2, Cl_2, Br_2, I_2, and At_2).

This is because by dimerizing (forming molecules composed of two atoms) each can gain a rare gas configuration:

$$\overset{\times\times}{\underset{\times\times}{\times F \times}}\!:\!\overset{\cdot\cdot}{\underset{\cdot\cdot}{F}}:$$

All are predominantly nonmetallic, although not to the same degree. Their nonmetallic characters decrease in the following order (the symbol > means "more than"):

$$F > Cl > Br > I$$

Each atom, by virtue of the fact that it is deficient only one electron in its quest to acquire a rare gas configuration, is an oxidizing agent. Elementary fluorine is the most powerful chemical oxidizing agent known. It will displace any of the other nonmetals from their binary compounds.

$$2\,F_2 + 2\,H_2O \rightarrow O_2 + 4\,HF$$

$$F_2 + 2\,KCl \rightarrow Cl_2 + 2\,KF$$

In a similar manner chlorine will displace bromine from bromides and iodine from iodides. Each halogen is capable of replacing those that are less powerful oxidizing agents. In total there are five halogens.

1. Fluorine ("to flow" in Latin). It acquired this name because fluorspar (**CaF_2**), a mineral containing fluorine, is used as a flux (a substance that combines with the gangue in a furnace to form a fusible slag) in metallurgy.
2. Chlorine ("green" in Greek). The name refers to the greenish-yellow color of the gas. The liquid is actually a golden yellow.
3. Bromine ("stench" in Greek). Its name speaks for itself. Actually, it is not any worse-smelling than iodine. It is a reddish-brown liquid at room temperature.
4. Iodine ("violet" in Greek). The name refers to the color of its vapor. The solid has a silvery metallic luster.
5. Astatine ("unstable" in Greek). It is the heaviest of all the halogens. All its known isotopes are short-lived; the most stable has a half-life of only 7.5 hours. Because of its rarity, it is basically a laboratory curiosity.

Note. Astatine will not be considered in the further discussions of the halogens. Even though there is good reason to believe that its properties fall in line with the other halogens, insufficient experimental data have been collected to assign numerical values to most of these properties.

As uncombined elements, the halogens present an example of poetry in chemistry. Their properties change in an orderly fashion from fluorine, the lightest, to iodine, the heaviest. (Table 11–7.)

Much of the variation in the halogen characteristics can be predicted on the bases of their electronic structures. The electronegativity is greatest for fluorine because it lacks the shielding effect of a completed inner electron shell. It is much less for chlorine because chlorine does have a completed inner electron shell. When a negatively charged inner electron shell shields the positively charged atomic nucleus, it results in the reduction in the ability of the nucleus to attract electrons. As can be seen from Table 11–7, once the effect of an inner electron shell has been established, successive inner shells have less pronounced effects. It is because of this that the electronegativity difference between fluorine and chlorine is much greater than that between chlorine and bromine or between bromine and iodine. The outer electrons of iodine, because they are relatively far from the nucleus and shielded by three completed inner shells, are less tightly held and therefore cause iodine to display some metallic properties. The oxidation states of the halogens and examples representative of them are listed in Table 11–8. From this table it can be seen that with the exception of +6 and +2 the halogens may have any oxidation number from −1 to +7.

All the halogens combine with hydrogen to produce hydrogen halides (**HF, HCl, HBr, HI**). These are typical covalent compounds that

TABLE 11–7 The Properties of the Halogens

	F_2	Cl_2	Br_2	I_2
Atomic number	9	17	35	53
Outer electron configuration	$2s^2 2p^5$	$3s^2 3p^5$	$4s^2 4p^5$	$5s^2 5p^5$
Density of the solid (grams/cc.)	1.3	1.9	3.4	4.9
Melting point (°C)	−223	−102	−7.3	114
Boiling point (°C)	−187	−34.6	58.8	183
Heat of vaporization (K cal./mole)	1.64	4.42	7.42	10.38
Covalent radius (angstroms)	0.72	0.99	1.14	1.33
Ionization potential (electron volts)	17.42	13.01	11.84	10.44
Electronegativity	4.0	3.0	2.8	2.5

11–8 Oxidation States of the Halogens

TABLE

Oxidation Number	Compounds of Fluorine	Compounds of Chlorine	Compounds of Bromine	Compounds of Iodine
+7		$HClO_4$, Cl_2O_7		HIO_4,
+6				
+5		$HClO_3$	$HBrO_3$, BrF_5	HIO_3, I_2O_7
+4		ClO_2		I_2O_4
+3		$HClO_2$	BrF_3	ICl_3
+2				
+1		Cl_2O, HclO, ClF	HBrO, BrCl	HIO_3, ICl, IBr
0	F_2	Cl_2	Br_2	I_2
−1	HF, OF_2, BrF, ClF_3, IF_5, IF_7	HCl, ICl_3, ICl, BrCl	HBr, IBr	HI

increase in stability going from **HI** to **HF**. Their covalent character produces low melting and boiling points and high resistance to the flow of electricity. Their bonds are polar bonds and their polarity increases in the following order (the symbol < means "less than"):

$$HI < HBr < HCl < HF$$

With the exception of hydrofluoric acid (**HF**), the aqueous solutions of all the hydrohalogen acids—hydrochloric (**HCl**), hydrobromic (**HBr**), and hydroiodic (**HI**)—are all strong acids and excellent conductors of electricity. The electrical conduction results from the reaction of the hydrogen halide with water to produce hydronium ions.

$$HX + H_2O \rightarrow H_3O^+ + X^-$$

(The symbol **X** indicates any halogen.) This reaction goes to completion in dilute solutions of all the hydrogen halides except hydrogen fluoride.

The chemical reactions of the hydrogen halides can be divided into two groups, those in which they act as acids and those in which they act as reducing agents. As acids they undergo typical acid-base neutralizations to produce a halide salt and water.

$$HBr + LiOH \rightarrow LiBr + H_2O$$

As reducing agents the strength of their properties increases in the following order:

$$HF < HCl < HB_2 < HI$$

Although hydrofluoric acid has no reducing strength to speak of, the strength of hydroiodic acid by contrast is so great that even in dilute solutions it is capable of reducing atmospheric oxygen. It is often used for the reduction of organic compounds.

$$4\,HI + O_2 \rightarrow 2\,H_2O + 2\,I_2$$

The hydrogen halide of greatest commercial importance is **HCl**. It is used in the manufacture of drugs, dyes, and soaps and in the surface treatment of metals prior to plating.

Hydrofluoric acid, the weakest of the hydrohalogen acids, possesses a very unusual property. It is capable of dissolving silicates and silicon dioxide.

$$SiO_2 + 4\,HF \rightarrow SiF_4 + 2\,H_2O$$

Since silicon dioxide and silicates are the main constituents in glass, hydrofluoric acid must be stored in either plastic or wax containers.

The halogens as a group form an extensive number of oxy-acids. The *hypo* and *per* prefixes used in naming these compounds indicate that the oxidation states of the halogen atoms are less than the halous or more than the halic acids. Chlorine, the most versatile of the halogens, forms four oxy-acids, one for each of the four odd-numbered oxidation states from +1 to +7. Their names and electronic structures are

H:Cl:	H:O:Cl:	H:O:Cl:O:	H:O:Cl:O: (with :O: above Cl)	H:O:Cl:O: (with :O: above and below Cl)
Hydrochloric acid	Hypochlorous acid	Chlorous acid	Chloric acid	Perchloric acid

Oxy-Halogen Acids

All are monobasic acids. Starting with hydrochlorous acid, all four oxy-acids of chlorine can be considered as the product of coordinating one, two, three, or four oxygen with the central chlorine atom. The stability and acid strength of these oxy-acids increase with the number of coordinated oxygen. It is the least for hypochlorous acid and the most for perchloric. The stability, however, even at its greatest, does not approach that of sulfuric or phosphoric acid. Perchloric acid, the most stable, will explode if heated in contact with oxidizable materials. The chlorate salts also present a hazard: in contact with strong acids they disproportionate (convert to higher and lower oxidation states), liberating explosive chlorine dioxide gas.

$$3\,KClO_3 + H_2SO_4 \rightarrow 2\,ClO_2 + KClO_4 + H_2O + K_2SO_4$$

In contrast to their acid strengths, the oxidizing abilities of the oxy-halogen acids decrease with increasing oxygen coordination. The ability to oxidize is greatest for hypochlorous acid and least for perchloric.

Let's hang in there. We're all one family.

A set of reactions peculiar to the halogens is the formation of *interhalogen compounds.* The stabilities of the diatomic halogen molecules suggest that structures similar to F_2, Cl_2, Br_2, and I_2 may also be achieved between atoms of more than one element. With the exception of **IF**, examples of all the possible diatomic interhalogen molecules have been synthesized. The compounds, however, are not restricted to diatomic molecules. Molecules have been prepared with the general composition $X(X')_n$. The notation indicates a halogen atom of one type (**X**) may be associated with as many as n halogen atoms of another type (**X′**); n may be any odd number from 3 to 7. The nature of the structure of these molecules and their bonds is a matter still awaiting elucidation. Chemically, the interhalogens behave similarly to the free halogens. They react with metals and nonmetals to produce the corresponding halides and usually hydrolyze according to the following equation:

$$XX' + H_2O \rightleftarrows H^+ + X'^- + HOX$$

The prime sign (′) indicates the more electropositive of the two elements; for example,

$$BrCl + H_2O \rightleftarrows HCl + HOBr$$

REVIEW

1. The chemistry of an inorganic molecule is a consequence of its parts. Although one group of atoms may serve as a component part of more than one molecule, that group, if it remains intact throughout a chemical reaction, will always possess the same chemical properties. Phosphate ion will give the same reaction regardless of whether it is a constituent of sodium, potassium, or ammonium phosphate. By the same token,

reactions characteristic of iron with an oxidation number of +2 can be produced by using either ferrous chloride, nitrate, or sulfate.

2. When comparing the chemical properties of atoms within a given group, the most marked differences exist between elements of the second and third periods. This is a result of the much smaller degree of nuclear shielding of the second-period valence electrons as compared to the valence electrons of higher periods. As the periods increase in number the degree of nuclear shielding becomes more extensive, causing the valence electrons to be less tightly held and giving the element more of a metallic character.

3. As a rule of thumb, and by no means hard and fast, when nonmetallic elements form multiatomic compounds—e.g., $BaSO_4$ and $KClO_4$—the oxidation states of the central atom tend to be even-numbered if the element is found in an even-numbered periodic group and odd-numbered if it is found in an odd-numbered periodic group. Sulfur in its compounds tends to have even oxidation numbers, whereas the halogens tend to display odd oxidation numbers. Exception to this rule are compounds in which the atom under consideration is bonded to another atom of the same type—e.g., Na_2S_2, N_2H_4—and compounds such as NO, NO_2, and ClO_2.

4. The two most common groupings are chosen as the basis of naming acids containing composite groups with the same central atom and different ratios of oxygen. That group in which the central atom has the higher of the two oxidation states is assigned an *ic* suffix. The group of lower oxidation state is assigned an *ous* suffix. If groups exist with oxidation states higher than the *ic* or lower than the *ous*, they are assigned respectively *per* and *hypo* prefixes. The salts of *ic* acids have *ate* suffixes, and those of *ous* acids *ite* suffixes. Simple acids containing no oxygen, such as hydrochloric acid (**HCl**), are named hydro______ic acids, where the abbreviated name of the appropriate atom or group is inserted in the space. The salts of these acids have *ide* suffixes—e.g., sodium bromide (**NaBr**), potassium cyanide (**KCN**).

5. If an element is capable of existing in more than one oxidation state in different compounds, it may function either as an oxidizing or a reducing agent. When the other oxidation state is lower than the state it possesses, it can act as an oxidizing agent. Conversely, if the other state is higher, it can act as a reducing agent. If oxidation states above and below the value of the compound under consideration exist, it will be capable of reacting either as an oxidizing or a reducing agent, depending of course on the nature of the compound with which it is reacting.

6. The oxidation state of an atom in a composite group, whether ion or molecule, can be determined if an oxidation number of −2 is assigned to each oxygen and +1 to each hydrogen in the group. The sum of the oxidation numbers of the atoms in the group must equal the charge on the group. Exceptions to this procedure are oxygen atoms in peroxides and hydrogen atoms in hydrides. The oxidation number is −1 for both hydrogen and oxygen in these compounds.

7. Oxygen, when incorporated in peroxidic compounds, is in a −1 oxidation state. Depending upon the materials it reacts with, it may be either reduced or oxidized. The former process produces a compound with the anticipated amount of oxygen, the latter produces free oxygen.

8. Nitrogen can exist in a large number of oxidation states. The state most frequently encountered when dealing with oxy-nitrogen compounds is the +5 state of nitric acid. Nitric acid is both a strong acid and a strong oxidizing agent. Because the +5 oxidation state is the maximum degree of oxidation that nitrogen can attain, compounds containing nitrate groups are excellent oxidizers and the basis for most chemical explosives.

   Nitrogen, element 7, contains an odd number of valence electrons. As a consequence of this, compounds of nitrogen and elements with even numbers of electrons are often colored. If the odd electron is paired off through reactions such as dimerization, the color disappears.

   Nitrogen can be made to combine with hydrogen directly to produce ammonia, the cornerstone of commercial nitrogen chemistry. Its oxidation state as a component of ammonia is −3.

9. The pride and joy of the sulfur community is sulfuric acid, a jack-of-all trades. It is a strong acid with a high boiling point that can be used to produce other acids from their salts. As a concentrated solution it is also an excellent oxidizing and dehydrating agent.

10. The oxy-phosphorous acids are each associated with three hydrogens, but only phosphoric acid is triprotic. Phosphorous and hypophosphorous acids are respectively diprotic and monoprotic acids. This is because only the hydrogens attached to the oxygens are ionizable. One of the hydrogens in phosphorous and two of the hydrogens in hypophosphorous acid are attached to the phosphorous directly. To augment the devious nature of the oxy-phosphorous acids, phosphoric acid, besides existing in the ortho form, also exists in a large number of condensed forms.

11. The halogens are components in a large variety of acids and salts. All the hydrohallic acids with the exception of **HF** are strong acids. The acid strengths and stabilities of the oxy-halogen acids increase with the number of coordinated oxygen atoms.

    In a unique approach to attaining a rare gas configuration, a halogen atom may combine with a second type of halogen to produce interhalogen compounds. Chemically, these compounds behave like the respective free halogens.

REFLEXIVE QUIZ

1. The one common product of reactions between $Ba(OH)_2$ + H_2SO_4, $Ba(NO_3)_2$ + $(NH_4)_2SO_4$, and $Ba(NO_3)_2$ + K_2SO_4 is ______.
2. Ammonium chloride ionizes according to the following equation:

$$NH_4Cl \rightarrow NH_4^+ + Cl^-$$

   The oxidation state of nitrogen in ammonium chloride is ____.
3. Of the compounds NH_3, PH_3, H_2S, and N_2O_5, the one with the greatest potential for oxidizing other compounds is ______.
4. Of the compounds **HF**, **HClO**, H_2S, $HClO_4$, the strongest acid is ______.
5. In the selection H_2SO_4, HNO_2, H_3PO_2, H_3PO_3, the diprotic acids are ____________.
6. Nitric acid is able to dissolve metals that are lower than hydrogen in the electromotive series because it is a good ____________ as well as a strong acid.
7. In the compound $KClO_3$ the chlorine atom has an oxidation state of ____.
8. The differences between ortho-phosphoric acids are dependent on the number of ______ molecules incorporated in the acid.
9. The process by which sulfuric acid can convert sugar to carbon is ____________.
10. N_2O_5 is the anhydride of ______ acid.

Answers: (1) $BaSO_4$ [173, 177, 208]. (2) −3 [183–185, 209]. (3) N_2O_5 [186, 190, 199, 208, 209]. (4) $HClO_4$ [206]. (5) H_2SO_4, H_3PO_3 [194, 197–202]. (6) Oxidizing agent [183, 209]. (7) +5 [204–206]. (8) Water [199–201, 209]. (9) Dehydration [193, 209]. (10) Nitric [182].

SUPPLEMENTARY READINGS

1. R. T. Sanderson. "Principles of Halogen Chemistry." *Journal of Chemical Education* 41 (1964): 361.
2. W. Haynes. *Brimstone: The Stone That Burns.* New York: Van Nostrand–Reinhold Corp., 1959.
3. E. O. Sherwin and G. J. Weston. *Chemistry of the Non-Metallic Elements.* New York: Pergamon Press, 1966.
4. C. J. Pratt. "Chemical Fertilizers." *Scientific American* 212 (1965): 62–72.
5. M. J. Bigelow. *The Representative Elements.* Croton-on-Hudson, N.Y.: Bogden & Quigley, Inc., 1970.
6. S. Y. Tyree, Jr., and K. Knox. *Textbook of Inorganic Chemistry.* New York: Macmillan Co., 1961.

Section 12

Organic Chemistry: Saturated Compounds

Section 12

Organic Chemistry: Saturated Compounds

It All Started with Urea.

Since 1828, when Friedrich Wöhler synthesized urea, the first man-made organic compound, chemists have produced in excess of one million distinctly different others. These, however, represent a mere drop in Mother Nature's bucket. Biologists estimate that the life forms on this earth contain something of the order of 10^{12} different types of organic molecules. Prior to Wöhler's discovery, it was thought that these compounds could be synthesized only by living organisms. For this reason, they were given the misnomer "organic." The single factor that makes possible the creation of so large a number of compounds from so small a number of elements is the unique properties of carbon.

Carbon, element 6, has the electronic configuration $1s^2 2s^2 2p^2$. In order for it to acquire a rare gas configuration, it must either gain or lose four electrons.

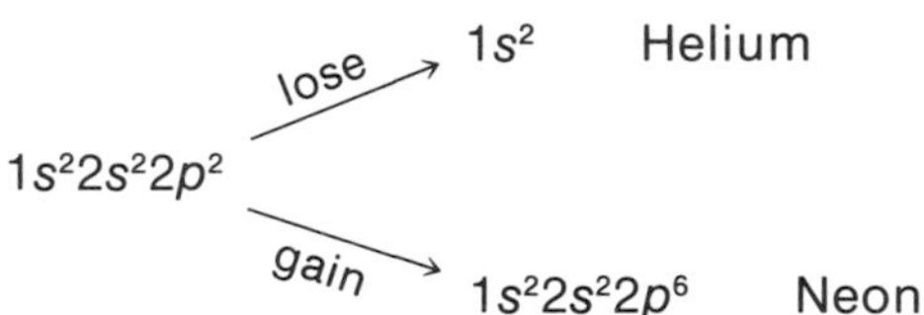

As pointed out earlier, four is an inordinately large number of electrons to be either gained or lost. The carbon atom therefore shares rather than transfers them. The result of this sharing is the formation of four covalent bonds. It might be expected from the electronic configuration of carbon that two types of covalent bonds could be formed (that is, those that were originally $2s$ and those that were $2p$ electrons). In actual practice, however, this does not happen.

What does happen is that these four orbitals *hybridize,* that is, form four new orbitals, each of which has one-quarter the properties of a $2s$ orbital and three-quarters the properties of a $2p$ orbital.* The notation for each of these new orbitals is sp^3, since they are each composed of one part s to three parts p. Each carbon still has four electrons with which to form covalent bonds, but because of hybridization they are indistinguishable. Hybridization does not change the number of

*The process is analogous to the production of the hybrid color purple. Purple is *not* a component of the spectrum of white light, or a rainbow, but must be produced by the *hybridization* of the spectrum colors red and blue.

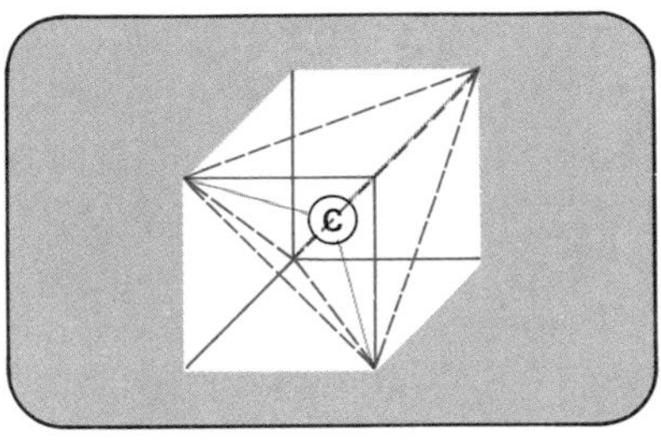

FIGURE

12–1 Orientation of sp^3 Orbitals in Carbon

orbitals, *only the geometry*. Before hybridization there were two non-directional 2*s* orbitals and two sausage-shaped 2*p* orbitals oriented along two of three coordinates (left–right, up–down, front–back). After hybridization the four equivalent orbitals are oriented so that they point to the four corners of a tetrahedron. This orientation can be pictured in terms of a cube with a carbon atom at its center and each of the sp^3 orbitals pointed to alternate corners. (Figure 12–1.) The figure formed by joining together these four points is a *tetrahedron*. It has four faces, four corners, and six edges as compared to a cube, which has six faces, eight corners, and twelve edges. The tetrahedral configuration produces stronger bonds and, through superior bond placement, minimizes the repulsion between adjacent bonds.

Paraffin Is Not Just Wax.

The simplest organic compounds are those composed solely of carbon and hydrogen joined by single bonds. They comprise a series known as the *alkanes* or paraffins, and they are serially named according to the number of carbon atoms each contains. Methane has one carbon; ethane has two, propane three, butane four, pentane five, and so on. The number of hydrogen atoms in each alkane is equal to two times the number of carbon atoms plus two (C_nH_{2n+2}). For example, in methane (CH_4),

$$\begin{array}{c} H \\ H:\ddot{\underset{\cdot\cdot}{C}}:H \\ H \end{array} \qquad \begin{array}{l} \text{Number of Hydrogens} = 2 \times \text{Number of Carbons} + 2 \\ \qquad\qquad 4 \qquad\qquad = 2 \times \qquad\qquad 1 \qquad\qquad + 2 \end{array}$$

Each of the four hydrogens in methane is covalently bonded to one of the carbon's sp^3 orbitals. As a result of this electron sharing, each hydrogen assumes the rare gas configuration of a helium atom, the carbon assumes the electronic configuration of a neon atom, and the entire molecule assumes a tetrahedral configuration. (Figure 12–2*a*.) In this configuration the angle formed by any two hydrogen atoms and the carbon atom is 109° 28′. (Figure 12–2*b*.) This sp^3 configuration is typical of the bonding in all alkanes.

The first three members of the alkane series are all colorless gases. They are unique among the alkanes because the carbon atoms can be joined to each other in only one fashion.

```
    H           H  H             H  H  H
    |           |  |             |  |  |
 H—C—H       H—C—C—H         H—C—C—C—H
    |           |  |             |  |  |
    H           H  H             H  H  H
```

Methane CH_4 | Ethane C_2H_6 | Propane C_3H_8

Note. For simplicity of notation, the two electrons that comprise each covalent bond are usually indicated by a single —, as here.

In alkanes containing four or more carbon atoms, however, the carbons can be joined to each other in different ways. Butane (C_4H_{10}), for example, can have its carbon skeleton connected in either of two possible configurations:

```
    H  H  H  H                 H  H  H
    |  |  |  |                 |  |  |
 H—C—C—C—C—H      or      H—C—C—C—H
    |  |  |  |                 |  |  |
    H  H  H  H                 H  |  H
                                  |
                               H—C—H
                                  |
                                  H
```

Normal butane | Iso butane

These two compounds, although they have the same chemical formula, are structurally different. They differ also with respect to physical and chemical properties, having different melting and boiling points and different degrees of solubility in water. The two compounds are *isomers* (molecules that have the same type and number of atoms arranged differently) of the butane molecule. The straight-chained compound is the "normal" isomer and the compound having a carbon branch is

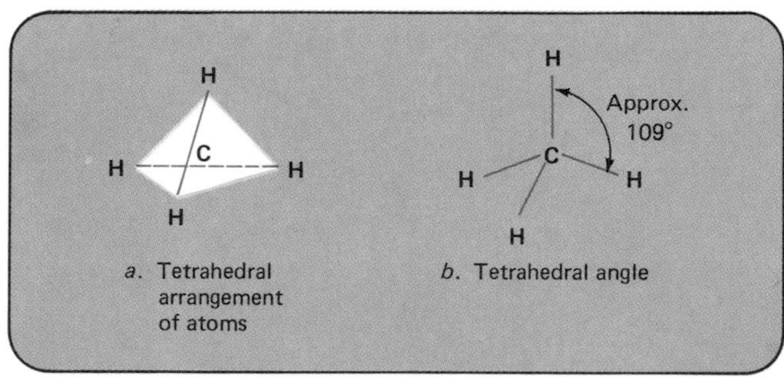

FIGURE

12–2 Configuration of Methane (CH_4)

the "iso" isomer. Since there is only one possible arrangement of the branch, there are only two isomers of butane. At first glance, it might seem that the carbon branch could be placed on either of the other two carbon atoms, but a moment's reflection will show that this would produce only a bent representation of normal butane. In normal butane the maximum number of carbons connected to any one carbon atom is two. In isobutane the maximum number is three.

To avoid confusion as to which compound is meant by the notation C_4H_{10}, the International Union of Pure and Applied Chemistry (IUPAC) has devised a naming system capable of distinguishing between molecular structures. Compounds are named after the longest carbon chain in the molecule. Smaller carbon chains are then considered as additions to this main chain. The location of these smaller chains is established by numbering the carbons in the main chain and citing the number of the carbon to which they are attached.

Normal butane contains only one chain, which is of necessity the longest chain in the molecule.

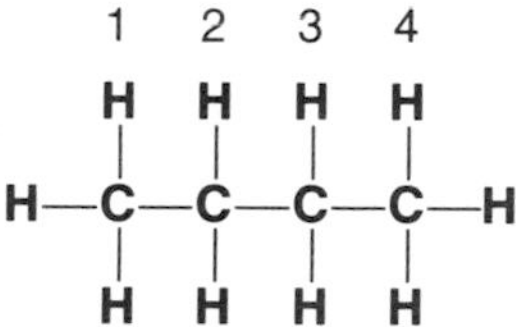

The term *normal* (abbreviated by the prefix n-) denotes the lack of any branching. The *butane* indicates that the longest chain contains four carbon atoms.

Unlike the chain in n-butane, the longest chain in isobutane contains only three carbons. This is so no matter which carbon atom is chosen as the starting point of the counting:

```
   1  2  3                  1  2                        2  3
   H  H  H                  H  H  H                  H  H  H
   |  |  |                  |  |  |                  |  |  |
H—C—C—C—H      or      H—C—C—C—H      or      H—C—C—C—H
   |  |  |                  |  |  |                  |  |  |
   H  |  H                  H  |  H                  H  |  H
      |                        |                        |
   H—C—H                  H—C—H                  H—C—H
      |                        |                        |
      H                        H                        H

                               3                        1
```

Isobutane can therefore be considered a modified propane molecule that has one methyl (**CH_3**) group attached to its second carbon atom. Because of this, it is named 2-methylpropane by the IUPAC system.

Naming organic compounds with greater numbers of carbon atoms by this system results in some elaborate but logical labels. Pentane, for example, has three isomers:

```
                                                           H
                                                           |
                                                        H—C—H
                                                           |
   H  H  H  H  H              H  H  H  H               H   |   H
   |  |  |  |  |              |  |  |  |               |   |   |
H—C—C—C—C—C—H          H—C—C—C—C—H          H—C——C——C—H
   |  |  |  |  |              |  |  |  |               |   |   |
   H  H  H  H  H              H  |  H  H               H   |   H
                                 |                         |
                              H—C—H                     H—C—H
                                 |                         |
                                 H                         H
```

Normal pentane | Isopentane | Neopentane

According to the IUPAC system, their names would be:

```
                                                    C
 1  2  3  4  5           1  2  3  4                 |
 C—C—C—C—C               C—C—C—C                 C—C—C
                            |                    1 2|
                            C                      3C
```

Pentane | 2-Methylbutane | 2,2-Dimethylpropane

Depending on which end of the chain counting begins, isopentane may be labeled either 2-methylbutane or 3-methylbutane.

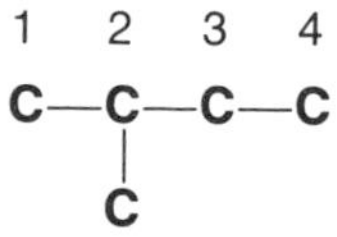

```
1  2  3  4           4  3  2  1
C—C—C—C      or      C—C—C—C
  |                    |
  C                    C
2-Methylbutane       3-Methylbutane
```

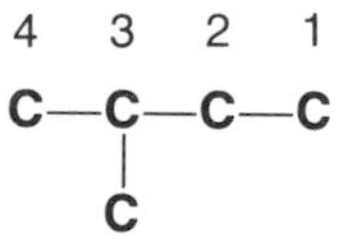

When such a choice exists, convention dictates counting from the nearer end of the chain so that the lower of the two possible position numbers is assigned to the compound. Isopentane is therefore called 2-methylbutane rather than 3-methylbutane. There is no choice in the case of neopentane; the designation must be 2,2-dimethylpropane. Working backward, you can easily see why: *propane,* the root name, indicates that the longest chain has three carbon atoms; *dimethyl* indicates two methyl groups; and *2,2* indicates that both the methyl groups are attached to the second (number 2) carbon of the propane chain.

Table 12–1 gives the name of the alkanes having from one to ten carbon atoms and the number of isomers associated with each. Although 75 may be thought a large number of isomers for a compound of only 10 carbon atoms, it is trivial by comparison to the 366,319 and 4.11×10^9 isomers that compounds of twenty and thirty carbon atoms respectively have.

When naming the side chains of organic molecules, the same root name is employed as in the naming of the longest chain. It is, however,

TABLE 12–1 Names and Numbers of Isomers of the First Ten Alkanes

Number of Carbons	Name	Number of Isomers*
1	Methane	1
2	Ethane	1
3	Propane	1
4	Butane	2
5	Pentane	3
6	Hexane	5
7	Heptane	9
8	Octane	18
9	Nonane	35
10	Decane	75

* Note: Cyclic isomers are not included.

TABLE 12–2 Common Alkyl Groups

Group	Name
CH_3—	Methyl
CH_3CH_2—	Ethyl
$CH_3CH_2CH_2$—	n-Propyl

CH_3CH—
      |
    CH_3
	Isopropyl

$CH_3CH_2CH_2CH_2$—	n-Butyl

CH_3CHCH_2—
      |
    CH_3
	Isobutyl

CH_3CH_2CH—
          |
        CH_3
	sec-Butyl

CH_3
      |
CH_3C—
      |
    CH_3
	tert-Butyl (or t-Butyl)

$CH_3CH_2CH_2CH_2CH_2$—	n-Pentyl (Alternate name: n-Amyl)

$CH_3CHCH_2CH_2$—
      |
    CH_3
	Isoamyl

modified by replacing the *ane* ending with *yl*. Table 12–2 lists some common side-chain (alkyl) groups and their names. In this table, two new prefixes are introduced: *sec-* and *tert-*, standing for *secondary* and *tertiary*. They indicate that the carbon atom that is the point of attachment of the alkyl group is attached additionally to two or three other carbon atoms. When a molecule contains more than one side chain, they are listed in alphabetical order.

Example. Name the following decane according to the IUPAC system.

```
      H   H   H   H   H   H   H
      |   |   |   |   |   |   |
  H — C — C — C — C — C — C — C — H
      |   |   |   |   |   |   |
      H   H   |   H   H   H   H
              |   H   H
              |   |   |
          H — C — C — C — H
              |   |   |
              H   H   H
```

The longest continuous chain contains eight carbon atoms; therefore, the root name is octane. The side chain contains two carbon atoms; therefore, it is an ethyl group. Counting from the nearer end of the main chain, so that the ethyl group will have the lower possible position number, reveals that it is attached to the carbon in position 4. Put them all together, and they spell 4-ethyloctane, the name of the compound.

The naming of

```
    H   H   H   H   H
    |   |   |   |   |
H — C — C — C — C — C — H
    |   |   |   |   |
    H   H   H   |   H
            H   |   H
            |   |   |
        H — C — C — C — H
            |   |   |
            H   |   H
            H — C — H
                |
                H
```

is a bit more tricky. Although at first glance one might describe it as a tertiary butyl derivative of pentane, the compound is more simply named 2,2,3-trimethylhexane.

```
6   5   4   3
C — C — C — C — C
           2|
        C — C — C
           1|
            C
```

The lesson to be learned from this example is that the longest chain is not always apparent at first glance.

Spaghetti Has No Top or Bottom. Frosted Doughnuts Do.

The alkanes can exist in ringed forms as well as in straight-chained forms. Ringed alkanes containing as few as three or as many as twelve carbon atoms have been synthesized. *Cycloalkanes,* as ringed alkanes are called, have several interesting features, three of which are germane here.

1. Cycloalkanes possess two less hydrogens than the normal alkanes because they join to themselves.

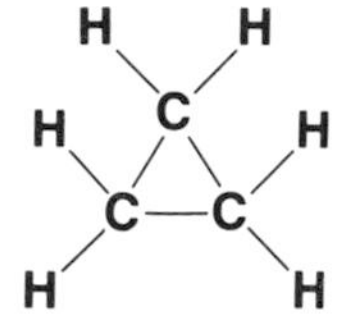

```
    H   H
     \ /
 H    C    H
  \  / \  /
   C --- C
  /       \
 H         H
```

Cyclopropane (C_3H_6)

```
   H   H   H
   |   |   |
H— C — C — C —H
   |   |   |
   H   H   H
```

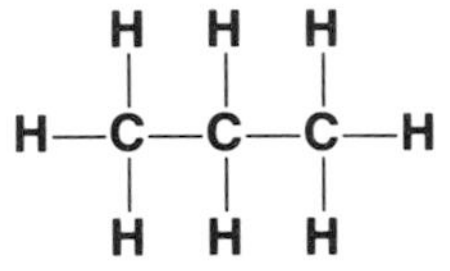

n-Propane (C_3H_8)

2. Cycloalkanes have strained bonds because the cyclic configuration restricts the free movement of any one carbon atom about its neighbor.

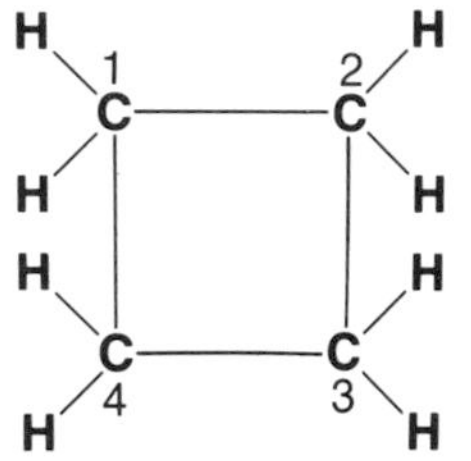

```
H             H
 \1         2/
  C ------- C
 /|         |\
H |         | H
H |         | H
 \|         |/
  C ------- C
 /4         3\
H             H
```

Cyclobutane (C_4H_8)

```
   H   H   H   H
   |   |   |   |
H— C — C — C — C —H
   |   |   |   |
   H   H   H   H
```

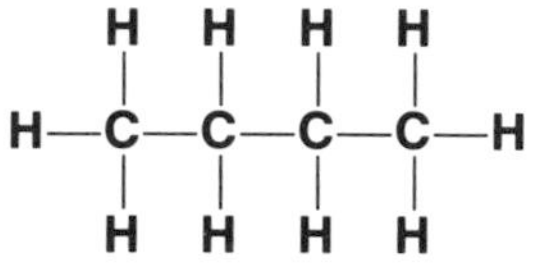

n-Butane (C_4H_{10})

For ease of visualization, the carbon atoms in alkanes may be considered attached to their neighbors through ball-and-socket joints. In the "straight-chained" alkanes, such joints allow the free rotation of each atom about its neighbor. In cyclic structures, however, movement is greatly restricted because of the lack of free ends. Before moving, each atom must consider not only its position but also the position of the two atoms to which it is attached and how they are joined to all the other atoms in the molecule. If an atom attempts to make an extreme movement, strain will be placed on at least one of the two bonds that join it to the rest of the molecule; thus it will be forced to move in a more orthodox fashion.

Example. The 4 carbon in n-butane can easily orient itself such that it is midway between the 2 and 3 carbon atoms.

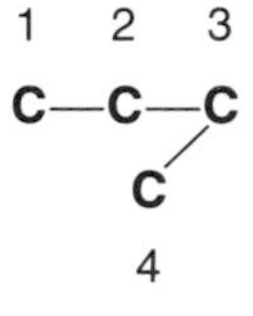

```
1   2   3
C — C — C
       /
      C
      4
```

This is not possible in cyclobutane without introducing great strains into the bond between the 1 and 4 carbon atoms.

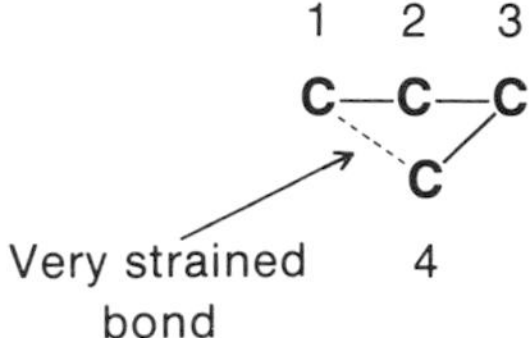

Note. In this explanation and at subsequent points in this book the hydrogen atoms have been deliberately omitted from molecular representations so that they may not obscure the argument being stressed. All carbon atoms in organic molecules, however, unless specifically indicated to the contrary, may be considered to form four molecular bonds each. If a carbon atom is depicted as having less than this number of bonds, it is to be understood that the difference between four and that number of bonds should be interpreted as **C—H** bonds. The notation **C—C** should not, therefore, be interpreted as C_2 but as C_2H_6. Since one of each carbon atom's bonds has been expressly assigned to the other carbon atom, the other three bonds that each carbon possesses must of necessity be attached to hydrogen atoms.

```
    H  H
    |  |
 H—C—C—H
    |  |
    H  H
```

3. The carbon atoms in cyclic compounds, because they are all joined one to another, create a surface similar to a doughnut, to which a top and a bottom may be assigned. Both sides of a doughnut appear identical, but in terms of properties they can be distinctly different. This is apparent as soon as the doughnut is frosted.

If two dollops of frosting are to be spread on a doughnut, they may be applied either both to the same side or one to each side. By the same token, when groups other than hydrogen atoms are attached to cyclic carbon atoms, they may be oriented such that they extend from one or the other side of the molecular surface. The determination of which side is which becomes significant only when two or more groups protrude. With the advent of the second group the question arises as to whether it is protruding from the same side as the first (the *cis* form) or the other side (the *trans* form). Those molecules that differ only in terms of group orientations are known as *geometrical isomers*. Unlike *structural isomers* such as butane and isobutane, geometrical isomers have identical carbon skeletons; they differ merely in the orientation of their groups.

Example. *Cis-* and *trans*-1,2-dimethylcyclohexane are geometrical isomers.

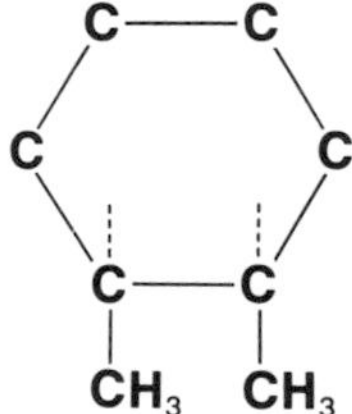

Cis-1,2-dimethylcyclohexane

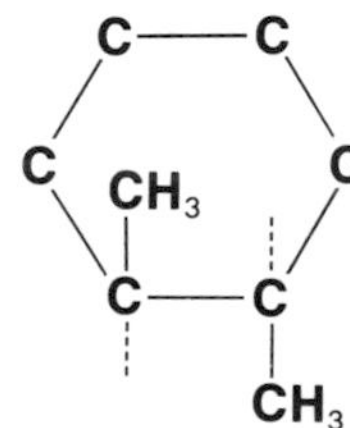

Trans-1,2-dimethylcyclohexane

In the *cis* isomer the methyl groups are both protruding from the same side of the molecule, whereas in the *trans* compound they protrude from different sides. In straight-chained compounds this geometric isomerism is not possible because the methyl groups have free rotation around the chain and therefore do not allow the establishment of a top or bottom side.

```
C—C—C—C—C—C
  |  |
  C  C
```

As Different as Righthanded and Lefthanded Screws

At this point it should be apparent to the reader that organic molecules as a group are inherently more complex than inorganic. Although the empirical formula (the formula that expresses only the simplest numerical ratio of atoms in the compound) of an inorganic compound is often sufficient data to identify it, this is seldom true for an organic molecule. Most organic molecules possess one or more structural isomers, and where restricted rotation exists, geometric isomers add to the number of possibilities that an empirical formula may represent. To complicate the situation further, there exists in addition to these two types of isomers yet a third, the *optical isomer.*

Optical isomers, because of the particular arrangement of the four groups attached to a given carbon atom, can interact with light in more than one way.

Note. Light as we normally encounter it travels in an infinite number of planes. If we took a telescope and covered all of the front lens except an ultrathin slot, we would still see the same light no matter how the telescope was rotated while looking through it. That is, the amount of light seen at every position of rotation would be the same. (Figure 12–3.) If, however, the light source were polarized so that it emitted in only one plane, then the amount of light seen would depend on the orientation of the slot in the telescope. (Figure 12–4.)

Such a condition can be visualized by imagining an opaque wall with a single ultrathin slit set between the light source and the telescope. Because the slit in the wall is oriented in one particular direction, the maximum amount of light will be visible in the telescope only

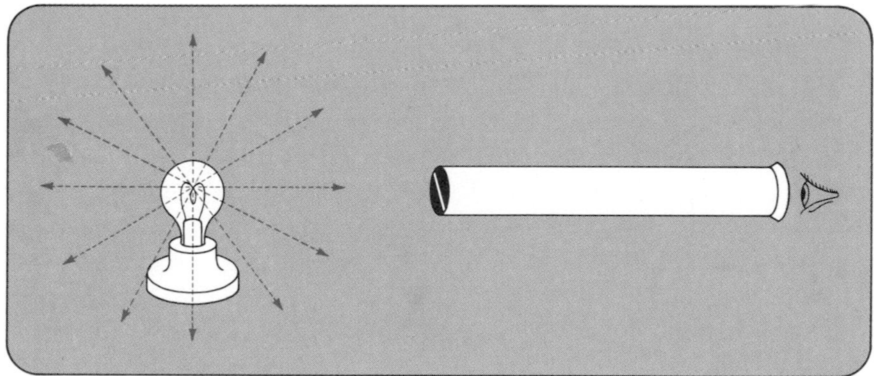

FIGURE 12–3 Nonpolarized Light

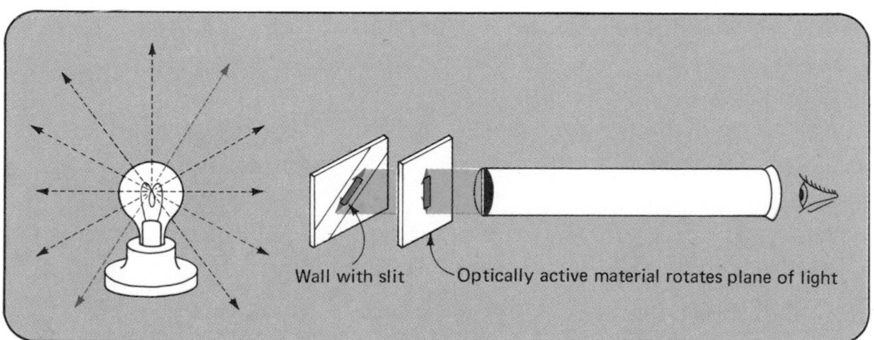

FIGURE 12–4 Plane Polarized Light

when its slit is oriented in the same direction. The light after passing through the slit in the wall can be considered plane polarized, because the slit forced it to travel in only one plane. Had the light not been oriented in the same manner it would have been blocked. If a material placed between the slit and the telescope causes a change in the direction of polarization of the light, then the material is said to be optically active. To allow the maximum amount of light to pass through a properly oriented telescope after an optically active material has been placed in the light beam, it is necessary to reorient the telescope to compensate for the change in direction produced by the active material. If the telescope is rotated to the right to make this accommodation, then the material is said to be *dextro-rotatory* or $d(+)$ rotary (*dexter* in Latin means *right*). If the accommodation is made by a left, or counterclockwise, rotation, then the material is said to be *levo-rotatory* or $l(-)$ rotary (*laevus* in Latin means *left*).

The above description of rotating polarizers is merely a schematic one. In actual practice it is not feasible to construct slits capable of polarizing light; instead, crystals of herapathite or calcite that display this property are employed. The effect, however, is the same.

Any molecule capable of rotating polarized light must be asymmetric. This means that it and its mirror image are not the same. They differ in the same manner as a left glove differs from a right. Although each glove has five fingers and there is a one-to-one correspondence between the number and size of these fingers, there is no way, save by turning it inside out, that a left glove can be worn on a right hand. Organic compounds, because of their tetrahedral carbon bonding, can be considered analogous to four-fingered gloves. If each of the bonds at the four corners of the tetrahedron are attached to four different groups of atoms, then the mirror image of the molecule will differ from the original.

To determine whether a molecule and its mirror image are identical it is necessary merely to superimpose one on the other. If superpositioning is possible, then the molecules are identical and they will have no effect on polarized light.

Example. 3-Methylhexane possesses optical isomers; 3-ethylhexane does not.

The reason for this is that the 3 carbon in 3-methylhexane is attached to four dissimilar groups, whereas the 3 carbon in 3-ethylhexane is attached to only three. If we picture both molecules staring into a mirror with arms outstretched, the difference becomes immediately apparent. When 3-ethylhexane looks in the mirror, the reflected left hand is the same as the right and the reflected right hand is the same as the left; no difference is noticeable because both groups are identical. (Figure 12–5.) This is not the case when 3-methylhexane looks into the mirror. (Figure 12–6.) 3-Methylhexane has one arm shorter than the other. In the real molecule it is the left arm. In the reflection it is the right.

```
                                                     H
                                                     |
          H                                       H—C—H
          |                                          |
       H—C—H                                      H—C—H
  H  H    |   H  H  H                       H  H    |   H  H  H
  |  |    |   |  |  |                       |  |    |   |  |  |
H—C—C—C—C—C—C—H                       H—C—C—C—C—C—C—H
  |  |  |  |  |  |                          |  |  |  |  |  |
  H  H  H  H  H  H                          H  H  H  H  H  H
   3-Methylhexane                              3-Ethylhexane
```

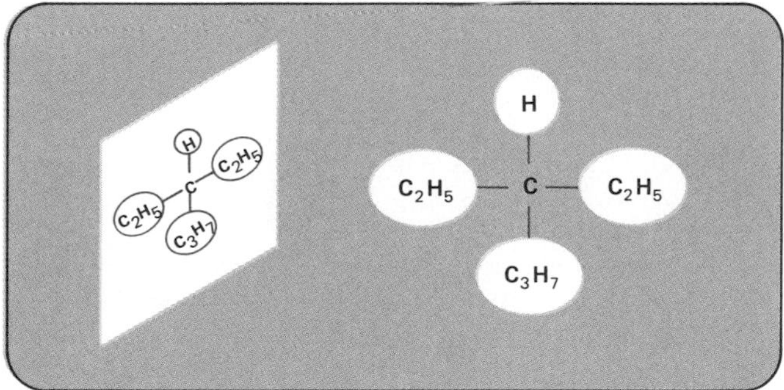

12–5 3-Ethylhexane FIGURE

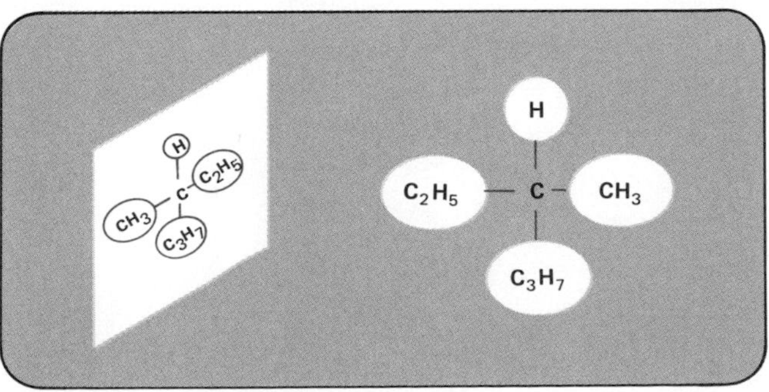

12–6 3-Methylhexane FIGURE

Two isomers that are related in this way are enantiomers (*enantio* in Greek means "opposite," *meros* means "part"). Because of the strong resemblance for each other, many of the properties of enantiomers are identical. They have the same boiling and melting points and are equally soluble in the same solvents. They do differ, however, with respect to the manner in which they interact with living systems and plane polarized light. The dextro form will rotate the plane of polarized light exactly the same number of degrees in the clockwise direction as the levo will rotate it in the counterclockwise direction for the same concentration of compound. When equal amounts of both enantiomers are present in a mixture, their effects oppose and cancel each other. A mixture of this type is known as a racemic mixture (*racemus* in Latin means "a bunch of grapes").

The separation of a racemic mixture into its component enantiomorphs is known as resolutions. Louis Pasteur in 1848 was the first man to perform such a feat. Working at the Ecole Normale in Paris, he became interested in the chemistry of wine making. He found that the sodium ammonium salt of tartaric acid, a substance produced during fermentation, crystallized into two forms that were mirror

images of each other. With a hand lens and a pair of tweezers he separated them, just as one would separate left shoes from right. These crystals when separately dissolved in the same amount of water rotated the plane of polarized light equally but in opposite directions. When equal amounts of both crystals were combined, the resultant solution was inactive. From these observations Pasteur correctly assumed that the optical properties of this compound were characteristics of the molecule and not the crystal.

Tartaric acid, as a compound exhibiting optical isomerism, has two distinctive features:

1. It possesses more than one asymmetric carbon atom.
2. Both of the asymmetric carbon atoms have the same four groups attached to them.

```
        O                    O
        ‖                    ‖
        C—OH                 C—OH
        |                    |
     H—C*—OH            HO—C*—H
        |                    |
    HO—C*—H              H—C*—OH
        |                    |
        C—OH                 C—OH
        ‖                    ‖
        O                    O
   Dextro (+)            Levo (−)
```

Each of the carbon atoms with an asterisk is asymmetric. Its mirror image cannot be superimposed upon it. Each molecule is the mirror image of the other.

Although either of the two asymmetric carbon atoms would be sufficient reason for tartaric acid to be optically active, the presence of two such atoms on one molecule produces additional complications. These two optically active carbon atoms can align themselves such that they produce a symmetrical, non-optically-active mesomolecule. What happens is that the optical rotation of one asymmetric group is equally and oppositely counterbalanced by the other. A mirror placed across the center of the molecule between the two asymmetric carbon atoms will produce a reflected image that is exactly the same as its real counterpart. Because the molecule itself possesses both components of a racemic mixture, it is optically inactive. When more than one optically active, dissimilar carbon atom exist within a compound,

the resultant molecule is not inactive but rather has an activity dependent upon the orientation of the individual asymmetric atoms.

The reason that Pasteur did not encounter the meso form in his separation was due to his fortuitous use of the sodium ammonium salt rather than the free tartaric acid. The salt, lacking the symmetry of the acid, does not exist in the meso state.

```
        O
        ‖
        C—O—NH₄
        |
   HO—C—H
- - - - - - - - - - - - -
   HO—C—H
        |
        C—O—Na
        ‖
        O
```

When depicting the meso form of tartaric acid, the significant point to remember is that both the hydrogen and hydroxide groups must be paired on the same side of the molecule. The choice of sides is immaterial since rotating the molecule 180° will show both forms to be identical.

The importance of optical isomers does not reside in their optical characteristics but rather in the configuration of atoms that causes them. Most living organisms will contain only one of two possible enantiomorphs. Microorganisms such as molds will preferentially consume one enantiomorph over another. Enantiomorphs will taste and smell different. Enzymes of one configuration will be biologically active, whereas those of the other configuration will not.

What It All Means So Far.

The alkanes are the simplest organic compounds, and yet, for all their simplicity, they are capable of existing in structural, geometric, and optical isomeric forms, each with its own unique set of properties. In the forthcoming discussion of organic compounds and their reactions, it should be kept in mind that all of the compound types to be discussed are equally capable of existing in these three isomeric forms.

Little Affinities—Little Action.

Alkanes or paraffins (*parum* in Latin means "little," *affinis* means "affinity") are the least reactive of the organic compounds. Normally unaffected by strong acids and bases, they react only with strong oxidizing agents under forced conditions. All are nonpolar; all have the general formula C_nH_{2n+2} for the noncyclic compounds; and all burn in air to produce carbon dioxide and water.

$$C_nH_{2n+2} + \left(\frac{3n+2}{2}\right) O_2 \rightarrow n\,CO_2 + (n+1)\,H_2O + \text{Heat}$$

$$C_4H_{10} \quad + \quad 7\,O_2 \rightarrow 4\,CO_2 + 5\,H_2O + \text{Heat}$$

Petroleum deposits provide the widest natural selection of these compounds. The first four, methane, ethane, propane, and butane, are fuel gases. Propane and butane under pressure are sold in rural communities as LPG or liquefied petroleum gas for cooking. In recent

TABLE

12–3 Petroleum Distillation Fractions

Name	Composition (percent)	Molecular Size	Boiling Range (°C)	Uses
Gases	2	C_1–C_5	0	Fuel
Petroleum ethers	2	C_5–C_7	30–110	Solvents
Gasoline	32	C_6–C_{12}	30–200	Auto fuel
Kerosene	18	C_{12}–C_{15}	175–275	Jet and Diesel fuel
Fuel oil	20	C_{15}——	250–400	Heating Fuel
High-temperature residue	26	C_{19}——	300——	Lubricants, wax, and asphalt

years, butane has become familiar to the consumer as the fuel in his cigarette lighter. Gasoline is a mixture of alkanes with formulas ranging from C_5H_{12} to $C_{12}H_{26}$. The mixture in kerosene ranges from $C_{10}H_{22}$ to $C_{15}H_{32}$. With increased molecular weight the melting and boiling points also increase. This gives the alkanes with higher molecular weights the properties of greases and waxes and allows them to be separated from the crude petroleum by fractional distillation. (Table 12–3.)

Fractional distillation consists of heating the crude petroleum until it is vapor and then allowing it to cool gradually. As the vapor cools, the component with the highest boiling point condenses first, followed in order by materials with successively lower boiling points. If the cooling is so controlled that the highest boiling component can be removed before the second highest condenses, then the remaining mixture will have one less component than the original vapor. With the absence of this component the second highest becomes the highest boiling component and the process is repeated. In practice, properly designed and constructed fractionating columns are capable of simultaneously and continuously separating crude oil into its desired component mixtures. Figure 12–7 is a schematic of such a column. As the vapors rise, they cool, allowing the various components to be removed from different levels of the column.

Out of Gas? Make Your Own Alkanes.

Specific alkanes can be produced in the laboratory without concern for their abundance, if any, in crude oil. This can be accomplished by either of two synthetic procedures. The first involves the substitutions of a hydrogen atom for an already existing functional group. The second involves the combination of two alkyl halides to produce a combined alkane. In reactions that are not dependent upon one particular alkyl group, the generalized notation **R** will be used to indicate that the reaction can be applied to any alkyl group (i.e., methyl, ethyl, etc.). The remainder of the molecule may consist of any of a number of functional groups. Table 12–4 is a listing of these groups and the functional nature that they confer upon the alkyl group. They will be dealt with individually in forthcoming pages.

Alkanes from Acids Can Be as Easy as Breathing Out.

Alkanes can be synthesized from organic acids by removing the equivalent of a carbon dioxide molecule (CO_2) from the carboxylic acid group $\left(-\overset{\overset{\large O}{\|}}{C}-OH\right)$. It is necessary when synthesizing alkanes by this method that the starting compound contain one more carbon atom than the desired product. The method is of value only if the starting acid can

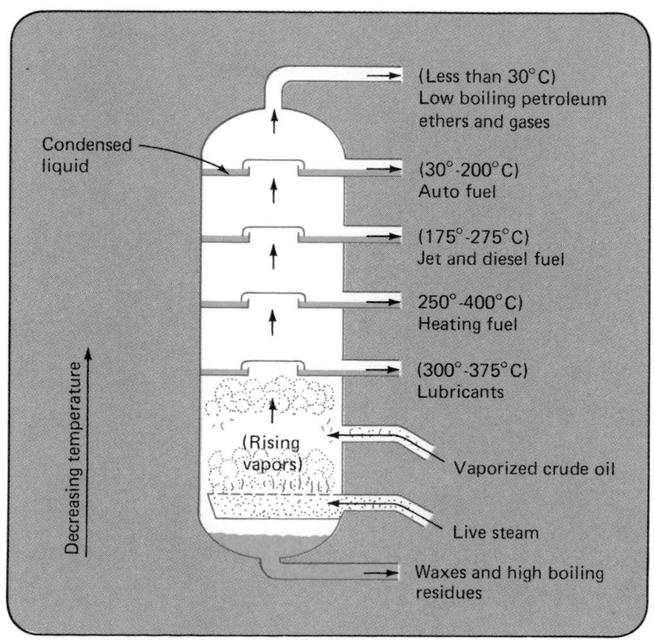

FIGURE 12–7 Fractionating Column for Distillation of Petroleum

TABLE

12–4 Functional Groups of Organic Compounds

Functional Group	Class Name	Example*	Example Name
—H	Alkanes	$CH_3CH_2CH_3$	Propane
—C=C—	Alkenes	$CH_3CH{=}CH_2$	Propene
—C≡C—	Alkynes	$CH_3C{\equiv}CH$	Propyne
—OH	Alcohols	$CH_3CH_2CH_2OH$	Propanol
$-\overset{H}{\overset{\vert}{C}}{=}O$	Aldehydes	$CH_3CH_2\overset{H}{\overset{\vert}{C}}{=}O$	Propanal
$-\overset{O}{\overset{\Vert}{C}}-$	Ketones	$CH_3\overset{O}{\overset{\Vert}{C}}-CH_3$	Propanone
$-\overset{O}{\overset{\Vert}{C}}-OH$	Acids	$CH_3CH_2\overset{O}{\overset{\Vert}{C}}-OH$	Propionic acid
$-NH_2$	Amines	$CH_3CH_2CH_2NH_2$	Propyl amine
—C—O—C—	Ethers	$CH_3OCH_2CH_3$	Methyl ethyl ether
—X	Halides	$CH_3CH_2CH_2Cl$	Propyl chloride

*Double bonding (=) is discussed in detail starting on page 243, and triple bonding (≡) is discussed starting on page 248.

be obtained as a cheap industrial by-product or is readily available in nature.

$$RCOOH + NaOH \xrightarrow[\text{Lime}]{\text{Soda}} RH + Na_2CO_3$$

$$\underset{\text{2-Methylbutyric acid}}{CH_3CH_2CH\overset{CH_3}{\overset{\vert}{C}}OOH} + NaOH \rightarrow \underset{\text{Butane}}{CH_3CH_2CH_2CH_3} + Na_2CO_3$$

Butane can also be prepared by reacting its iodide with hydrogen iodide or hydrolyzing its magnesium iodide with water:

$$RI + HI \rightarrow RH + I_2$$

$$CH_3CH_2CH_2CH_2I + HI \rightarrow CH_3CH_2CH_2CH_3 + I_2$$

$$RMgI + HOH \rightarrow RH + Mg\begin{matrix} I \\ OH \end{matrix}$$

$$CH_3CH_2CH_2CH_2MgI + HOH \rightarrow CH_3CH_2CH_2CH_3 + Mg\begin{matrix} / I \\ \backslash OH \end{matrix}$$

The alkyl magnesium halide is not listed in Table 12–4 because it is not a stock organic compound but is normally prepared specifically as a reagent for the synthesis of other organic compounds. It is known as a Grignard reagent in honor of the chemist who discovered its properties.

Grow Bigger Alkanes. Use Alkyl Halide instead of Water.

The Grignard reagent that was hydrolyzed to produce butane can also be used to produce higher numbered alkanes. This can be accomplished by substituting an alkyl halide for the water and running the reaction in an ether solution.

$$RMgX + XR' \rightarrow R - R' + MgX_2$$

$$\underset{\text{Butyl magnesium bromide}}{C_4H_9MgBr} + \underset{\text{2-Bromopropane}}{BrCH(CH_3)CH_3} \rightarrow \underset{\text{2-Methylhexane}}{C_4H_9—CH(CH_3)CH_3} + MgBr_2$$

The **R′** in the generalized reaction indicates that this alkyl group may be the same as or different from the **R** alkyl group. When the alkyl groups are identical, the molecule produced is symmetrical.

Symmetrical alkanes can also be produced by the reactions of alkyl halides with sodium. This procedure produces an alkane with twice the number of carbon atoms as the original halide.

$$RX + 2\,Na + XR \rightarrow R - R + 2\,NaX$$

$$2\,CH_3CH_2CH_2CH_2Br + 2\,Na \rightarrow CH_3CH_2CH_2CH_2 \vdots CH_2CH_2CH_2CH_3 + 2\,NaBr$$

The method is useful only for the production of symmetrical compounds. If two different alkyl halides were employed in this synthesis, three products rather than one would be produced. This is because each halide, in addition to being able to react with the other type halide, is also capable of reacting with one of its own kind. The net effect of these additional reactions is to reduce the amount of desired product that is produced.

$$CH_3CH_2CH_2CH_2Br + CH_3Br + 2\,Na \rightarrow CH_3CH_2CH_2CH_2{\vdots}CH_3 + 2\,NaBr$$

Butyl bromide Methyl bromide Pentane

$$2\,CH_3CH_2CH_2CH_2Br + 2\,Na \rightarrow CH_3CH_2CH_2CH_2{\vdots}CH_2CH_2CH_2CH_3 + 2\,NaBr$$

Butyl bromide Octane

$$2\,CH_3Br + 2\,Na \rightarrow CH_3{\vdots}CH_3 + 2\,NaBr$$

Methyl bromide Ethane

Example. List all the bromides that can be reacted with **HI** to produce 2-methylbutane.

1-Bromo-2-methylbutane $CH_2Br—C(CH_3)(H)—CH_2—CH_3$

2-Bromo-2-methylbutane $CH_3—C(CH_3)(Br)—CH_2—CH_3$

3-Bromo-2-methylbutane $CH_3—C(CH_3)(H)—CHBr—CH_3$

4-Bromo-2-methylbutane
1-Bromo-3-methylbutane $CH_3—C(CH_3)(H)—CH_2—CH_2Br$

Each of the above compounds will yield 2-methylbutane when a hydrogen atom is substituted for the existing bromine.

Example. Starting with ethyl bromide, write a reaction for the syntheses of 3-methylpentane using a Grignard reagent.

$$CH_3CH_2Br + BrMg\overset{\overset{\displaystyle CH_3}{|}}{C}H{-}CH_2CH_3 \rightarrow CH_3CH_2{-}\underset{\underset{\displaystyle H}{|}}{\overset{\overset{\displaystyle CH_3}{|}}{C}}{-}CH_2CH_3 + MgBr_2$$

Secondary butyl magnesium bromide | 3-Methylpentane

In order to produce 3-methylpentane the Grignard functional group must be attached to either of the two secondary butyl carbon atoms. If it is attached to a primary carbon, normal hexane rather than 3-methylpentane is produced.

Like the proverbial cat that can be skinned in more than one way, these examples illustrate that, more often than not, organic products can be produced in more than one way from more than one kind of reactant.

When They Shouted "Charge!" the Radical Said "No."

The preceding reactions have all been concerned with producing alkanes from other chemicals. Once produced, however, the reactions of alkanes are limited. For the most part they are restricted to oxidations. A typical oxidation reaction is the halogenations of methane by chlorine. This reaction is thought to proceed by a free radical mechanism. Free radicals are atoms or groups of atoms with unshared electrons. They are usually highly reactive and combine readily with other atoms or atomic groupings that can pair their unshared electrons.

In organic chemistry, molecules react by either ionic or free radical mechanisms. Ionic mechanisms should be familiar from the discussions of inorganic reactions. They are the method by which most substitution and neutralization reactions in aqueous solution occur. They are due to the interaction of ionic charges. Organic compounds may also form ions. The positive ones are called carbonium ions and the negative ones carbanions. They correspond respectively to inorganic cations and anions:

$CH_3CH_2CH_2^+$
Carbonium ion

$CH_3CH_2CH_2^-$
Carbanion

Free radicals by contrast do not normally possess charges. Their radical nature is indicated by a dot signifying the unpaired electron.

$$CH_3CH_2CH_2\cdot$$

They are produced by subjecting molecular bonds to sufficient thermal or light energy so that a bond is broken into two fragments, each of which has one half of an electron pair. It is by means of these odd electron fragments that chlorine and methane react.

$$:\ddot{\mathrm{Cl}}—\ddot{\mathrm{Cl}}: \xrightarrow[\text{light}]{\text{heat or}} 2\,:\ddot{\mathrm{Cl}}\cdot$$

Chlorine molecule — Chlorine atom (Free radical)

By the action of heat or light the chlorine molecule absorbs sufficient energy to break into free radical atoms. These free radical atoms, in order to pair their odd electrons, abstract a hydrogen atom from a methane molecule.

$$\mathrm{H{:}\overset{H}{\underset{H}{\ddot{C}}}{:}H} + :\ddot{\mathrm{Cl}}\cdot \rightarrow \mathrm{H{:}\overset{H}{\underset{H}{\ddot{C}}}}\cdot + \mathrm{H}{:}\ddot{\mathrm{Cl}}:$$

Methane — Chlorine radical — Methyl radical — Hydrogen chloride

This abstraction produces a methyl radical that then interacts with another chlorine molecule to abstract a chlorine atom to satisfy its unpaired electron.

$$\mathrm{H{:}\overset{H}{\underset{H}{\ddot{C}}}}\cdot + :\ddot{\mathrm{Cl}}{:}\ddot{\mathrm{Cl}}: \rightarrow \mathrm{H{:}\overset{H}{\underset{H}{\ddot{C}}}}{:}\ddot{\mathrm{Cl}}: + :\ddot{\mathrm{Cl}}\cdot$$

Methyl radical — Methyl chloride

This abstraction results in the production of a methyl chloride molecule and another chlorine atom with which the free radical reaction chain reaction may be propagated. Because most free radical reactions proceed by a self-generating process, only a small concentration of radicals is often sufficient to make a reaction go. The chain nature of these reactions is reminiscent of nuclear reactions where the neutron yield is one. The overall reaction can be summarized by

$$Cl_2 + CH_4 \xrightarrow[\text{light}]{\text{Heat or}} CH_3Cl + HCl$$

It tells what happened. The mechanism tells how.

If an excess of chlorine is present, this free radical chlorination need not stop with methyl chloride but may continue by the same procedure to abstract additional hydrogens to produce the di-, tri-, and tetrachlorine derivatives of methane.

$$\underset{\text{Chloromethane}}{CH_3Cl + Cl_2} \rightarrow \underset{\text{Dichloromethane}}{CH_2Cl_2 + HCl}$$

$$\underset{\text{Dichloromethane}}{CH_2Cl_2 + Cl_2} \rightarrow \underset{\substack{\text{Trichloromethane} \\ \text{(Chloroform)}}}{CHCl_3 + HCl}$$

$$\underset{\text{Trichloromethane}}{CHCl_3 + Cl_2} \rightarrow \underset{\substack{\text{Tetrachloromethane} \\ \text{(Carbon tetrachloride)}}}{CCl_4 + HCl}$$

In actual practice this halogenation reaction always produces some of each type of chlorinated methyl derivatives. Only their relative amounts change with the chlorine concentration. Knowledge of reaction mechanisms is often helpful in providing a systematic understanding of the reasons a reaction will produce one set of products as opposed to another. It provides a network for organizing the many otherwise isolated facts encountered in chemistry.

REVIEW

1. The simplest class of organic compounds are the alkanes (paraffins).
2. They are composed exclusively of carbon and hydrogen atoms.
3. Each carbon atom in an alkane forms four covalent bonds. These bonds are between the carbon atom and four other atoms, either carbon or hydrogen.
4. The general formula of an alkane is $\mathbf{C_nH_{2n+2}}$. The number of hydrogen atoms in a molecule equals twice the number of carbon atoms plus two.

5. All the carbon orbitals involved in the bonding together of an alkane molecule are sp^3 hybrids, which are arranged tetrahedrally about each carbon atom.
6. In most organic compounds component atoms may attain a complete complement of valence shell electrons (rare gas configuration) through more than one arrangement of atoms and bonds.
7. Molecules having different arrangements of the same number of atoms and bonds are known as isomers.
8. Although isomers possess the same general formula, they display significantly different physical and chemical properties.
9. The three major types of isomers are structural, geometrical, and optical. These classifications are applicable to other classes of organic compounds as well as alkanes.
10. Structural isomers contain the same number and type of atoms but differ in the manner in which these atoms are bonded to each other.

    Geometrical isomers contain the same number and type of atoms but differ in terms of the spatial orientation of their groups.

    Optical isomers contain the same number and type of atoms but differ in the manner in which they rotate the plane of polarized light. The ability to rotate the plane of polarized light is due to the presence within the molecule of one or more asymmetric carbon atoms.
11. In the naming of compounds possessing more than one isomer it is necessary to specify by notations such as "levo," "dextro," "n," "iso," "*cis*," or "*trans*" the isomer of interest.
12. When naming alkanes it is first necessary to determine the number of carbons in the longest chain. This number specifies the name of the compound. All other shorter chains are referred to this chain length.
13. When organic compounds react they do so by either an ionic or a free radical reaction mechanism. Positive and negative organic ions are known respectively as carbonium ions and carbanions.
14. Ionic organic reaction mechanisms are a consequence of the charges of the reactant ions.
15. Free radical organic compounds are not charged. The mechanism by which they react is governed by the process of electron pairing.

REFLEXIVE QUIZ

1. The names of the three structural isomers of C_5H_{12} are ______________, ______________, and ______________.
2. There are no __________ or __________ butane isomers.
3. *Cis*-1,2-ethylpropylcyclohexane is an example of a __________ isomer.
4. "Straight-chained" alkanes do not possess __________ isomers.
5. If isomers of a compound rotate the plane of polarized light, they are _________ isomers.
6. The plane of polarized light is rotated in a clockwise fashion by the _________ optical isomer.
7. Mixtures of equal amounts of enantiomers are known as ________ mixtures.
8. The method of separating compounds by means of differences in their boiling points is known as ___________________________.
9. Reaction mechanisms that are dependent upon the unpaired electrons in some of their reactants are known as ______________ reactions.
10. An organic ion that possesses a positive charge is called a __________.

Answers: (1) Normal pentane, isopentane, neopentane; or pentane, 2-methylbutane, 2,2-dimethylpropane [216]. (2) Geometrical, optical [216–222]. (3) Geometrical [222, 238]. (4) Geometrical (222, 238]. (5) Optical [222–229, 238]. (6) Dextro (+) [226–228]. (7) Racemic [226]. (8) Fractional distillation [230–231]. (9) Free radical [235–238]. (10) Carbonium ion [235].

SUPPLEMENTARY READINGS

1. M. Orchin. "Determining the Number of Isomers from a Structural Formula." *Chemistry* 42 (1969): 8.
2. F. D. Rossini. "Hydrocarbon in Petroleum." *Journal of Chemical Education* 37 (1960): 554.
3. G. A. Mills. "Ubiquitous Hydrocarbons." *Chemistry* 44 (1971): 12.
4. T. O. Lipman. "Wöhler's Preparations of Urea and the Fate of Vitalism." *Journal of Chemical Education* 41 (1964): 45.
5. R. T. Wendland. "Petrochemicals." New York: Anchor Books—Doubleday and Co., 1969.
6. D. F. Mowery, Jr. "Criteria for Optical Activity." *Journal of Chemical Education* 46 (1969): 269.

Section 13

Organic Chemistry: Unsaturated Compounds

Section 13

Organic Chemistry: Unsaturated Compounds

When There's Not Enough Hydrogen to Go Round

The alkanes are indeed fortunate in having only single bonds and a hydrogen for each noncarbon–carbon bond. Since their carbon skeletons are completely attached to hydrogen atoms they are known as *saturated compounds.* When insufficient hydrogen atoms exist to satisfy a carbon skeleton's needs, the compound is *unsaturated*, and it must resort to multiple-bond formation to complete its structure. Depending on the ratio of carbon to hydrogen atoms, this feat is accomplished through the formation of one or more *double* or *triple bonds.*

Molecules containing double bonds are known as *alkenes.* They have the general formula C_nH_{2n}. Those containing a triple bond are *alkynes.* They have the general formula C_nH_{2n-2}.

Example. Ethane (C_2H_6) is a two carbon saturated compound, an alkane. The two carbon unsaturated compounds are ethene and ethyne.

	Alkane	Alkene	Alkyne
General formula	C_nH_{2n+2}	C_nH_{2n}	C_nH_{2n-2}
Name	Ethane	Ethene	Ethyne
Configuration	$CH_3—CH_3$	$CH_2═CH_2$	$CH≡CH$

Large molecules may contain more than one multiple bond. When this occurs, it is possible to have certain alkenes with generalized formulas that are identical to those of alkynes and vice versa.

For example, each of the three compounds below has a general formula C_nH_{2n-2} and a specific formula C_4H_6:

```
H  H  H  H                 H  H           H           H
|  |  |  |                 |  |           |           |
C═C—C═C      H—C≡C—C—C—H     H—C—C≡C—C—H
|        |                 |  |           |           |
H        H                 H  H           H           H
 Butadiene        1-Butyne              2-Butyne
```

Butadiene and both butynes contain the same number of bonds, carbon atoms, and hydrogen atoms. In spite of this, their chemistries are not identical. This is because the chemistry of each is dependent upon its respective functional groups (double or single bonds) and configuration. Although each molecular configuration provides each of its component atoms with a rare-gas-type valence shell, the three different approaches produce three different molecules. In each approach, however, the electron requirements of four carbons and six hydrogens are satisfied with only twenty-two electrons.

```
  H  H H  H           H H          H      H
H:C::C:C::C:H   H:C:::C:C:C:H   H:C:C:::C:C:H
                      H H          H      H
```

$CH_2CHCHCH_2$ $\quad$ $CHCCH_2CH_3$ $\quad$ CH_3CCCH_3

In Section 12 we saw that the single bonds of the four carbon atoms of an alkane are arranged tetrahedrally and are equivalent one to another. This tetrahedral arrangement, the result of sp^3 hybridization of orbitals, although the most favorable for single bonds, is energetically poor for multiple bonds. If two sp^3 carbon atoms are to be joined so that there are multiple bonds between them, all but the initial bond would have to bend back on themselves in order that the additional bonds might interact. Such bending would produce great strain in these bonds and make them energetically unstable. (Figure 13–1.)

If one *s* and two rather than three *p* orbitals are hybridized, the resulting three equivalent hybrid orbitals are more suitable for double bond formation. Such a hybridization produces a configuration where three sp^2 orbitals are all in one plane and the remaining *p* orbital extends out perpendicular to that plane. The results have the appear-

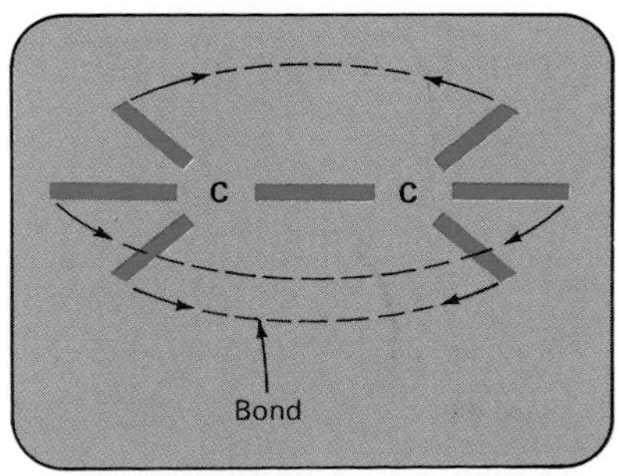

FIGURE 13–1 Bond Strains Which Would Result from sp^3 Multiple Bonds

ance of a three-position turnstile where each arm represents one of the three sp^2 hybrid orbitals and the shaft the remaining unsullied p orbital. (Figure 13–2.)

The p orbital extends out equally above and below the plane of the turnstile. When a double bond is formed between two carbon atoms, one sp^2 orbital from each carbon overlaps one sp^2 orbital from the other to form a bond. Such a bond joining two atoms along their centers is known as a sigma (σ) bond. (Figure 13–3.)

When oriented properly, the p orbitals of the two bonding carbons also overlap. This results in the formation of a second bond above and below the σ bond. A bond of this type, constructed from two p electrons, is known as a pi (π) bond. Although the electron density of the π bond as pictured in Figure 13–4 appears above and below the σ bond, it is only one additional bond because it involves only two

OPTOMETRIST
FORMING MULTIPLE BONDS WITH SP3 CARBONS IS LIKE SHAKING HANDS WITH YOUR PALM OUT.

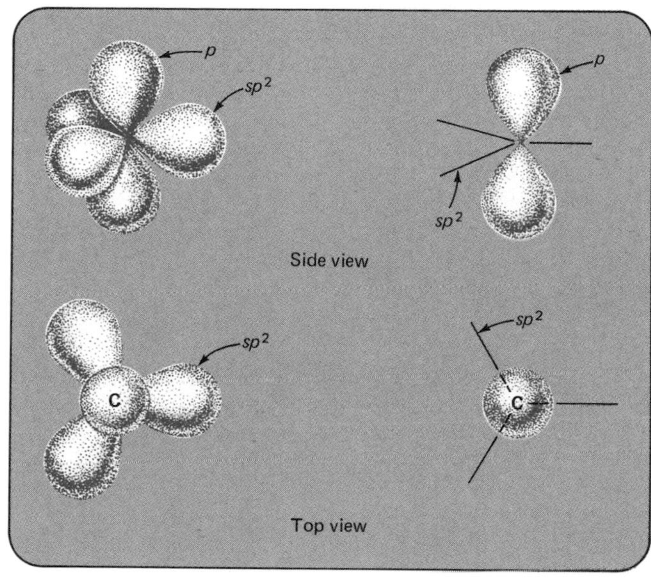

FIGURE 13–2 sp^2 Orbital Hybridization

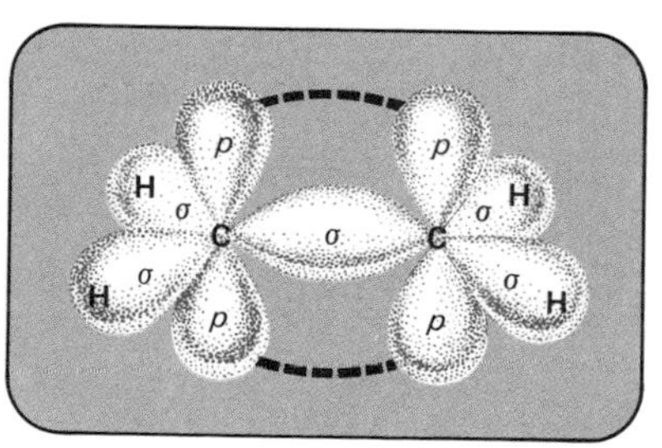

FIGURE 13–3 σ Bonding in Ethene

additional electrons. Double bonds consist, therefore, of one σ bond and one somewhat weaker π bond. The atoms joined by them are more closely spaced and more tightly bound than equivalent atoms held together by only a single bond. Each carbon, in addition to contributing two electrons for the formation of the double bond, also possesses two sp^2 hybrid orbitals that lie in a plane passing through the σ bond and

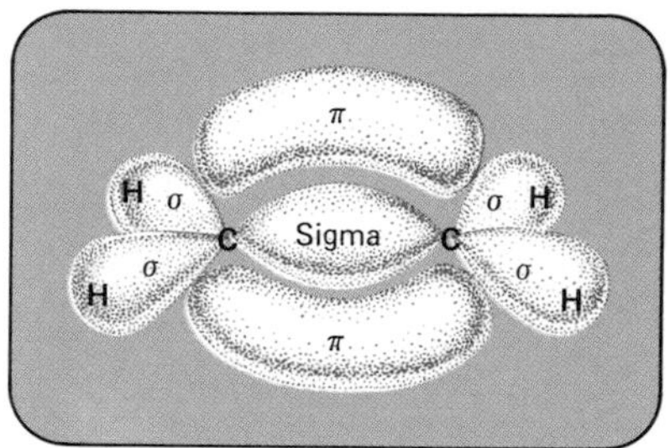

FIGURE 13–4 π Bonding in Ethene

perpendicular to the plane created by the two sausage-shaped π bonds. (Figure 13–4.)

Because of this planar nature, these orbitals are capable of forming geometrical isomers. This may be readily seen in the 1,2-dichloro isomers of ethene. Although there is only one type of ethene (ethylene) molecule,

```
H       H
 \     /
  C═C
 /     \
H       H
```

there are two dichloroethylene isomers, each with its own set of physical properties:

```
Cl      Cl              H       Cl
  \    /                 \     /
   C═C                    C═C
  /    \                 /     \
 H      H              Cl       H
```

Cis-1,2-dichloroethene   *Trans*-1,2-dichloroethene

Cis		*Trans*
−80.5°C	Melting point	−50°C
60.3°C	Boiling point	47.5°C
1.284 g/ml	Density	1.256 g/ml

These geometric isomers are only possible because the π bond effectively prevents the free rotation of the carbon atoms about the σ bond joining their centers. The molecule that has adjacent chlorine atoms is the *cis* form; in the *trans* form, the chlorine atoms are at opposite edges of the molecular plane. (Reminder: *Cis* in Latin means "on this side," *trans* means "on the other side.") Geometric isomers of 1,2-dichloroethane do not exist because no π bond is present to restrict the rotation of the carbon atoms and cause all the functional atoms to align themselves in one plane.

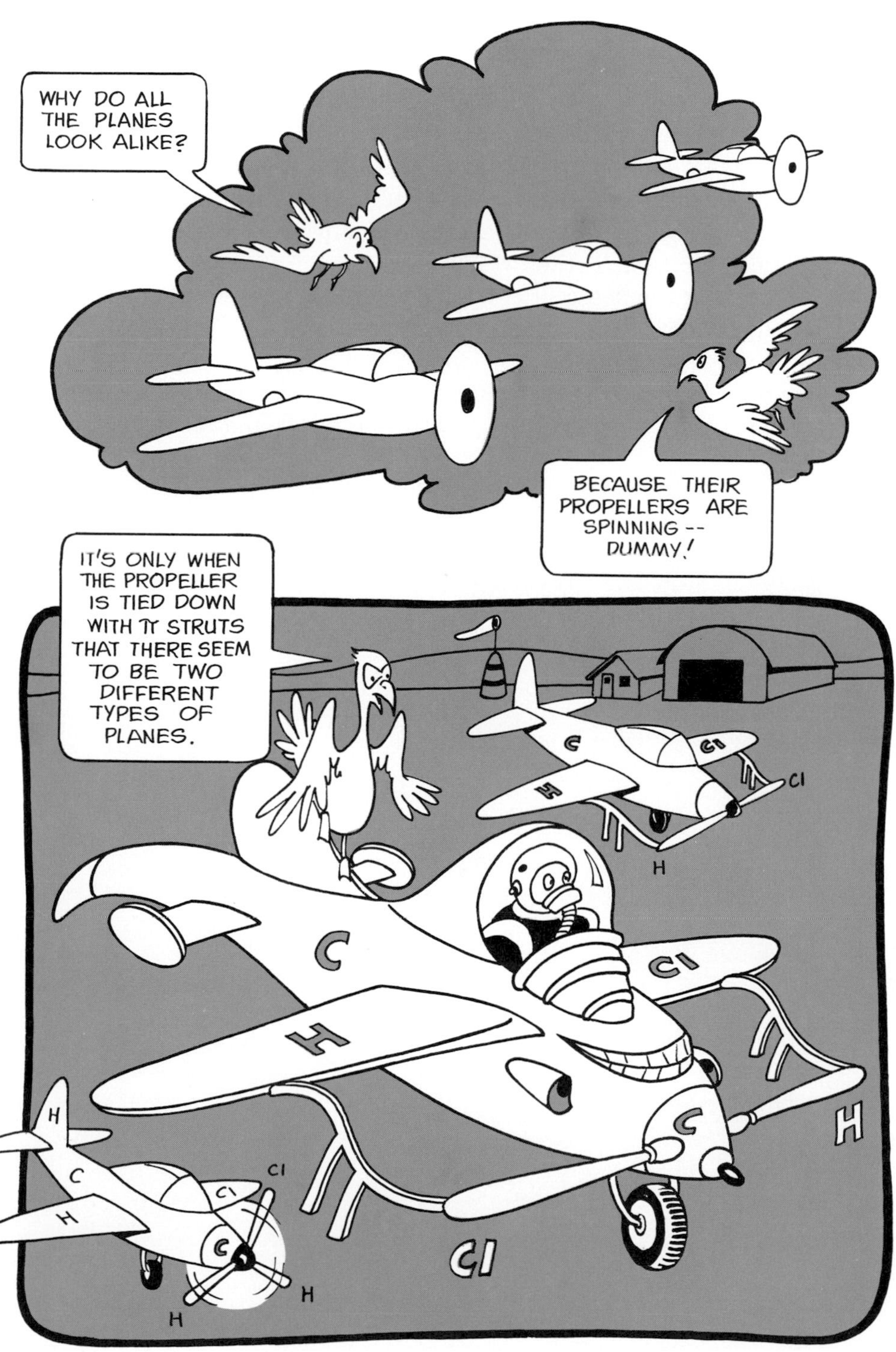
WHY DO ALL THE PLANES LOOK ALIKE?
BECAUSE THEIR PROPELLERS ARE SPINNING -- DUMMY!
IT'S ONLY WHEN THE PROPELLER IS TIED DOWN WITH π STRUTS THAT THERE SEEM TO BE TWO DIFFERENT TYPES OF PLANES.
C
Cl
H
Cl
H
C
Cl
H
Cl
H
H
C
H
C
Cl
H
H
C
Cl
Cl
H

Triple Bonds Make Tight Little Molecules.

Derivatives of ethyne (**C_2H_2**, acetylene), like those of ethane, do not have geometric isomers. Simple compounds of triple-bonded carbon atoms are linearly oriented. This is the result of *sp* hybridization. When one *s* and one *p* carbon orbital hybridize, the resulting atom acquires two *sp* hybrid orbitals, which are oriented back to back. The two remaining *p* orbitals are oriented perpendicular to these *sp* hybrids and to each other. (Figure 13–5.)

Ethyne is composed of two such carbons and two hydrogens. (Figure 13–6.) The σ bond along the center line of the molecule is formed by the overlap of one *sp* hybrid from each carbon. The two additional bonds necessary for the formation of the triple bond are π bonds. They are formed by the overlapping of the *p* orbitals and result in the pro-

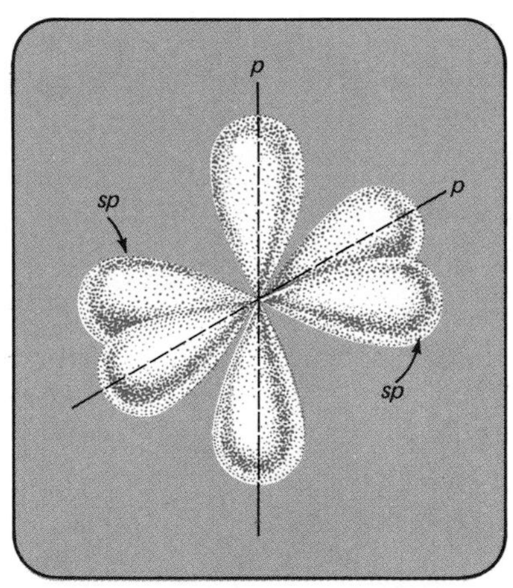

FIGURE 13–5 ***sp* Orbital Hybridization**

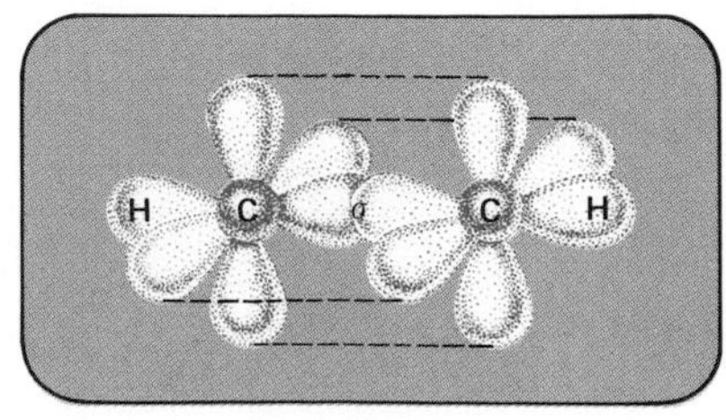

FIGURE 13–6 **Bonding in Ethyne**

duction of π sausages above and below and in front and back of the σ bond. The net effect of these four sausages is to create a hollow superbaloney containing two carbon atoms with hydrogen atoms protruding from each end. The hydrogen atoms are joined to the carbons by the remaining *sp* hybrid orbitals.

Alkynes occur in only one isomeric form, since all four atoms are oriented in a straight line because of *sp* hybridization. The triple bond is shorter and stronger than either the single or double bond.

Where, Oh Where, Have the Multiple Bonds Gone?

Alkenes occur widely in nature. These double-bonded carbon compounds are intermediates in the biosyntheses and metabolism of fats and carbohydrates and the miracle ingredient in polyunsaturated margarines. Alkenes at room temperature may be either gases, liquids, or solids, depending on their molecular weight. When they are composed only of carbon and hydrogen, they are all insoluble in water and all have a density of less than 1. Unlike alkanes, they are easily attacked by many reagents and may serve as useful intermediates in the preparation of other compounds.

Alkynes do not generally occur in nature; however, plants of the composital family such as daisies and sunflowers do contain triple-bonded molecules of the type.

$$CH_3C{\equiv}C{-}C{\equiv}C{-}C{\equiv}C{-}\overset{\displaystyle H}{\overset{|}{C}}{=}\overset{\displaystyle H}{\overset{|}{C}}{-}\overset{\displaystyle O}{\overset{\|}{C}}{-}OCH_3$$

Industrially, ethyne (acetylene) is the only alkyne of significance. It was discovered in 1836 by Sir Humphry Davey, who prepared it by adding water to a metallic carbide. Commercial producers of acetylene utilize calcium carbide for this purpose because it can be produced cheaply by reacting calcium oxide with coke.

$$CaO + 3\,C \xrightarrow{3000°C} \underset{\text{Calcium carbide}}{CaC_2} + CO$$

$$CaC_2 + 2\,H_2O \longrightarrow \underset{\text{Acetylene}}{C_2H_2} + Ca(OH)_2$$

If They Can't Be Found, They'll Have to Be Made.

Alkenes can be prepared commercially by the dehydration of alcohols or, in the cases of ethene, propene, and butene, by the catalytic cracking of the appropriate alkane.

Dehydration

$$\mathrm{R{-}\underset{H}{\overset{H}{C}}{-}\underset{OH}{\overset{H}{C}}{-}R \xrightarrow[\text{Dehydrating agent}]{H_2SO_4} R{-}\overset{H}{C}{=}\overset{H}{C}{-}R + H_2O}$$

The sulfuric acid removes the components of water from vicinal (adjacent) carbon atoms. The optimum temperature for the process is dependent on the particular alcohol. With ethanol, a temperature of 160°C is employed.

$$\mathrm{CH_3CH_2OH \xrightarrow[160^\circ]{H_2SO_4} \underset{Ethene}{CH_2{=}CH_2} + H_2O}$$

The dehydration of ethanol is quite straightforward. If, however, more than one possible product can be produced by the dehydration of an alcohol, the product formed in greatest abundance will be the most highly substituted alkene. The dehydration of 2-butanol will produce both 1-butene and 2-butene.

$$\mathrm{CH_3\underset{OH}{C}HCH_2CH_3 \xrightarrow{H_2SO_4} \underset{\substack{\text{2-Butene}\\ \text{(Major product)}}}{CH_3CH{=}CHCH_3} + \underset{\text{1-Butene}}{CH_2{=}CHCH_2CH_3} + H_2O}$$

2-Butene, which has two methyl groups substituted for two hydrogens on the double-bonded carbons, is more highly substituted than 1-butene, in which only one hydrogen is replaced by an ethyl group. Although this rule is empirical and not always completely accurate, it is often of value in predicting the products of dehydration and dehydrohalogenation of alcohols and alkyl halides. It is commonly called the Saytzev Rule, after its discoverer.

Cracking

At elevated temperatures, gaseous alkanes break into smaller molecules:

$$CH_3CH_2CH_3 \xrightarrow{500°C} CH_4 + CH_2{=}CH_2$$

The process usually occurs so that the electrons of a broken bond are distributed equally between the two fragments that are produced. Since each possesses an odd electron where once there was an electron pair, both fragments are free radicals.

$$CH_3CH_2CH_3 \rightarrow CH_3\cdot + \cdot CH_2CH_3$$

These free radicals are relatively unstable and short-lived. They react with one another to produce longer-lived, more conventional molecules in which each atom has a rare gas configuration. In the case of the methyl and ethyl free radical fragments, the methyl abstracts a hydrogen atom from the ethyl to produce methane and ethene.

4 atoms		7 atoms		5 atoms		6 atoms
$CH_3\cdot$	+	$\cdot CH_2CH_3$	→	CH_4	+	$CH_2{=}CH_2$
7 electrons		13 electrons		8 electrons		12 electrons

It should be recognized that the numbers of atoms and electrons on both sides of the equation are the same. The reaction merely rearranges atoms and electrons so that each atom acquires a rare gas configuration.

This cracking reaction is also accompanied by a dehydrogenation that produces propene:

$$CH_3CH_2CH_3 \xrightarrow{500°C} CH_3CH{=}CH_2 + H_2$$

The mechanism by which the dehydrogenation reaction occurs is thought to be similar to that occurring in the cracking process, with the exception that the elevated temperature breaks a carbon–hydrogen bond rather than a carbon–carbon bond.

Dehydrohalogenation

Alkenes and alkynes can also be synthesized from alkylhalides and dihalides. The procedure is seldom used for the formation of alkenes, however, because the alkylhalides that might be utilized for this purpose usually have to be produced from alcohols. The alkenes are therefore usually produced directly from the same alcohols, without the intermediate step. If cheap alkylhalides are available as by-products of industrial processes, however, they may indeed be used for this purpose. The reaction is the abstraction by alcoholic base of the components of a hydrohalogen acid from vicinal carbon atoms. Alcohol is used as a solvent because both reactants are soluble in it. **KOH** is usually the base of choice.

$$R{-}\overset{\overset{\large R}{|}}{\underset{\underset{\large X}{|}}{C}}{-}\overset{\overset{\large H}{|}}{\underset{\underset{\large H}{|}}{C}} + KOH \xrightarrow{C_2H_5OH} R{-}\overset{\overset{\large R}{|}}{C}{=}\overset{\overset{\large H}{|}}{C}{-}R + KX + H_2O$$

$$CH_3\underset{\underset{\large Br}{|}}{C}HCH_3 + KOH \xrightarrow[\text{Heat}]{C_2H_5OH} CH_3CH{=}CH_2 + KBr + H_2O$$

2-Bromopropane
(A secondary halide)

The reaction gives best results with tertiary halides, least satisfactory results with primary halides, and intermediate results with secondary halides.* Typically, as much as 100 percent of the tertiary halides can be converted to the alkene, whereas the conversion of primary halides is seldom greater than 20 percent efficient. The production of ethers, a second product, is the reason for the reduced alkene production.

$$CH_3CH_2CH_2Br \xrightarrow[\text{Heat}]{\text{Alcoholic } \mathbf{KOH}} \underset{20\%}{CH_3CH{=}CH_2} + \underset{60\%}{CH_3CH_2CH_2{-}O{-}CH_2CH_2CH_3}$$

When the dehydrohalogenation of an alkylhalide can produce more than one alkene, the Saytzev Rule may once again be employed to determine which of the two possible products will be produced in greater abundance.

Alkylene dihalides can be used to produce either alkenes or alkynes. The deciding factor is the reagent employed to produce the multiple

*The terms *primary, secondary,* and *tertiary* when applied to a halide indicate that the number of alkyl groups attached to the carbon atom that is bonded to the halide (—**X**) is, respectively, one, two, and three.

CH_3CH_2X	$CH_3\underset{\underset{\large X}{|}}{C}HCH_3$	$(CH_3)_3CX$
Primary	Secondary	Tertiary

bond structure. If the reagent is a metal, then a metallic halide and an alkene will be produced.

$$\underset{\quad\ \ X\qquad X}{R{-}CH{-}CH{-}R} + Zn \rightarrow R{-}CH{=}CHR + ZnX_2$$

$$\underset{\qquad\quad Cl\qquad Cl}{CH_3CH{-}CH{-}CH_3} + Zn \rightarrow CH_3CH{=}CHCH_3 + ZnCl_2$$

If the reagent is an alcoholic base, then where possible it may be expected that two hydrohalogen acids will be removed from each molecule and an alkyne, two water, and two metal halide molecules will be produced. The production of alkynes by this procedure is merely an extension of the process for the production of alkenes. Because the dihalide contains the components of two hydrohalogen acids, their removal by an alcoholic base produces a triple-bonded structure rather than a double-bonded one.

$$R{-}\underset{X}{\overset{H}{C}}{-}\underset{X}{\overset{H}{C}}{-}R + 2\,KOH \rightarrow R{-}C{\equiv}C{-}R + 2\,KX + 2\,H_2O$$

$$CH_3\underset{Cl}{CH}{-}\underset{Cl}{CH}{-}CH_3 + 2\,KOH \rightarrow CH_3C{\equiv}CCH_3 + 2\,KCl + 2\,H_2O$$

Multiple Bonds Are Just Voids Waiting to Be Filled.

As a group, the reactions of unsaturated molecules are just the reverse of those necessary to produce them. Compounds with multiple bonds are synthesized by the removal of hydrogen, halogens, hydrohalogen acids, or water from vicinal atoms in singly bonded molecules. The acquisition of new atoms to eliminate these multiple bonds is the most common type of reaction that these unsaturated compounds undergo.

The addition of atoms to a double bond is more often more complicated than the removal of the same two atoms from the parent single-bonded compound. This is because there can be only one kind of double bond produced by the removal of two dissimilar atoms from the parent alkene, but the return of these same two atoms can be accomplished in either of two ways. 2-Bromopropane, when reacted with alcoholic **KOH**, produces propene.

$$CH_2CHBrCH_2 + KOH \rightarrow CH_2CH{=}CH_2 + KBr + H_2O$$

Propene, when reacted with hydrogen bromide, however, can, depending on conditions, produce either 1-bromopropane or 2-bromopropane.

$$CH_3CH{=}CH_2 + HBr \rightarrow \begin{cases} \nearrow CH_3CHBrCH_3 \\ \searrow CH_2BrCH_2CH_3 \end{cases}$$

The decision as to which product is produced in greater abundance is a function of the mechanism by which the molecules react. The reaction sequence by which dry hydrogen bromide adds to a double bond may be visualized in terms of the following equation:

$$H_2C{::}CHCH_3 + H{:}\ddot{\underset{..}{Br}}{:} \rightarrow \left[\begin{matrix} & \ddot{CH_3} \\ CH_3{:} & C \\ & \ddot{H} \end{matrix} \right]^{+} + {:}\ddot{\underset{..}{Br}}{:}^{-}$$

The hydrogen bromide reacts with the propene to produce a positively charged carbonium ion, a charged carbon ion in which the carbon possesses a sextet of electrons. This carbonium ion is electrophilic (attracted to negative charges) in nature and readily combines with the bromide to form bromopropane. The mechanism is ionic. The π electrons of the alkene encourage the dissociation of the hydrogen bromide into a negative bromide ion and a hydrogen, which combines with propene to form a positively charged carbonium ion. The carbonium ion that is produced may be either primary or secondary.

$CH_3CH_2CH_2^+$	$CH_3\overset{+}{C}HCH_3$
Primary	Secondary
(Ethyl carbonium ion)	(Dimethyl carbonium ion)

The factor that determines which product will be produced in greater abundance is the stability of the precursor carbonium ion. This stability decreases in going from tertiary to primary carbonium ions. (Tertiary > secondary > primary. Tertiary carbonium ions have the form R_3C^+.) As a result, the addition of **HBr** to propene yields more 2-bromopropane than 1-bromopropane. When writing the equations of the reaction, for clarity it is customary to indicate only the major reaction product.

$$CH_3CH{=}CH_2 + HBr \rightarrow CH_3\overset{+}{C}HCH_3 + Br^- \rightarrow CH_3CHBrCH_3$$

HI and **HCl** add to a double bond in the same manner as **HBr**. The ease of addition decreases in going from **HI** to **HCl** (**HI** > **HBr** > **HCl**). This is because chlorine is a more reactive element than iodine. It forms a stronger bond with hydrogen and is therefore more reticent to allow its hydrogen to be incorporated into a carbonium ion.

The Russians Do It Differently: Free Radical Addition

The production of 2-bromopropane rather than 1-bromopropane by the addition of **HBr** to propene is due to the ionic character of the reaction mechanism. When the mechanism of chemical interaction changes, so too do the products. If there is no reason for chemicals to react in a particular way, they don't. The major reason for the production of 2-bromopropane is the greater stability of the secondary carbonium ion.

The characteristics of this free radical mechanism were first described by Markownikov, a Russian chemist, in 1870. Markownikov's rule states that when an asymmetrical reagent (**HBr**) adds to an asymmetrical double-bonded molecule (propene), the negative part of the adding reagent (bromine) will bond to the carbon with the smaller number of hydrogen atoms. If these compounds react without the production of carbonium ion, these restrictions do not hold and 1-bromopropane then becomes the predominant product.

Such is the case in free radical reactions. When the reaction is conducted in the presence of peroxides, **Br·** free radicals are produced. These atoms combine with the terminal double-bonded carbon atom to produce an organic free radical.

$$CH_3CH{=}CH_2 + :\ddot{\underset{..}{Br}}\cdot \rightarrow CH_3\underset{\cdot}{C}HCH_2Br$$

Once formed, the organic free radical can then interact with other **HBr** molecules to produce 1-bromopropane and regenerate the bromine atom free radical.

$$CH_3\underset{\cdot}{C}HCHBr + H:\ddot{\underset{..}{Br}}: \rightarrow CH_3CH_2CH_2Br + :\ddot{\underset{..}{Br}}\cdot$$

Because of the chain nature of this process, only a small amount of free-radical-inducing peroxide is necessary to alter the reaction products from 2-bromopropane to 1-bromopropane. As evidenced by this discussion, knowledge of the reaction mechanism is extremely useful in predicting reaction products.

Alkenes, in the presence of metal catalysts such as platinum and palladium, can acquire additional hydrogens and be converted to alkanes.

$$CH_3CH{=}CH_2 + H_2 \xrightarrow[\text{catalyst}]{\text{Pt}} CH_3CH_2CH_3$$

In hydrogenation reactions, because of the symmetry of the substance being added, the product that results is always the same, irrespective of the mechanism of addition. The speed of the reaction can be markedly influenced, however, by changes in reaction conditions that may result in a change in reaction mechanism.

Hydrogenation reactions are used commercially to convert unsaturated liquid vegetable oils into solids. The process is attractive because the resultant alkane derivative is more stable and therefore less likely to go rancid or produce smoky decomposition products when heated. Shortenings such as Spry and Crisco and many margarines are produced in this manner. In recent years, concern that the ingestion of saturated fatty acids may be linked with heart disease has caused food producers to advertise the unsaturated or partially saturated content of their products. Corn and cottonseed oils used to be hardened by hydrogenation to produce margarine, but these fatty acids are now only partially hydrogenated, and other chemicals are then added to make them suitable for margarine. Linolenic acid is a typical polyunsaturated acid found in linseed and fish-liver oil. Stearic acid, which is found in plant and animal fats, is its saturated analog.

$$CH_3CH_2{-}CH{=}CH{-}CH_2{-}CH{=}CH{-}CH_2{-}CH{=}CH{-}(CH_2)_7COOH$$

Linolenic acid (Polyunsaturated)

$$CH_3(CH_2)_{16}COOH$$

Stearic acid (Saturated)

Alkenes End to End: Plastics

One of the most important reactions of simple and polyunsaturated alkenes is polymerization, the process of joining together many small molecules to form very large ones. The resulting compound is a *polymer* (*poly* in Greek means "many," *meros* means "parts"). The individual molecules that make it all possible are the *monomers*. If the monomer is ethylene (ethene), for example, then its polymeric form is polyethylene:

$$n\,\mathbf{CH_2{=}CH_2} \xrightarrow[\text{and pressure}]{\mathbf{O_2}\text{, heat,}} \sim\!\sim\mathbf{CH_2{-}CH_2{-}CH_2{-}CH_2}{-}\sim\!\sim \text{ etc.}$$

$$\text{or } (\mathbf{{-}CH_2{-}CH_2{-}})_n$$

Polyethylene

where *n* equals a large number. Polyethylene is the plastic most familiar in packaging films, flexible bottles, and children's toys. If substituted ethylenes are polymerized, the resulting polyethylene polymer will contain the substituted group attached at more or less regular intervals along its length. Polyvinylchloride is the most familiar of the substituted ethylene polymers. It is used in phonograph records, plastic raincoats, and a number of enamel paint preparations.

$$n\,\mathbf{CH_2{=}\underset{\underset{Cl}{|}}{CH}} \xrightarrow{\text{Peroxide}} \sim\!\sim\mathbf{CH_2{-}\underset{\underset{Cl}{|}}{CH}{-}CH_2{-}\underset{\underset{Cl}{|}}{CH}{-}CH_2{-}\underset{\underset{Cl}{|}}{CH}}{-}\sim\!\sim \text{ etc.}$$

$$\text{or } \left(\mathbf{{-}CH_2{-}\underset{\underset{Cl}{|}}{CH}{-}}\right)_n$$

Vinyl chloride

Polyvinylchloride

Ethylene containing substituted $\mathbf{{-}C_6H_5}$, $\mathbf{{-}CN}$, or $\mathbf{{-}COOCH_3}$ groups as well as the halides can be employed in the formation of polymers. The polymerization process for all of these is basically the same. However, each polymer will have different properties and uses depending upon its particular side chains. Polymerization usually requires the presence of an initiator to start the chain formation. Typically, this is a peroxide that converts some of the monomer into free radicals. These free radicals combine with other monomer molecules to initiate the formation of radical polymer chains. These increase in size by further combination with other monomer molecules. As long as the chain retains its radical properties, it is capable of combining with additional monomers and growing still larger. Once the radical properties are removed, the growth of that particular molecule is terminated. Since this termination is unique for each molecular chain, polymers usually have a range of sizes rather than one specific molecular weight.

Chain initiation:

$$\text{Peroxide} \rightarrow 2\ \text{Radical}\cdot\ (\mathbf{R}\cdot)$$

$$\mathbf{R}\cdot + \mathbf{CH_2{=}CH(R)} \rightarrow \mathbf{R{-}CH_2{-}CH(R)}\cdot$$

Chain propagation:

$$\mathbf{R{-}CH_2{-}CH(R)}\cdot + \mathbf{CH_2{=}CH(R)} \rightarrow \mathbf{R{-}CH_2CH(R){-}CH_2CH(R)}\cdot \text{ etc.}$$

Fifty or more monomer molecules may be incorporated into a single chain by a succession of chain-propagating steps. The growth of the chain is terminated when a single covalent bond is formed from the unpaired radical electron. The size and nature of the two radicals that combine to form this covalent bond determine the size and nature of the completed molecule.

Chain termination:

$$\mathbf{R{-}CH_2{-}CH(R){-}\left(CH_2CH(R)\right)_n}\cdot + \cdot\mathbf{\left(CH(R){-}CH_2\right)_n{-}CH(R){-}CH_2{-}R} \rightarrow$$

$$\mathbf{R{-}CH_2{-}CH(R){-}\left(CH_2CH(R)\right)_{2n}{-}CH(R){-}CH_2{-}R}$$

Polymerizations are not restricted to free radical mechanisms. In ionic environments they can proceed by ionic means. The catalyst for such polymerizations is often boron trifluoride ($\mathbf{BF_3}$). Low temperatures are usually employed if products of maximum molecular weights are desired. Polymers of isobutylene ($\mathbf{CH_2{=}C(CH_3)_2}$), which are useful in the manufacture of auto tires, can be prepared in this manner. Like free radical polymerization, ionic polymerization can be divided into three steps.

Chain initiation:

$$\mathbf{F{:}\ddot{B}(F)(F)} + \mathbf{{:}\ddot{O}(H){:}H} \rightarrow \mathbf{F_3B{:}\ddot{O}(H){:}H} \rightleftarrows \mathbf{F_3B{:}\ddot{O}{:}H^-} + \mathbf{H^+}$$

$$\mathbf{H^+} + \mathbf{CH_2{=}C(CH_3)_2} \rightarrow \mathbf{H_3C{-}C(CH_3^{\,+})(CH_3)}$$

The Lewis acid, boron trifluoride, functions as a catalyst by reacting with water to produce a hydrogen ion. This hydrogen ion, when combined with the isobutylene, produces a trimethyl carbonium ion, which is capable of further reaction to produce a polymer.

$$(CH_3)_3C^+ + (n+1)CH_2{=}C(CH_3)_2 \rightarrow (CH_3)_3C{-}[CH_2C(CH_3)_2]_nCH_2\overset{\overset{CH_3{}^+}{|}}{\underset{\underset{CH_3}{|}}{C}} \quad \text{Chain propagation}$$

The growth of the chain is terminated by the expulsion of a proton from the carbonium ion and its recombination with the anion that was created by the reaction of boron trifluoride and water.

$$(CH_3)_3C[CH_2C(CH_3)_2]_nCH_2\overset{\overset{CH_3{}^+}{|}}{\underset{\underset{CH_3}{|}}{C}} \rightarrow (CH_3)_3C[CH_2C(CH_3)_2]_n\overset{\overset{H}{|}}{C}{=}\overset{\overset{CH_3}{|}}{\underset{\underset{CH_3}{|}}{C}} + H^+ \quad \text{Chain termination}$$

$$H^+ + F_3BOH^- \rightleftarrows F_3BOH_2$$

Natural rubber, another possible component of tires, is formed from isoprene (1,3-butadiene).

$$H_2C{=}\underset{\underset{CH_3}{|}}{C}{-}CH{=}CH_2$$

The molecular weight of its molecules is usually about a million, which means that each polymeric molecule contains approximately fourteen thousand isoprene units.

Aromatic Chemistry: Dream Cultures and Monkey Business

The planning of a country's course based upon the dreams of its statesmen is well established in antiquity. Genesis 41 tells of Pharaoh, the king of Egypt, who dreamed about seven lean cows devouring seven fat ones. Since this was not his everyday, garden variety of dream, he was disturbed. Joseph, one of his prisoners, interpreted it as meaning that Egypt would have seven years of prosperity followed by seven years of rough going. In order to soften the depression, Pharaoh mobilized Egypt to store grain in anticipation of the predicted famine.

When seven years later that region of the world suffered a general famine, Egypt alone had sufficient grain.

The most famous animal dream to be found in the chemical literature concerns Kekulé's snakes. Friedrich August Kekulé von Stradovitz, architect turned chemist, while riding a London bus in the year 1865, dozed off and dreamed of snakes. The snakes were wildly dancing chains of carbon atoms. Suddenly the head end of one attached itself to the tail and began to spin. The British had not scheduled a famine for that year so, rather than let a good dream go to waste, Kekulé used it to explain the structure of benzene.

Benzene and its related coal tar compounds were so fashionable at the time that *Punch,* the English humor magazine, waxed poetic:

> There's hardly a thing that a man can name
>   Of use or beauty in life's small game
> But you can extract in alembic or jar
>   From the "physical basis" of black coal tar
> Oil and ointment and wax and wine
>   And lovely colors called aniline;
> You can make anything from salve to a star,
>   If you only know how, from black coal tar.

Benzene is a queer duck; although its formula of C_6H_6 would lead one to believe that it was either an alkene or an alkyne, it does not behave as such.

$HC{\equiv}C{-}C{\equiv}C{-}CH_2{-}CH_3$

$HC{\equiv}C{-}CH_2{-}C{\equiv}C{-}CH_3$

$HC{\equiv}C{-}CH_2{-}CH_2{-}C{\equiv}CH$

$HC{\equiv}C{-}CH{=}CH{-}CH{=}CH_2$

$H_3C{-}C{\equiv}C{-}C{\equiv}C{-}CH_3$

$H_3C{-}C{\equiv}C{-}CH_2{-}C{\equiv}CH$

$H_2C{=}CH{-}C{\equiv}C{-}CH{=}CH_2$

Each of these compounds has the empirical formula of benzene but none behaves chemically like benzene.

Alkenes and Alkynes	*Benzene*
1. Undergo addition reactions	**1.** Doesn't
2. Are easily oxidized	**2.** Isn't
3. Produce several products by monosubstitution	**3.** Produces only one product by monosubstitution

The kicker in this comparison is that there is *only one* monosubstituted benzene product.

$$C_6H_6 + Br_2 \xrightarrow[\text{Catalyst}]{\text{Iron}} C_6H_5Br + HBr$$

Bromobenzene

If this same substitution were made on any of the listed straight-chained compounds, more than one product would be produced. If bromine were substituted for one of the hydrogens in

$$\overset{1}{HC}{\equiv}\overset{2}{C}{-}\overset{3}{CH}{=}\overset{4}{CH}{-}\overset{5}{CH}{=}CH_2$$

five different products could be produced, depending on which of the five hydrogens was replaced.

The snakes in Kekulé's dream told him that if he stayed with a straight-chained representation of benzene, the problem of molecular structure would always be confused. If, however, he let one end of the molecule join with the other, then not only would each know what the other was doing, but the resulting ring structure would make all the carbon atoms equivalent. Kekulé, like Eve, took the snake's advice. Unlike Eve, however, he found the results rewarding.

On the strength of the snake's suggestion, he proposed two equivalent structures for the benzene molecule.

```
          H                     H
          |                     |
          C                     C
       /     \\              //     \
  H—C          C—H      H—C          C—H
     ||         |          |         ||
  H—C          C—H      H—C          C—H
       \     //              \\     /
          C                     C
          |                     |
          H                     H
         (a)                   (b)
```

Kekulé suggested that the alternate double bonds were not fixed but moved at great speed within the molecule, constantly converting it from form *a* to form *b* and back. The properties of the molecule as a result were not of either form but actually those produced by the constant electron resonance between the two forms. This electron movement prevented the localization of single or double bonds between any two specific carbon atoms and, as a consequence, the molecule as a whole did not display properties typical of an unsaturated compound.

Considering the state of atomic and molecular theories at the time,

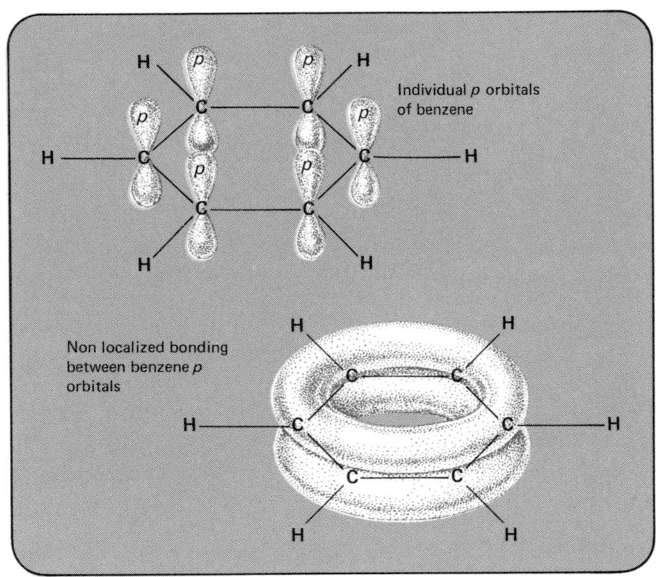

FIGURE 13–7 **Bonding in Benzene**

Kekulé's depiction of benzene was highly revolutionary. Today we believe that the benzene molecule is a resonance hybrid of these two Kekulé structures. That is, each carbon in benzene is considered bound to its two neighboring carbons and one hydrogen by sp^2 hybrid orbitals. In addition, the *p* orbital of each carbon, which contains one electron and is perpendicular to the surface of the ring, overlaps neighboring *p* orbitals to form π bonds between adjacent carbon atoms. Since each carbon atom is equidistant from its two neighbors, it shows no tendency to overlap one to the exclusion of the other. Because of this equivalency, what in fact does happen is the formation of two *p* electron π bond shadows above and below the benzene plane. These shadows produce a carbon–carbon bonding that is neither single nor double but uniquely different. Its length is intermediate between that of a single and a double bond. (Figure 13–7.)

The π bonds in benzene, because they are not localized as in alkenes and alkynes, act in concert to stabilize the molecule. Chemical reagents that normally would react with isolated π bonds are ineffective against this network. The result of this stability is to produce a network of bonds that remain intact through nearly all the reactions of benzene. The single form of a benzene molecule may therefore be more correctly depicted by

$$C_6H_6 \text{ (ring of six C—H units with inner circle)}$$

The inner circle indicates that the π electrons are delocalized and mutually shared among all the carbons. It should be evident from this structure that all the carbons are equivalent and that there is only one monobromobenzene.

$$C_6H_6 + Br_2 \rightarrow C_6H_5Br + HBr$$

"Let us learn to dream, gentlemen, then perhaps we shall find the truth . . . but let us beware of publishing our explanations before they have been put to proof by working understanding."—Kekulé.

There's Only One One, But There Are Three Two's.

There is only one isomeric form of all monosubstituted benzenes. When the substitution is greater than one, however, this uniqueness disappears. Toluene is a monosubstituted benzene where one methyl has replaced one hydrogen.

$$C_6H_5CH_3$$

If two hydrogen are each replaced by a methyl, the resultant product is xylene, which exists in three isomeric forms.

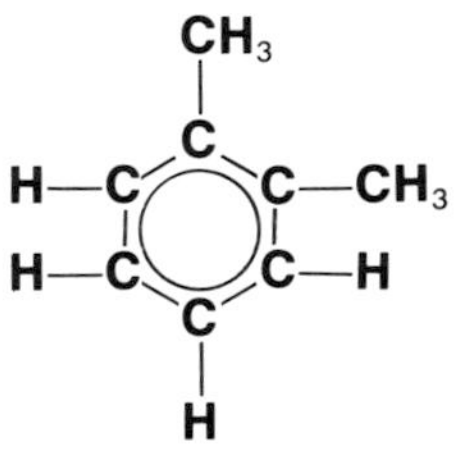

Ortho-xylene 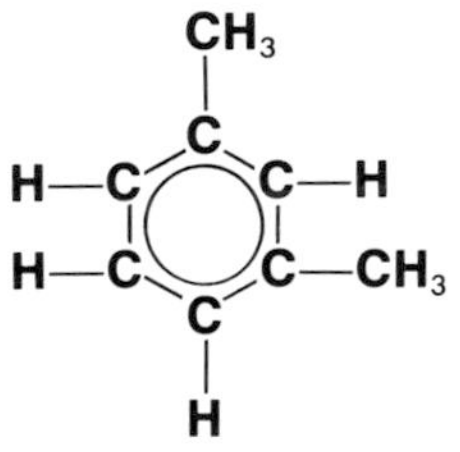 Meta-xylene 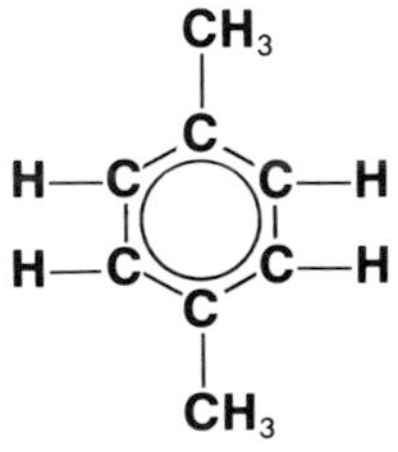 Para-xylene

If xylene is considered the product of two successive methyl substitutions of benzene, then it may be said that the first substitution, which produced toluene, removed the equivalency from the remaining five unsubstituted carbon atoms. This is because the first substitution establishes a particular carbon atom as a point of reference from which all future substitutions can be measured. The substitution of the second group can be described as occurring 1, 2, or 3 carbon atoms away from the first substituted atom. The first substitution cannot be so described because all six carbons in the unsubstituted benzene are identical.

Starting at the first substituted carbon atom and counting around the ring, chemists have labeled the positions in the benzene ring as *ortho, meta,* and *para* ("straight," "after," and "beyond," respectively, in Greek). Thus, 1,2-dimethylbenzene is known as ortho-xylene; 1,3-dimethylbenzene is known as meta-xylene; and 1,4-dimethylbenzene is known as para-xylene. The ortho-meta-para notation is commonly used to describe disubstituted benzene derivatives. The specific isomeric form is usually denoted by a lowercase *o, m,* or *p* preceding the name of the compound, as in the sentence "One of the active ingredients in mothballs is *p*-dichlorobenzene."

Just as methane is the simplest aliphatic molecule, benzene is the simplest aromatic.* Coal tar, whose aromatic praise has been previously sung, contains, in addition to the xylenes, aromatic compounds in which the benzene ring is fused with other rings. (Figure 13–8.) These condensed ring systems are generally more reactive than benzene. The larger of them possess, in addition to aromatic characteristics, olefinic (double-bond) properties. It should be noted that although the alternating double-bond structure has been discarded as a method of

*Organic molecules that possess benzene-like structures, or properties, are called *aromatic* compounds. All other organic compounds are called *aliphatic* compounds.

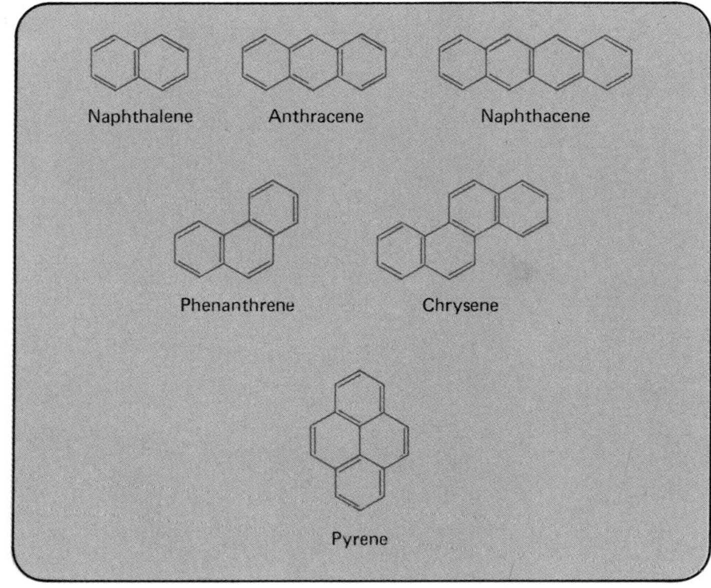

FIGURE 13–8 Fused Ring Components of Coal Tar

depicting benzene, it has been retained in the case of its condensed-ring homologs. This is because all the positions in these molecules are not equivalent. In the case of phenanthrene, one position in particular has more olefinic character than the others. This may readily be demonstrated by writing the five equivalent Kekulé representations of the molecule.

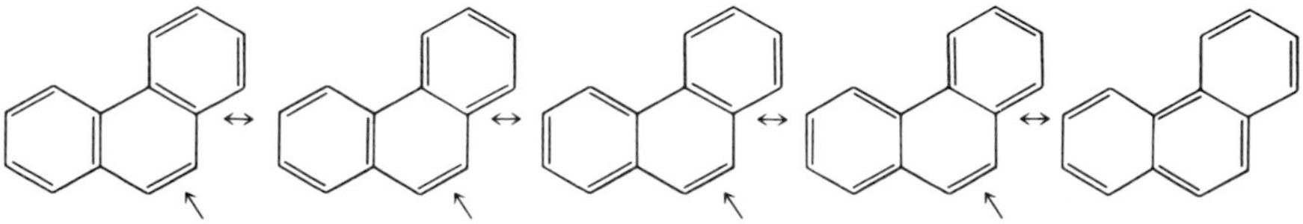

In four of the five representations the double bond is shown to exist between the 9 and 10 carbons. At no other position in the molecule does this occur. The implication of this fact is that of all the bonds in the molecule this one will display the highest olefinic character. Experimentally this has been found to be true.

The physical properties of benzene and its homologs are similar in many ways to nonaromatic hydrocarbons. Like alkanes they are soluble in organic solvents and insoluble in water. Because of their planar structure and delocalized π electrons, which produce an increase in intermolecular forces of attraction, their boiling points are slightly higher than comparable nonaromatic alkanes. In the case of benzene itself, its symmetrical structure allows for better crystal packing than normal hexane and, as a result, its melting point is much higher. (Benzene m.p. $= +5.5°C$; n-hexane m.p. $= -95°C$.)

You May Substitute My Hydrogen, but Touch Not the Carbon in My Ring.

Substitutions rather than additions are the characteristic reactions of aromatic hydrocarbons. In these reactions one or more of the ring hydrogen atoms are replaced by the desired functional group. Typical substitutions are halogenations, nitrations, sulfonations, and alkylations.

Halogenations

The halogenation of aromatic rings is most successfully accomplished with chlorine or bromine and the respective Lewis acid iron halide catalyst.

$$\text{Benzene} + \mathbf{Br_2} \xrightarrow{\mathbf{FeBr_3}} \text{Bromobenzene} + \mathbf{HBr}$$

Benzene   Bromobenzene

Note. To simplify the representation of aromatic compounds the carbon and hydrogen atoms composing the basic benzene ring will no longer be explicitly depicted. It is to be understood that they are contained in the hexagon-circle symbol for benzene.

In multihalogenated aromatic compounds the introduction of the first halogen is the easiest to effect. As the number of halogen atoms already present on the ring increases, subsequent substitutions become more difficult. This is because of the electrophilic ("electron loving" in Greek) nature of the substituting halogen species.

The bromine molecule may be considered to be composed of a positive ion ($\mathbf{Br^+}$) and a negative ion ($\mathbf{Br^-}$). This molecule interacts with the ferric bromide catalyst to produce an ionic super-halide complex.

$$\mathbf{Br_2 + FeBr_3 \rightarrow Br(FeBr_4)}$$

Once formed, this complex can ionize to produce a positively charged bromonium ion.

$$\mathbf{Br(FeBr_4) \rightarrow Br^+ + FeBr_4^-}$$

It is this positively charged, electron-seeking bromonium ion that attaches the benzene ring to produce a short-lived intermediate compound.

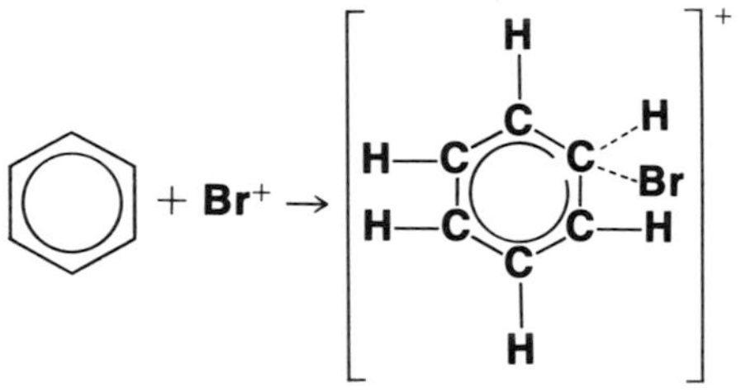

$$\text{C}_6\text{H}_6 + Br^+ \rightarrow [\text{H—C—C—C(H)(Br)—C—H ring intermediate}]^+$$

The intermediate compound represents the conflict between a proton and a bromonium ion for the electrons at that ring position.

In such a case there is really no contest. The needs of the bromonium ion are greater. The hydrogen as a result is released and the bromine is substituted.

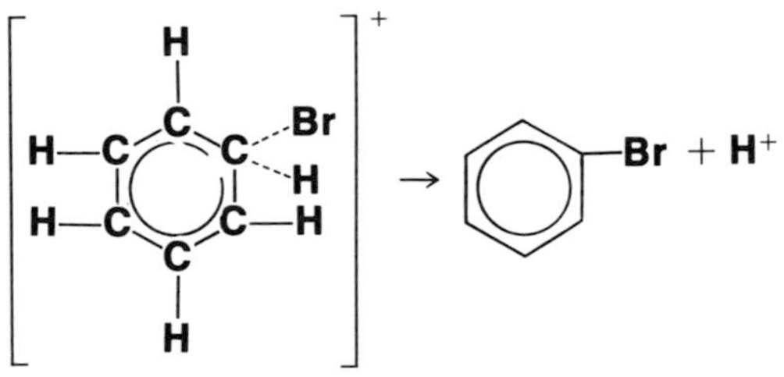

$$[\text{ring intermediate with Br and H}]^+ \rightarrow \text{C}_6\text{H}_5\text{—Br} + H^+$$

Once released, the proton reacts with the super-halogen anion to produce hydrohalogenic acid and regenerated catalyst, which may be reused for additional halogenations.

$$H^+ + FeBr_4^- \rightarrow HBr + FeBr_3$$

Nitration, sulfonation, and alkylation are all reactions that are similarly dependent upon the electrophilic character of the substituting group.

Nitration

Concentrated nitric acid is dehydrated by concentrated sulfuric acid to produce nitronium bisulfate and water.

$$HNO_2 + H_2SO_4 \rightleftarrows H_2O + NO_2(HSO_4)$$

This complex is capable of ionizing to produce the electrophilic nitronium ion (**NO_2^+**), which nitrates the ring.

$$NO_2(HSO_4) \rightarrow NO_2^+ + HSO_4^-$$

$$C_6H_6 + NO_2^+ \rightarrow C_6H_5NO_2 + H^+$$

$$H^+ + HSO_4^- \rightarrow H_2SO_4$$

As in the halogenation, the acid catalyst is regenerated at the end of the reaction. Mechanistically, the bases $FeBr_4^-$ and HSO_4^- assisted in the removal of the benzene proton concurrent with the addition of the substituting nitro group.

Sulfonation

The sulfonation of benzene can be produced through reaction with concentrated sulfuric acid at elevated temperatures or fuming sulfuric acid ($H_2SO_4 \cdot SO_3$) under ordinary conditions. In either case the SO_3 molecule, most probably, is the active electrophilic specie. In contrast to the cationic intermediates present in the other electrophilic aromatic substitutions discussed, the sulfonation intermediate is neutral.

$$C_6H_6 + SO_3 \rightleftarrows [C_6H_6 \cdots SO_3]$$

$$[C_6H_6 \cdots SO_3] \rightarrow C_6H_5SO_3^- + H^+$$

The benzene sulfonate ion exists in equilibrium with the reasonably strong benzene sulfonic acid.

$$C_6H_5SO_3^- + H^+ \rightleftarrows C_6H_5SO_3H$$

Alkylation

Carbonium ions (R^+) will react with benzene to produce alkyl derivatives. The desired carbonium ion can be prepared by reacting its alkyl halide with an aluminum halide. The aluminum tetrahalide basic ion that results reacts with the benzene ring to remove a proton and allow the alkyl carbonium ion to be incorporated.

$$RX + AlX_3 \rightarrow R(AlX_4) \qquad (X = \text{halogen})$$

$$R(AlX_4) \rightarrow R^+ + AlX_4^-$$

$$C_6H_6 + R^+ \quad C_6H_5R + H^+$$

Although this may seem an ideal method for substituting specific alkyl groups onto a benzene ring, life is never that simple. In many cases there is a rearrangement of the alkyl-group carbon skeleton on substitution that results in the production of an isomer of the intended compound. If benzene, for example, is reacted with n-propylchloride and aluminum chloride, the expected n-propylbenzene is not produced. The product is, instead, the isopropyl isomer.

$$\text{C}_6\text{H}_6 + CH_3CH_2CH_2Cl \xrightarrow{AlCl_3} \text{C}_6\text{H}_5\text{—}CH(CH_3)_2 + HCl$$

In all the aromatic substitutions discussed, the driving force is the electrophilic nature of the substituting group. Once the substitution has been accomplished, it is this same force that makes the substituted benzene ring less susceptible to additional electrophilic substitution. The substituted group, because of its electronegativity, channels some of the ring's original attractiveness to itself and away from the ring proper. The effect is known as ring deactivation. A partial listing of ring deactivating groups can be found in Table 13–1.

Just as there are ring deactivating groups, there are also ring activating groups. If a substituted benzene ring contains a ring activating group, it will undergo electrophilic substitution more readily than an unsubstituted benzene molecule. A partial listing of these groups appears in Table 13–2.

TABLE 13–1 Ring Deactivating Groups in Electrophilic Substitutions Reactions

Nitro	$—NO_2$
Sulfonic Acid	$—SO_3H$
Formyl	$—C(=O)—H$
Carboxylic acid	$—C(=O)—OH$
Cyano	—CN
Halo	—F, —Cl, —BR, —I

TABLE 13–2 Ring Activating Groups in Electrophilic Substitution Reactions.

Alkyl	—R (e.g., $—CH_3$, $—C_2H_5$, etc.)
Amino	$—\ddot{N}H_2$
Hydroxy	$—\ddot{O}H$
Thiol	$—\ddot{S}H$

Is It 5 to 1 against Predicting Where the Second Group Will Add?

The cardinal rule in determining where in an aromatic ring the second group will add is first to determine where the first group is. Once this appraisal has been made, the choice of location for the second group can be restricted to either of two possibilities. These possibilities are ortho-para or meta. If the first substituent on the ring is an ortho-para directing group, then the second substituent will add to the ring in either the ortho or para position. If, on the other hand, the first substituent is a meta directing group, then the second group will add in a position meta to the first. By and large, all deactivating groups with the exception of the halogens **Cl**, **Br**, and **I** are meta orienting groups. Those that are ring activating groups are ortho-para orienting. The operation of these rules can be demonstrated by comparing the nitration reactions of toluene with that of nitrobenzene.

$$C_6H_5CH_3 + HNO_3 \xrightarrow{H_2SO_4} o\text{-}CH_3C_6H_4NO_2 + p\text{-}CH_3C_6H_4NO_2$$

Toluene — Ortho-nitrotoluene — Para-nitrotoluene

$$C_6H_5NO_2 + HNO_3 \xrightarrow{H_2SO_4} m\text{-}C_6H_4(NO_2)_2$$

Nitrobenzene — Meta-nitrobenzene

The methyl group is a ring activator; therefore, both ortho-nitrotoluene and para-nitrotoluene are produced. The nitro group, on the other hand, is a ring deactivator; therefore, dinitrobenzene is meta substituted. These rules predict the major reaction products. They are, however, not the only products. Smaller amounts of the other isomers are also produced.

When considering the reactions of alkyl substituted aromatic molecules, one should keep in mind that these molecules are composed of two distinctly different types of carbon bonds. Because of this it is possible to adjust reagents and reaction conditions so that only the aromatic or the aliphatic portion of the molecule will react. The chlorinations of toluene by both ionic and free radical mechanisms is an example of the selective effect of reaction conditions on the type of product produced.

$$C_6H_5CH_3 + Cl_2 \xrightarrow[\text{Free radical}]{\text{UV light}} C_6H_5CH_2Cl + HCl$$

Toluene

Benzylchloride (Note: $C_6H_5CH_2$ is known as the benzyl group.)

$\downarrow$ Cl_2, $FeCl_3$

Ortho-chlorotoluene + Para-chlorotoluene

REVIEW

1. In compounds where the number of hydrogen atoms is less than the number of carbon atoms plus two (C_nH_{2n+2}), it is necessary to form multiple bonds in order that the valence shells in each of the component atoms be completely filled.
2. The sp^3 tetrahedral carbon orbital configuration is a poor one for the formation of multiple bonds. sp^2 and sp hybrid orbitals are used respectively for double and triple bonds.
3. The multiple bonds that join two carbon atoms in a molecule can be subdivided into two basic components. The bond that joins the carbon atoms along the line of their centers is known as a sigma (σ) bond. The bond or bonds that join them above and below or in front and in back of this center line are known as pi (π) bonds.
4. π bonds are somewhat weaker than σ bonds between equivalent atoms.
5. Compounds that contain multiple bonds can possess geometric isomers because the multiple-bond structure restricts the rotation of the carbon atoms to which it is bonded.

6. Alkenes and alkynes as a group are more reactive than alkanes. They can participate in addition reactions with other reactants and in so doing saturate their multiple-bond structure (convert double and triple bonds to single bonds).
7. The orientation in the addition of an asymmetrical reactant to a double or triple bond is dependent upon the composition of the molecule on either side of the multiple bond and upon the mechanism by which the reactant adds to the multiple bond.
8. Multiple-bonded compounds often combine with themselves to reduce the number of multiple bonds and produce long-chained macromolecules called polymers.
9. In addition to forming compounds composed mostly of single-bonded atoms interspersed with discretely isolated multiple bonds, carbon also forms unsaturated "aromatic" compounds. The quality of aromatic compounds that makes them unique is that the unsaturated bonds in ringed aromatic compounds act in concert with each other to produce a very different type of chemistry. Whereas the main chemistry of unsaturated "straight-chained" compounds is dependent on addition reactions, the chemistry of aromatic ring compounds is mostly dependent upon substitution reactions.
10. The position and nature of functional groups already substituted in an aromatic ring compound as well as the nature of the substituting specie are the factors that determine the position at which a substituting reactant will be incorporated in an aromatic compound.
11. If an organic compound is composed of both aromatic and aliphatic parts, each part will display its characteristic chemistry.

REFLEXIVE QUIZ

1. Organic compounds that contain double bonds are known as ________.
2. If a compound has the general formula C_nH_{2n-2}, it may contain either ____ double bonds or ____ triple bond.
3. ______ hybrid orbitals are necessary for the formation of double-bonded carbon compounds.
4. The π bonding in alkenes restricts rotation about the double bond and therefore permits the formation of ____________ isomers.
5. At elevated temperatures, alkanes crack by a free radical mechanism to produce __________.
6. ____________ charged carbonium ions are electrophilic.

7. The three component steps in a chain polymerization reaction are ____________, ____________, and ____________. Such reactions may occur by either an _______ or a ____________ mechanism.

8. _______ substituted aromatic molecules have functional groups on adjacent ring carbon atoms.

9. The unique chemistry of aromatic compounds is due to the delocalization of the ring's ___ electrons.

10. The three factors that determine the products of an aromatic substitution reaction are ________________, ________________, and ______________________________.

Answers: (1) Alkanes [242]. (2) 2, 1 [242]. (3) sp^2 [243, 273]. (4) Geometrical [245–247, 273]. (5) Alkenes [251]. (6) Positively [254]. (7) Initiation, propogation, termination; ionic, free radical [256–259]. (8) Ortho [264, 266]. (9) π [262, 266]. (10) Nature of the reactant, reaction mechanism, nature and location of group already substituted on the ring [264–274].

SUPPLEMENTARY READINGS

1. E. A. Walters. "Models for the Double Bond." *Journal of Chemical Education* 43 (1966): 134.
2. T. F. Rutledge. *Acetylenes and Allenes.* New York: Reinhold Book Corp., 1969.
3. D. P. Craig. "The Changing Concept of Aromatic Character." *Education in Chemistry* 1, (1964): 136.
4. B. F. Greek. "Ethylene." *Chemical Engineering News,* Feb. 22, 1971: 16.
5. C. A. Russell. "Kekulé and Benzene." *Chemistry in Britain* 1 (1965): 141.

Section 14

Organic Chemistry: Hetero-atom Compounds

Section 14

Organic Chemistry: Hetero-atom Compounds

Wine Tastes Better Than Vinegar.

The exploits of the chemist Saccharomyces Cerevisiae are found in the earliest writings of the ancients. Although having come from extremely humble surroundings and having never learned to read or write, Saccharomyces with the help of his friends developed a process for converting fruit juice into wine. By the process of fermentation they converted sugars into ethyl alcohol. Drinks prepared by this conversion were the rage of the ancient world. "Noah was the first tiller of the soil, He planted a vineyard and he drank of the wine and became drunk" (Genesis 9:20). Pharaohs of Egypt before they died had particularly good recipes put on the walls of their tombs so that they might use them in the afterlife.

Flushed with the success due to the immediate acceptance of their product, Saccharomyces and associates enlisted the aid of bacteria to carry the fermentation one step further. Their overzealous bacteria, however, converted the ethyl alcohol into acetic acid and the wine to vinegar. (Vinegar derives its name from the French *vin aigre*, meaning "sour wine.") What occurred in this transformation was the oxidation of an alcohol. Both alcohols and acids are stages in the oxidation of alkanes. Alcohols and ethers are the lowest stage, acids are the highest. Between these two there also exists an intermediate stage composed of aldehydes and ketones. The interrelationships between these compounds can be displayed through the use of progressive oxidation half-reactions.

Stage one in the oxidation of ethane produces ethanol. Ethanol is an *alcohol* because of its —**OH** functional group.

$$\underset{\text{Ethane}}{CH_3CH_3} + H_2O \rightarrow \underset{\text{Ethanol}}{CH_3CH_2OH} + 2\,H^+ + 2\,e$$

Ethyl ether, the ether used in hospital operating rooms, can be produced by dehydrating two molecules of ethanol. Ethyl ether is at the same stage of oxidation as ethanol since the reaction involves only the removal of a molecule of water and does not produce any electron change.

$$CH_3CH_2\boxed{OH + H}OCH_2CH_3 \rightarrow \underset{\text{Diethyl ether}}{CH_3CH_2{-}O{-}CH_2CH_3} + H_2O$$

When ethane undergoes stage two of its oxidation, the reaction product is an *aldehyde.*

$$\mathbf{CH_3CH_2OH} \rightarrow \mathbf{CH_3\overset{\overset{\textstyle O}{\|}}{C}H} + 2\,\mathbf{H}^+ + 2\,e$$

Ethanol — Ethanal (Acetaldehyde)

The $\mathbf{-\overset{\overset{\textstyle O}{\|}}{C}H}$ functional group of ethanal distinguishes it as an aldehyde. Similarly, other alcohols may be oxidized in the same manner to produce aldehydes. When, however, the alcohol does not have its **OH** functional group on a terminal carbon, a *ketone* rather than an aldehyde is produced. The oxidation of 2-propanol illustrates this point:

$$\mathbf{CH_3{-}\underset{\underset{\textstyle H}{|}}{\overset{\overset{\textstyle OH}{|}}{C}}{-}CH_3} \rightarrow \mathbf{CH_3{-}\overset{\overset{\textstyle O}{\|}}{C}{-}CH_3} + 2\,\mathbf{H}^+ + 2\,e$$

2-Propanol (Isopropyl alcohol) — 2-Propanone (Acetone)

Acetone is a ketone because of its $\mathbf{R{-}\overset{\overset{\textstyle O}{\|}}{C}{-}R}$ arrangement of atoms. The three-carbon aldehyde equivalent of 2-propanone is propanal ($\mathbf{CH_3CH_2\overset{\overset{\textstyle O}{\|}}{C}H}$). It differs from the ketone in that it has attached to the carbon atom with the double-bonded oxygen one alkyl group and a hydrogen rather than two alkyl groups. The state of oxidation of both compounds, however, is the same.

Stage three in the oxidation of an alkane is the production of an acid.

$$\mathbf{CH_3\overset{\overset{\textstyle O}{\|}}{C}H} + \mathbf{H_2O} \rightarrow \mathbf{CH_3\overset{\overset{\textstyle O}{\|}}{C}{-}OH} + 2\,\mathbf{H}^+ + 2\,e$$

Ethanal (Acetaldehyde) — Ethanoic acid (Acetic acid)

In this third oxidation stage, as in the preceding two, there is a two-electron loss in the process of going from the lower to the higher of the two oxidation states. There is, therefore, a total of six electrons lost in going the complete route from an alkane to an acid. Oxidizing and reducing agents similar to those utilized in the redox reactions described in earlier sections can be employed to convert these organic molecules from one stage of oxidation to another. Although alkanes can only be oxidized and acids can only be reduced, the intermediate alcohol and aldehyde states can be either oxidized or reduced, depending upon the appropriate choice of reagent. The complete equation for the oxidation of 2-propanol, a secondary alcohol,* to acetone by potassium permanganate can be balanced using the same rules as those employed for inorganic redox reactions.

$$5\,(CH_3CHOHCH_3 \rightarrow CH_3COCH_3 + 2\,H^+ + 2\,e)$$

$$2\,(8\,H^+ + MnO_4^- \rightarrow Mn^{++} + 4\,H_2O + 5\,e)$$

$$5\,CH_3CHOHCH_3 + 16\,H^+ + 2\,MnO_4^- \rightarrow 5\,CH_3COCH_3 + 2\,Mn^{++} + 10\,H^+ + 8\,H_2O$$

$$5\,CH_3CHOHCH_3 + 6\,H^+ + 2\,MnO_4 \rightarrow 5\,CH_3COCH_3 + 2\,Mn^{++} + 8\,H_2O$$

$$5\,CH_3CHOHCH_3 + 3\,H_2SO_4 + 2\,KMnO_4 \rightarrow 5\,CH_3COCH_3 + 2\,MnSO_4 + K_2SO_4$$

Functional Groups Have Dual Personalities.

The functional oxy-groups are not restricted to alkyl compounds. They may also be associated with aryl (aromatic) compounds. Typical compounds are

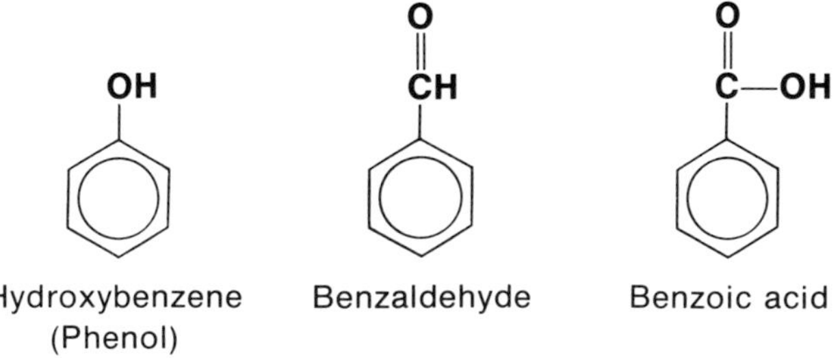

Hydroxybenzene (Phenol) | Benzaldehyde | Benzoic acid

* The terms *primary*, *secondary*, and *tertiary* when applied to an alcohol indicate that the number of alkyl groups attached to the carbon atom bonded to the —**OH** is respectively one, two, and three. The terminology is identical to that used to describe alkyl halides in Section 13.

CH_3CH_2OH	$CH_3CH(OH)CH_3$	$(CH_3)_3COH$
Primary	Secondary	Tertiary

The properties of these aryl derivatives differ appreciably from their alkyl counterparts. These differences are most marked in the hydroxy derivatives.

Hydroxy alkanes are known as alcohols; hydroxy aromatics are *phenols.* A comparison of their respective properties constitutes Table 14–1. Both alcohols and phenols can be prepared by the hydrolysis of halides.

$$\underset{\text{Ethyl chloride}}{CH_3CH_2Cl} + NaOH \xrightarrow{H_2O} \underset{\text{Ethanol}}{CH_3CH_2OH} + NaCl$$

$$\underset{\text{Chlorobenzene}}{C_6H_5\text{—}Cl} + 2\,NaOH \xrightarrow[\text{and pressure}]{\text{High temperature}} \underset{\text{Sodium phenoxide}}{C_6H_5\text{—}ONa} + NaCl + H_2O$$

$$2\,C_6H_5\text{—}ONa + CO_2 \xrightarrow{H_2O} 2\,C_6H_5\text{—}OH + Na_2CO_3$$

The synthesis of an alcohol by this procedure is particularly significant since it is thought to proceed through a *nucleophilic* mechanism. Such mechanisms are dependent upon the reaction of the alkyl halide with a substance that is capable of supplying the electron pair necessary to form a covalent bond. In addition to being able to supply an electron pair, nucleophilic reagents must be of the proper configuration and orientations to allow effective electron transfer and bond formations. If the reagents contain inappropriately placed functional groups, bond formations may be prevented or, at best, impeded. The nucleophilic reagent can be either negatively charged or neutral. The process of nucleophilic substitutions is driven by forces that are the reverse of those encountered in electrophilic substitutions. Groups that will activate an electrophilic reaction will deactivate a nucleophilic reaction, and vice versa. In electrophilic substitutions, such as the halogenations of benzene, the driving force is the greater electronegativity of the halogen atoms as compared to hydrogen.

In the nucleophilic substitution of a hydroxide for a halogen it is the high electron density of the attacking hydroxide ion that facilitates the removal of the halogen and allows the subsequent hydroxide substitution. The hydrolysis of 2-bromobutane illustrates this point. The hydroxide ion combines initially with 2-bromobutane to produce a temporary transition complex.

$$OH^- + H\text{–}\overset{Br}{\underset{CH_3}{C}}\text{–}C_2H_5 \rightarrow \left[HO\cdots\overset{\overset{H}{|}\ \delta^+}{\underset{CH_3}{C}}(C_2H_5)\cdots\overset{\delta^-}{Br} \right]^-$$

Transition complex

The combining of hydroxide ion with butyl bromide is facilitated by the polarity of the carbon–bromine bond. Bromine, because of its higher electronegativity, draws the bond electrons away from the carbon, making it relatively positive. With the arrival of the electron-rich hydroxide ions, the carbon atom can tolerate an even greater shift of bromine-bond electrons toward the bromine. This electron shift produces an even more highly polarized carbon–bromine bond. As the new carbon–oxygen bond is formed, the carbon–bromine bond becomes so highly polarized that it breaks. The bromine, having acquired the necessary number of electrons to form a bromide ion, exits gracefully. The butyl group, still no worse for electrons than when it was associated with the halogen, combines with the hydroxide to produce 2-butanol.

$$\left[HO\cdots\overset{\overset{H}{|}}{\underset{CH_3\ \ C_2H_5}{C}}\cdots Br \right]^- \rightarrow HO\text{—}\overset{\overset{H}{|}}{\underset{\underset{C_2H_5}{|}}{C}}\text{—}CH_3 + Br^-$$

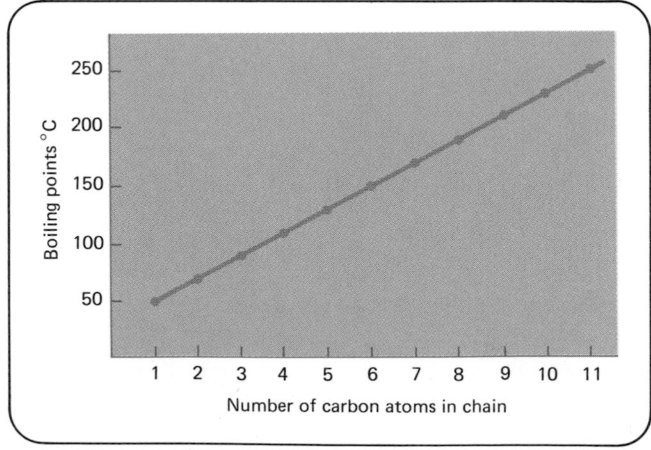

FIGURE 14–1 Boiling Points of Straight-Chained Alcohols

14–1 Comparison of Alcoholic and Phenolic Properties **TABLE**

Alcohols

The —**OH** group causes alcohols to be polar molecules. The alcohols with lower molecular weights are quite soluble in water, since they are structurally equivalent to a water molecule in which a hydrogen has been replaced by an alkyl group. As the size of the alkyl group increases, the physical properties of the alcohols become less like water and more like their alkane analogs. This results in a reduced water solubility.

The boiling points of alcohols are abnormally high by comparison to the hydrocarbon containing the same number of carbon atoms. (Figure 14–1.) The reason for this, as in the case of water, is hydrogen bonding.*

Alcohols are less acidic than water. Their ionization constants K_a are of the order of 10^{-18} ($K_a \approx 10^{-18}$). They react with both organic and inorganic acids to form esters and water.

$$\mathbf{CH_3C(=O)—OH + CH_3CH_2OH \rightarrow CH_3C(=O)—OCH_2CH_3 + H_2O}$$

Acetic acid — Ethanol — Ethyl Acetate (an Ester)

Phenols

Phenol (carbolic acid), the simplest hydroxybenzene compound, is a solid with a low melting point. It is only slightly soluble in water but very soluble in organic solvents. Phenols are more acid than water. Their ionization constants K_a are of the order of 10^{-10} ($K_a \approx 10^{-10}$). They will ionize in water to produce hydrogen ion and will react with bases such as sodium hydroxide to produce sodium phenoxide.

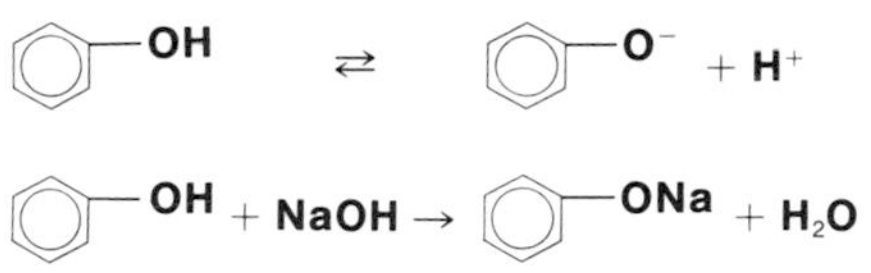

$$⌬\mathbf{—OH} \rightleftarrows ⌬\mathbf{—O^-} + \mathbf{H^+}$$

$$⌬\mathbf{—OH} + \mathbf{NaOH} \rightarrow ⌬\mathbf{—ONa} + \mathbf{H_2O}$$

In addition to the reactions of the phenol hydroxy group, phenols undergo substitution reactions characteristic of their aromatic ring. Most of these are electrophilic aromatic substitution. The presence of the hydroxyl group activates the aromatic ring and makes it more amenable to electrophilic substitution than either benzene or toluene. The substitution is ortho-para because the ring is activated.

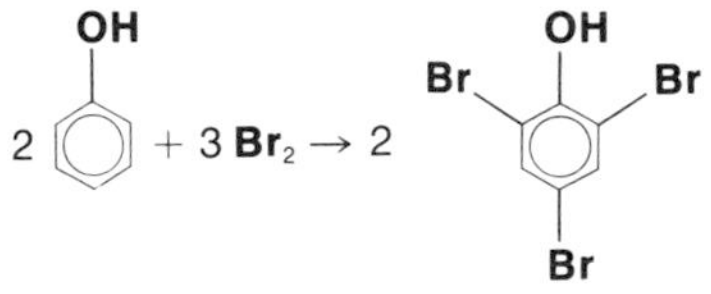

$$2\ ⌬\text{—OH} + 3\,\mathbf{Br_2} \rightarrow 2\ (\text{OH, Br, Br, Br})⌬$$

2,4,6-Tribromophenol

Dilute phenol solutions, because of their germicidal properties, were the first surgical antiseptics. Such use of them, however, has been discontinued in recent years in favor of less corosive chemicals.

---

*Hydrogen bonds are weak electrostatic bonds between molecules in which a component hydrogen atom serves as the bonding group between highly electronegative atoms such as oxygen, nitrogen, and fluorine.

An interesting sidelight to this reaction is that the arrangement of atoms of the butyl group after this substitution is the mirror image of their arrangement before the substitution took place. The effect is similar to what would happen if these groups were each attached to the spines of an umbrella and the wind turned the umbrella inside out. This inversion was first noted by Paul Walden and has been named in his honor.

Compounds, such as phenols and alcohols, that contain hydroxy groups do not have similar properties because the electrical environment of these groups is drastically altered when the compound is changed from an aromatic to an aliphatic one.

The alkyl aldehydes, ketones, and acids are much closer in chemical behavior to their aryl (aromatic) counterparts. This is because the chemistry of substituted aromatic compounds is the product of the chemistries of both the substituted rings and attached alkyl groups. In most cases, with the exception of phenols, halides, nitrates, and sulfonates, the functional substituted group contains its own carbon atom or atoms. These carbon atoms are not components of the aromatic π-ring structure and therefore serve to buffer the effect of the ring electrons on the functional group. Aldehydes, ketones, and carboxylic acids have, therefore, very similar chemistry, irrespective of whether they are aromatic or aliphatic. In all molecules, and in aromatics in particular, it must be kept in mind that a reagent in a reaction has the option of interacting with either the functional group or the rest of the molecule. If the functional group exerts an electron-activating or an electron-deactivating effect on the rest of the molecule, this will be reflected in how the molecule interacts with a given reagent.

Oxyorganics We Have Come to Know and Sometimes Love

We come in contact with many alcohols, ethers, aldehydes, ketones, and acids in our everyday living. The liquor we drink in cold weather to warm the body, the antifreeze we put in the car radiator to prevent the coolant from freezing, and the glycerin cough drops and candies we take for the cold we caught because the liquor didn't do an effective warming job, all contain alcohols. They contain respectively one, two, and three hydroxy alcohols.

CH_3CH_2OH	CH_2OHCH_2OH	$CH_2OHCHOHCH_2OH$
Ethanol	Ethylene glycol	Glycerin
(Wine, liquor)	(Antifreeze)	(Cough drops, candies, cosmetics)

The number of hydroxyl groups in each of these compounds does not correlate with either its antifreeze or its intoxicating properties.

Aldehydes are familiar as solvents, flavors, and fragrances. Dimethyl ketone $\left(CH_3{-}\overset{O}{\overset{\|}{C}}{-}CH_3\right)$, commonly known as acetone, is used in varnishes, lacquers, and plastics. Methyl ethyl ketone $\left(CH_3{-}\overset{O}{\overset{\|}{C}}{-}C_2H_5\right)$ is the wonder ingredient in nail polish remover. Cinnamaldehyde $\left(C_6H_5{-}CH{=}CHCHO\right)$ is the dominant flavor of cinnamon. Vanillin (benzene ring bearing OH, OCH_3 and $HC{=}O$) gives vanilla its characteristic flavor and odor. Eugenol (benzene ring bearing OH, OCH_3 and $CH_2CH{=}CH_2$)

produces the flavor in oil of cloves and provides a curious example of the wonders of chemistry: it can be converted to vanillin, an entirely dissimilar flavor and fragrance, merely by rearrangement and oxidation.

$$\text{Eugenol (4-}CH_2CH{=}CH_2\text{, 2-}OCH_3\text{-phenol)} \xrightarrow[\text{Alcohol}]{\text{KOH}} \text{Isoeugenol (4-}CH{=}CHCH_3\text{, 2-}OCH_3\text{-phenol)} \xrightarrow{\text{Oxidation}} \text{Vanillin (4-}HC{=}O\text{, 2-}OCH_3\text{-phenol)}$$

Eugenol — Isoeugenol — Vanillin

Carboxylic acids ionize in water to produce protons and carboxylate anions.

$$R{-}\overset{\displaystyle O}{\overset{\|}{C}}{-}OH \rightarrow R{-}\overset{\displaystyle O}{\overset{\|}{C}}{-}O^- + H^+$$

Carboxylic acid — Carboxylate anion

They are in general fairly weak acids and usually have K_a values of approximately 10^{-4} or 10^{-5}. The acid strengths may, however, be increased by substituting electrophilic atoms or groups on the carbon immediately adjacent to the carboxyl group. The aromatic acids as a class are slightly more acidic than their aliphatic counterparts. Sub-

MMMMM...
GLYCERIN HAS THREE TIMES AS MANY HYDROXIDES AS ETHANOL. IT MAKES A...
Gasp!
Ugh!
Brack!
...Terrible Martini!
SOCK!

stitution of electrophilic groups into the aromatic ring will further increase this acidity.

Acetic acid CH_3COOH — $K_a = 2 \times 10^{-5}$

Benzoic acid C_6H_5COOH — $K_a = 6 \times 10^{-5}$

Chloroacetic acid $CH_2ClCOOH$ — $K_a = 2 \times 10^{-3}$

p-Chlorobenzoic acid ClC_6H_4COOH — $K_a = 1 \times 10^{-4}$

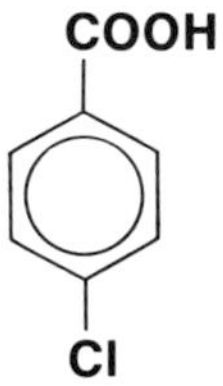

Like alcohols, carboxylic acids are capable of forming hydrogen bonds with other acid molecules and water. The nature of the process is usually dimerization. It results in abnormally high boiling points for these compounds compared to compounds of similar molecular weight and different functional groups.

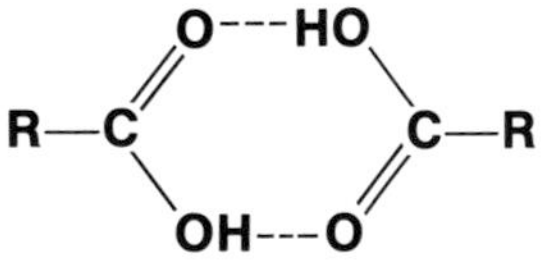

```
       O---HO
      //      \
  R—C          C—R
      \       //
       OH---O
```

Carboxylic acid dimer

Three common, everyday acids are acetic, citric, and acetylsalicylic. Acetic acid (**CH_3COOH**) in 4 to 5 percent solutions is sold as vinegar. Citric acid—or 2-hydroxy-1,2,3-propane tricarboxylic acid, as the IUPAC calls it—is a component of fruit juices, candies, and pharmaceuticals:

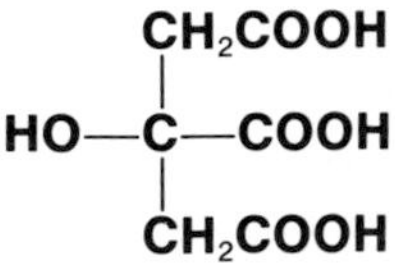

```
       CH2COOH
        |
  HO—C—COOH
        |
       CH2COOH
```

Acetylsalicylic acid, the proper scientific name for aspirin, is a useful medicine for most painful discomforts, with the notable exception of stomach ulcers. Besides being an acid, it is also an ester of acetic and salicylic acid.

$$C_6H_4(OH)COOH + CH_3C(=O){-}OH \rightarrow C_6H_4(O{-}C(=O){-}CH_3)COOH$$

Salicylic acid   Acetic acid   Aspirin

Putrescene Smells Terrible: Amines

Just as alcohols and phenols are alkyl and aryl derivatives of water, amines are alkyl and aryl derivatives of ammonia. Depending upon how many of the ammonia hydrogens have been replaced by organic functional groups, the compound can be either a primary, secondary, or tertiary amine. IUPAC naming of these compounds is seldom employed. More often than not, compounds are named in terms of the groups that substitute for the ammonia hydrogens.

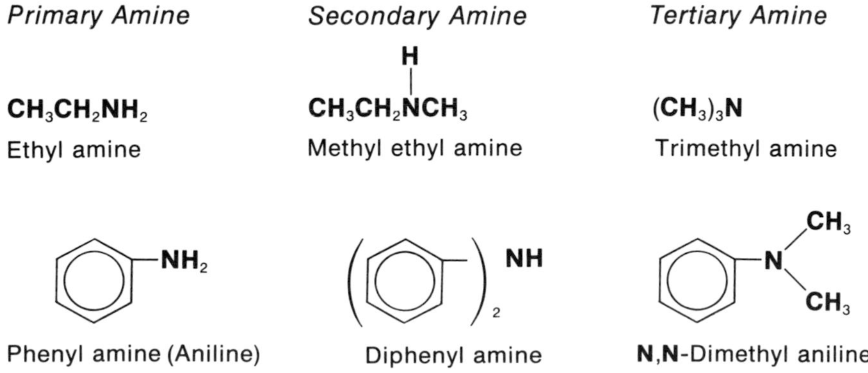

Note. The term *phenyl* is used to signify the benzene ring less one hydrogen (**C_6H_5**). The **N**,**N** notation signifies that both the groups are substituted on the amine nitrogen rather than on the phenyl ring.

Like ammonia, the amines with lower molecular weights are gases that dissolve in water to produce alkaline solutions. The alkalinity of these solutions is dependent upon the number and types of substituted groups attached to the nitrogen. Alkyl groups make the amine more basic than ammonia, whereas aryl groups tend to have the reverse effect. Aniline is a much weaker base than ammonia. Diphenyl amine, which contains two substituted phenyl groups, is only slightly basic. Triphenyl amine, the triple phenyl derivative of ammonia, does not display any significant basic properties.

Nearly all of the volatile alkyl amines have odors strongly reminiscent of decaying fish. The diamine derivatives of butane and pentane—$NH_2(CH_2)_4NH_2$ and $NH_2(CH_2)_5NH_2$—are so overwhelming in this respect that they are referred to commonly as putracine and cadaverine.

The unshared electron pair on the nitrogen atom of amines makes their chemistry very much like the parent ammonia molecule. Amines form salts with acids, can act as basic catalysts, and with the exception of tertiary amines can react with esters or acid anhydrides to produce amides.

Salt formation:

$$C_6H_5{-}\ddot{N}H_2 + HCl \rightarrow \left(C_6H_5{-}NH_3\right)^+ \; Cl^-$$

Aniline — Aniline hydrochloride

The product is known as a quaternary amine salt because there are four groups associated with the nitrogen.

Amide formation:

$$CH_3{-}\overset{O}{\overset{\|}{C}}{-}OCH_2CH_3 + C_2H_5NH_2 \rightarrow CH_3\overset{O}{\overset{\|}{C}}NHC_2H_5 + CH_3CH_2OH$$

Ethyl acetate — Ethyl amine — **N**-Ethyl acetamide — Ethanol

In the reactions of aromatic amines and their derivatives that do not involve the amino group directly, the ring activating effect of the amine functional group causes the substitution of additional electrophilic reagents to occur in either the ortho or the para positions.

$$\text{Acetanilide} + HNO_3 \xrightarrow[20°C]{H_2SO_4} p\text{-Nitroacetanilide (79\%)} + o\text{-Nitroacetanilide (19\%)} + m\text{-Nitroacetanilide (2\%)}$$

Acetanilide ($C_6H_5NHCOCH_3$); p-Nitroacetanilide (79%); o-Nitroacetanilide (19%); m-Nitroacetanilide (2%)

Synthetic methods for the preparation of amines vary depending upon whether a primary, secondary, or tertiary amine is the desired product. Primary amines can be produced by reacting the alkyl halide of the desired amine with potassium phthalimide and hydrolizing the product. If the desired product is isopropyl amine, then isopropyl bromide is a logical choice for a starting material.

$$(CH_3)_2CH{-}Br + KN(CO)_2C_6H_4 \rightarrow (CH_3)_2CH{-}N(CO)_2C_6H_4 + KBr$$

Isopropyl bromide; Potassium phthalimide

$$(CH_3)_2CH{-}N(CO)_2C_6H_4 + 2\,H_2O \rightarrow (CH_3)_2CH{-}NH_2 + C_6H_4(COOH)_2$$

Isopropyl amine; Phthalic acid

The potassium phthalimide substitutes an amine group for the halogen atom in the halide.

Secondary amines can be produced by reducing the condensation product of an aldehyde and a primary amine. Such a compound is known as a Schiff base.

$$\underset{\substack{\text{Acetaldehyde}\\\text{(Aldehyde)}}}{CH_3CH{=}O} + \underset{\substack{\text{Ethyl amine}\\\text{(Primary amine)}}}{C_2H_5NH_2} \rightarrow \underset{\substack{\text{Acetaldehyde ethyl imine}\\\text{(Schiff base)}}}{CH_3CH{=}NC_2H_5 + H_2O}$$

$$CH_3CH{=}NC_2H_5 + HCOOH \xrightarrow[\text{(Reducing agent)}]{\text{Formic acid}} CH_3CH_2NHC_2H_5 + CO_2$$

Note that in this synthesis the aldehyde is converted to an alkyl group with the same number of carbon atoms—that is, the acetaldehyde is converted to an ethyl group.

The reaction of alkyl halides with ammonia can be used to produce tertiary amines. This reaction actually results in a mixture of all types of amines; the tertiary product due to its higher boiling point, however, can usually be separated from this mixture. Since the process is merely the reaction of three alkyl halides with one ammonia molecule, it can be facilitated by using primary or secondary amines as the starting materials in place of ammonia.

$$CH_3Br + NH_3 \rightarrow CH_3NH_2 + HBr$$

$$CH_3Br + CH_3NH_3 \rightarrow (CH_3)_2NH + HBr$$

$$CH_3Br + (CH_3)_2NH \rightarrow (CH_3)_3N + HBr$$

When asymmetrical secondary amines are used as the starting materials, the final tertiary amine will possess more than one type of alkyl group.

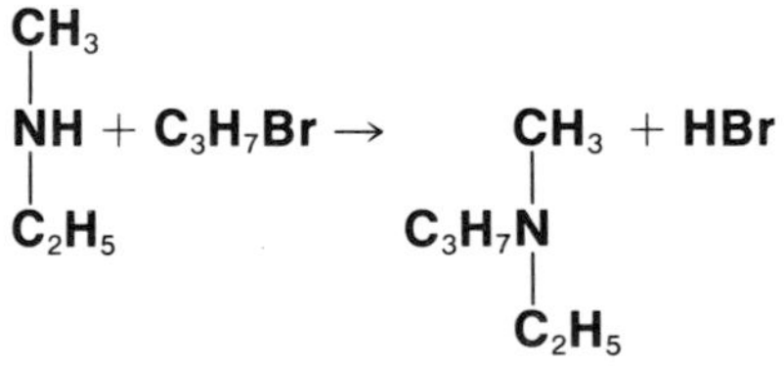

$$\begin{matrix} CH_3 \\ | \\ NH \\ | \\ C_2H_5 \end{matrix} + C_3H_7Br \rightarrow \begin{matrix} CH_3 \\ | \\ C_3H_7N \\ | \\ C_2H_5 \end{matrix} + HBr$$

Well-known and Sometimes Loved Amines

Adrenaline: 3,4-dihydroxy (1-hydroxy-2-methyl amino ethyl) ben-

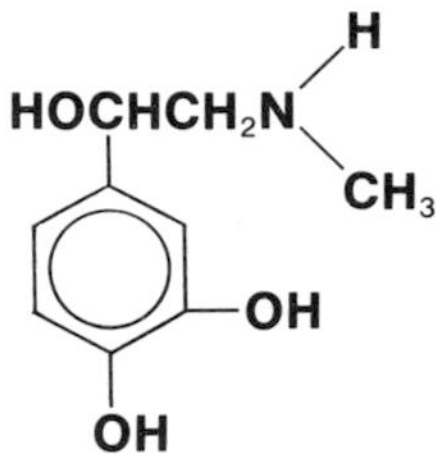

HOCHCH₂N(H)(CH₃) on a benzene ring bearing –OH at positions 3 and 4:

$$HOCHCH_2N\begin{matrix} H \\ CH_3 \end{matrix}$$

OH

OH

zene. If the complexity of this formula is enough to raise your blood pressure, adrenaline is the substance that is doing it. This is the substance secreted by the adrenal glands of animals, including *Homo sapiens*, in response to stress situations such as fear, rage, and physical exertion.

Phenacetin: p-ethoxyacetanilide.

$HNCOCH_3$

OC_2H_5

This compound is the mysterious P in the magical APC tablets that were prescribed for many years for "whatever ailed you." Like aspirin, it will ease pain and reduce fever. It is, however, appreciably more toxic.

PABA: p-amino benzoic acid.

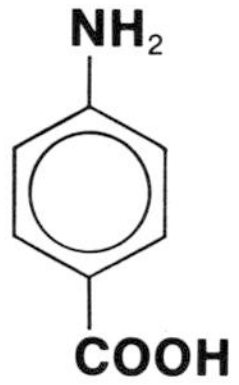

NH_2

COOH

Valued as a nutrient, this acid is one of the B vitamins. It is valuable, too, as an ingredient of suntan creams, as it can prevent burning. It can also be employed to prevent gray hair in rats, an effect of questionable value.

Novocain®: the hydrochloride of the 2-diethyl amino ethyl ester of *p*-amino benzoic acid.

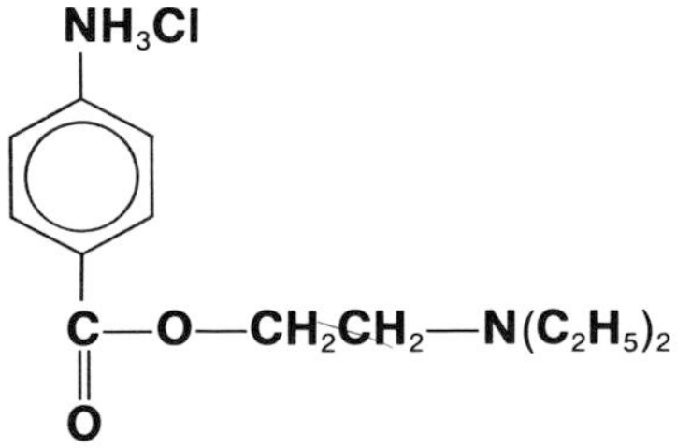

NH_3Cl

$C—O—CH_2CH_2—N(C_2H_5)_2$

O

Novocain (procaine hydrochloride) is a local anesthetic frequently used by dentists prior to drilling or extracting teeth. In recent years it has been fashionably employed, with dubious success, to stave off the perils of aging.

Butyl Mercaptans Don't Smell Too Great Either: Organo-Sulfur Compounds

Sulfur and oxygen are in the same group of the periodic chart. As a result it is not too surprising to find that there are sulfur compounds that are analogous to alcohols, acids, ketones, ethers, and esters. These compounds are thio alcohols or mercaptans (**RSH**), dithio acids (**RCSSH**), thioketones ($R_2C{=}S$), thio ethers (**R—S—R**), and thio esters (**RCSSR**).

Just as the alcohols are alkyl derivatives of water, the mercaptans or thio alcohols are alkyl derivatives of hydrogen sulfide.

HOH	**HSH**	CH_3CH_2OH	CH_3CH_2SH
Water	Hydrogen sulfide	Ethanol	Ethane thiol Ethyl mercaptan

Analogous to the production of alcohols through the reaction of alkyl halides with sodium hydroxide, thio alcohols can be produced by reacting alkyl halides with sodium hydrogen sulfide.

$$CH_3CH_2Cl + NaOH \rightarrow CH_3CH_2OH + NaCl$$

$$CH_3CH_2Cl + NaSH \rightarrow CH_3CH_2SH + NaCl$$

The alkyl derivatives of H_2S, like the parent molecule, are weakly acidic, easily oxidized, and marked by less than sachet odors. When treated with mild oxidizing agents, such as cupric salts, thiols are converted to disulfides, the more stable sulfur analogs of hydrogen peroxide.

$$2\,CH_3CH_2SH + 2\,CuCl_2 \rightarrow CH_3CH_2S{-}SCH_2CH_3 + 2\,CuCl + 2\,HCl$$

Just as alcohols by the addition of oxygen can be oxidized to carboxylic acids, thio alcohols by the addition of oxygen can be oxidized to a sulfonic acid.

UNCLE BUTYL TALES
BRER FOX, DO THE BRIAR PATCH SMELL FUNNY TO YOU?
BRER RABBIT, I WAS WONDERING WHEN YOU WERE GONNA ASK!
BRER SKUNK WAS HERE A WHILE BACK AND HE LEFT HIS CALLING CARD.
SINCE NONE OF US CAN READ NOR WRITE, HE SKUNKED THE WHOLE PATCH SO THERE'D BE NO MISTAKIN' HIM FOR NOBODY ELSE!
HE SAID THE MEMORY OF THE CROTYL MERCAPTAN, AN UNSATURATED BUTYL MERCAPTAN, WOULD STAY WITH US LONG AFTER WE FORGOT HIS FACE!

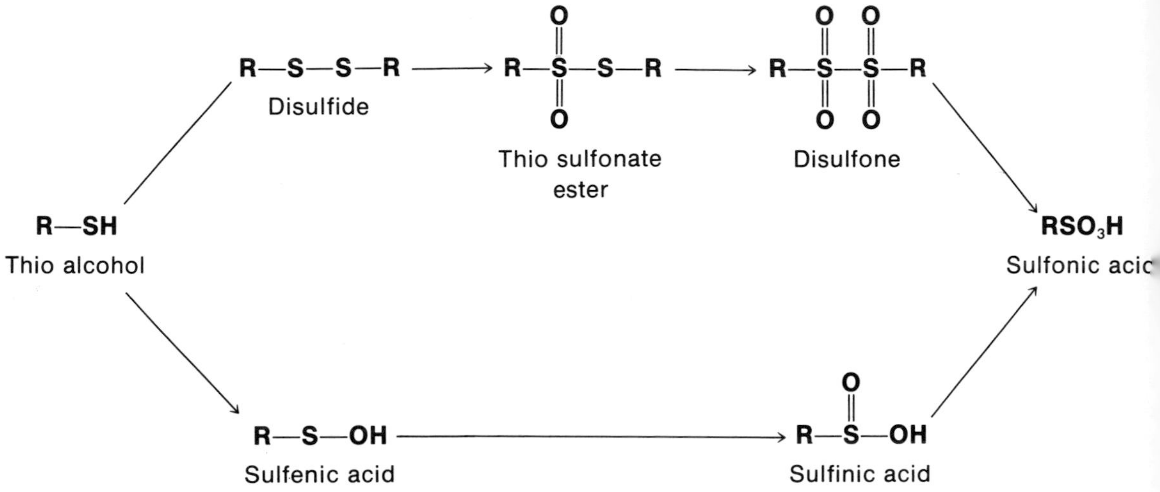

R—S—S—R → $R—S(=O)_2—S—R$ → $R—S(=O)_2—S(=O)_2—R$

Disulfide

Thio sulfonate ester

Disulfone

R—SH
Thio alcohol

RSO_3H
Sulfonic acid

R—S—OH → $R—S(=O)—OH$

Sulfenic acid

Sulfinic acid

Sulfur Compounds—Love 'Em or Leave 'Em.

The compound *p*-amino benzene sulfonamide, commonly called sulfanilamide, was the cornerstone of the many sulfa drugs which were developed in the late 1930s and early 1940s. Of the compounds based on the structure of sulfanilamide that were synthesized, only those containing the **SO_2NHR** group were found to be active against bacteria. From this finding it was deduced that bacteria assume that, because of its structure, the sulfa drug is PABA, one of their normal foods. For the bacteria, this was a bad mistake to make. Sulfa drugs cause their growth to be stunted and allow their potential victims the opportunity to destroy them with their antibodies.

The success of sulfa drugs when they were introduced was unprecedented. Although they had some minor drawbacks, these were small by comparison to their advantages. They reduced the death rate from streptococcal meningitis, for example, from 95 to 10 percent. It was not long before chemists discovered that by modifying the nature of the **R** in **SO_2NHR** the effects which these drugs had on patients could be similarly modified. Of the several hundred different sulfa drugs which were produced, only about thirty are currently in use. Figure 14–2 shows six of them.

Not until 1941, with the advent of penicillin, another sulfur-containing drug, did any drug in the history of mankind have so marked

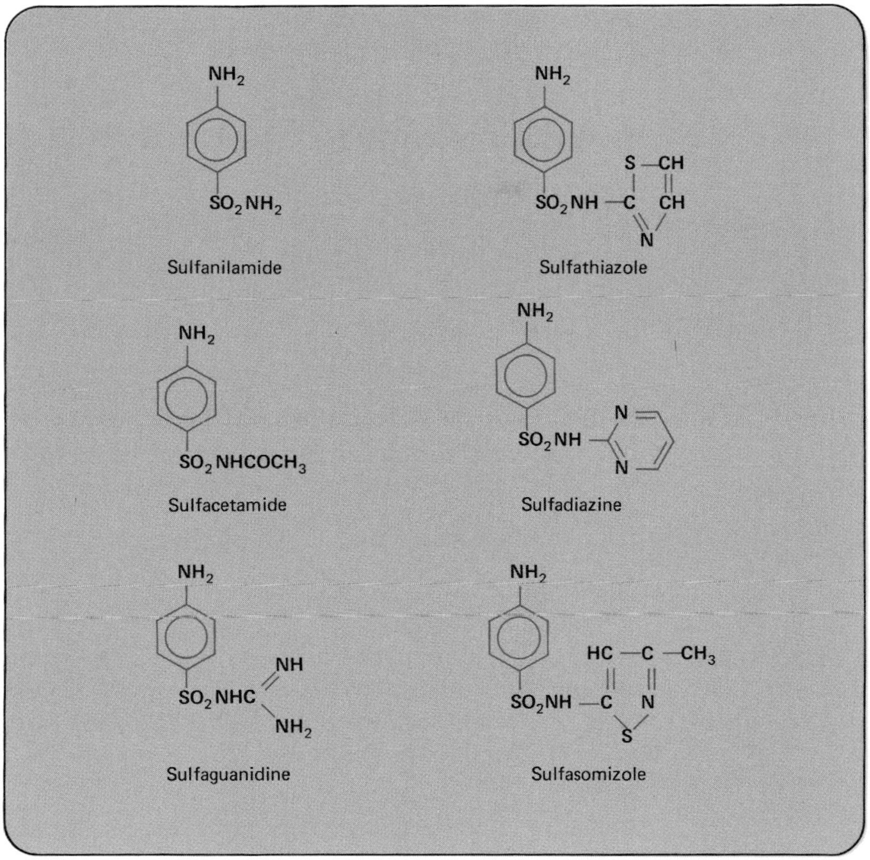

NH_2

SO_2NH_2

Sulfanilamide

NH_2

SO_2NH-C (S—CH, CH, N)

Sulfathiazole

NH_2

$SO_2NHCOCH_3$

Sulfacetamide

NH_2

SO_2NH (N, N)

Sulfadiazine

NH_2

SO_2NHC (NH, NH_2)

Sulfaguanidine

NH_2

SO_2NH-C (HC—C—CH_3, N, S)

Sulfasomizole

FIGURE

14–2 Representative Sulfa Drugs

an impact in the war against bacterial infection.

```
                       S     CH3
                     /   \  /
R—CONH—CH—CH        C
        |   |       |  \
        CO—N        |   CH3
            \       |  H
             \      | /
               C
                \
                 COOH
```

Penicillin

As in the case of sulfa drugs, the substitution of differing groups for **R** produces penicillins that have slightly different properties and side effects on patients.

REVIEW*

1. Carbon atoms have the ability to produce hybrid orbitals that, depending upon the number and type of component orbitals, display varying amounts of *s* and *p* characteristics. The superscripts following each orbital in the hybrid rotation indicate the relative proportions of each orbital comprising the hybrid.
2. The spatial orientation of hybrid orbitals differs from that of either of the parent component orbitals.
3. Once a particular type of hybrid orbital is formed, it is indistinguishable from another of the same type with respect to component orbital origin.
4. In alkanes, each carbon possesses four tetrahedrally oriented sp^3 orbitals.
5. The naming of noncyclic organic compounds is dependent upon the longest continuous carbon chain that the molecule possesses. Cyclic molecules are named in terms of the basic ring system from which they are derived.
6. Organic molecules containing chains of more than two carbon atoms may exist in more than one structural, geometric, or optical isomeric form.
7. Structural isomers differ one from the other in the manner in which the same number of carbon and associated atoms are joined together.

*The Review, Reflexive Quiz, and Supplemental Readings for Section 14 cover this section as well as the concepts developed in Sections 12 and 13 to provide a comprehensive review of organic chemistry.

8. Geometric isomers contain the same number of carbon atoms, each connected in an equivalent manner. However, one or more of the molecular side groups are oriented differently in space. In the *cis* isomer these groups are arranged on the same side of the molecule. In the *trans* isomer they are placed on opposite sides.

9. Optical isomers are compounds that contain asymmetric carbon atoms. Because of this asymmetry, they rotate the optical plane of light that passes through them. The isomeric forms differ each from the other only in the spatial configuration of the groups attached to the asymmetric carbon atom. The spatial relationship between the asymmetric groups of two optically active isomeric forms is equivalent to that of an object and its reflection. Equal amounts of each enantiomorph will rotate the plane of light in equal but opposite directions. The dextro form will rotate light to the right. The levo form will rotate it to the left.

10. Alkanes are a generally unreactive class of compounds. They can be prepared either by separation from natural products such as petroleum or chemically by saturation of multiple-bonded organic molecules to produce single-bonded ones.

11. Alkenes, unlike alkanes, contain double-bonded carbon atoms. For each double bond the molecule contains two less hydrogen atoms than the comparable alkane. Alkenes are more reactive than alkanes. They can undergo addition reactions with hydrogen and other molecules to remove their double bond. The double bond is shorter and stronger than the single bond. It is composed of one σ bond made up of two sp^2 hybrid orbitals and one π bond made up of p orbitals. If the double bond is converted to a single bond by the addition of atoms, then the sp^2 and p orbitals of those carbon atoms will be converted to sp^3 orbitals.

12. Carbon–carbon triple bonds are composed of sp and p orbitals. They are shorter and stronger than either single or double bonds. The chemistry of triple-bonded molecules is in many ways similar to its double-bonded cousins. Compounds containing triple bonds can be prepared by extensions of the procedures used to prepare double-bonded molecules.

13. Organic compounds can exist as neutral molecules, free radicals, positively charged carbonium, or negatively charged carbanions. The state in which molecules exist often determines the mechanisms by which they react with other molecules. When reaction conditions are altered so as to effect a change in reaction mechanism, the resulting reaction products are, as a rule, also altered.

14. The mechanisms by which organic reactions occur can be divided into three groups: electrophilic, nucleophilic, and radical reactions. The driving force of the first two is the desire, or lack of same, that reactants may have to be associated with an electron-rich environment. Radical reactions are driven by the reactant's desire to pair its electrons.

15. Aromatic molecules such as benzene contain carbon–carbon bonds that are intermediate in length between a single and a double bond. Unlike isolated double bonds in alkenes, which also are composed of sp^2 and p orbitals, the bonds in aromatic rings work together to stabilize the molecule, making it immune to the addition reactions characteristic of isolated double bonds.

16. The nature and location of atoms substituted on an aromatic ring determine both the feasibility and location of any future substitutions. Substituted atoms either activate or deactivate the ring toward additional substitutions. If a ring is activated in terms of an additional component substitution, then that component will add either ortho or para to the atom that caused the activation. If, however, the ring is deactivated, the additional substitution will occur in the meta position.

17. In composite molecules that contain both aromatic ring structures and straight-chained groups, the carbon atoms in each part of the molecule display chemistries that are essentially equivalent to that which each would have displayed had it not been connected to the other molecular section. The carbon atoms in close proximity to the juncture of the two molecular types, however, do have modified properties as a result of the presence of the second molecular group.

18. Organic molecules that contain atoms other than carbon and hydrogen may display acidic, basic, or redox properties. The atoms that most commonly combine with carbon to produce these properties are oxygen, nitrogen, and sulfur.

19. A large number of oxygen-containing organic molecules can be divided into three classes depending on their stage of oxidation. The lower oxidation stage encompasses the alcohols and ethers. The next higher stage contains the aldehydes and ketones. Acids are the highest stage of oxidation.

20. In nitrogen-containing amine compounds, one or more of the hydrogens of an ammonia molecule are replaced by either alkyl or aryl functional groups. If only one hydrogen is replaced, then a primary amine is produced. If two or three hydrogen are replaced, then secondary or tertiary amines are produced.

21. The chemical similarity between sulfur and oxygen allows the formation of many organic compounds where the sulfur atom is combined in a manner similar to oxygen in oxy-organic compounds. Analogous to the oxidation of alcohols to acids, thio alcohols can be oxidized to sulfonic acids.

REFLEXIVE QUIZ

1. In spite of the fact that the electronic configuration of carbon is $1s^22s^22p^2$, the four hydrogens in methane (**CH_4**) are equally bound. This is because the carbon atom has undergone _____ hybridization.
2. Compounds that have the same number and type of atoms yet differ in either physical or chemical properties are known as __________.
3. Noncyclic organic molecules are named in terms of the __________ continuous chain of carbon atoms in the molecule.
4. 2-Methylbutane and 2,2-dimethylpropane are both isomers of _________.
5. Cyclic compounds, because they have no ends, possess two less _______ atoms than the equivalent noncyclic compounds.
6. ___________ isomers can be found only in molecules which have two sides.
7. ___________ isomers possess asymmetric atoms that have the ability to rotate the plane of polarized light.
8. Enantiomers will rotate the plane of polarized light in __________ directions.
9. ___________ display very little chemical reactivity aside from being capable of oxidation.
10. ______________________ is the process of separating mixtures of compounds by means of their boiling point differences.
11. Atoms, molecules, or groups of atoms with unpaired electrons are ______________.
12. Organic reaction mechanisms are either radical or _________ in nature. Carbonium ions are ___________ charged organic ions. Carbanions are ___________ ones.
13. The chain nature of organic radical reactions makes them particularly suitable for the production of ___________.
14. Alkenes and the alkynes differ from alkanes in that they have multiple bonds that contain _____ and _____ hybrid orbitals.
15. In compounds containing double and triple bonds, the component bond joining these atoms along their centers is a _____ bond. The additional component bonds above and below or in front of and behind this fundamental bond are _____ bonds.
16. ___________ isomers do not exist for single-bonded compounds because atoms are free to rotate about the single bond.
17. Multiple-bonded structures can be prepared by removal of simple mol-

ecules such as water, hydrogen, halogens, and hydrohalogen acids from ________ carbon atoms.

18. The addition of asymmetrical, simple molecules to multiple-bonded carbon molecules will produce different products, depending on the reaction ________.

19. The carbon-to-carbon bonding in benzene and other aromatic molecules is composed of ____ hybrid orbitals and *p* orbital ____ electron shadows above and below the plane of the molecule.

20. Aromatic bonds are intermediate in length between double and ______ bonds.

21. In electrophilic substitution reactions, highly electrophilic groups already present on the aromatic ring ________ it toward further substitution.

22. In order for an electrophilic reagent to substitute in either the ortho or para position of a substituted compound, the ring must be ________.

23. Organic molecules composed of only carbon and hydrogen have neither acidic nor basic properties. In order for them to have such properties a third element must be present. Elements capable of providing this function are ________, ________, and sulfur.

24. When aldehydes are oxidized, they are converted to ______. When they are reduced, they are converted to ________.

25. Aromatic —**OH** compounds are ________. Aliphatic —**OH** compounds are ________.

26. In nucleophilic substitution reactions, the bonds of the carbon atoms at the point of substitution become ________.

27. Amines are essentially ________ molecules in which alkyl or aryl groups have substituted for ________.

28. Mercaptans are sulfur analogs of ________.

Answers: (1) sp^3 [212–215, 298]. (2) Isomers [214–216]. (3) Longest [215–219, 298]. (4) Pentane [216]. (5) Hydrogen [219–222]. (6) Geometric [219–222, 299]. (7) Optical [222–229, 299]. (8) Opposite [222–229, 299]. (9) Alkanes [229–231, 299]. (10) Fractional distillation [231]. (11) Free radicals [235]. (12) Ionic [235–237]. (13) Polymers [256–258]. (14) sp^2, sp [242–249, 299]. (15) σ, π [242–249, 299]. (16) Geometric [242–247, 299]. (17) Vicinal [250–253, 299]. (18) Mechanism [253–256, 299]. (19) sp^2, π [259–263, 300]. (20) Single [259–263, 300]. (21) Deactivate [267–273, 300]. (22) Activated [267–273, 300]. (23) Oxygen, nitrogen [267–271, 300]. (24) Acids, alcohols [278–280, 300]. (25) Phenols, alcohols [280–283]. (26) Inverted [282–285]. (27) Ammonia, hydrogen [289–294, 300]. (28) Alcohols [295–297, 300].

SUPPLEMENTARY READINGS

1. E. A. Walters. "Models for the Double Bond." *Journal of Chemical Education* 43 (1966): 134.
2. W. H. Saunders, Jr. *Ionic Aliphatic Reactions.* Englewood Cliffs, N.J.: Prentice-Hall, Inc., 1965.
3. W. R. Pryor. *Introduction to Free Radical Chemistry.* Englewood Cliffs, N.J.: Prentice-Hall, Inc., 1966.
4. W. R. Dolbier, Jr. "Electrophilic Additions to Olefins." *Journal of Chemical Education* 46 (1969): 342.
5. N. Isenberg and M. Grdinic. "A Modern Look at Markonekov's Rule and the Peroxide Effect." *Journal of Chemical Education* 46 (1969): 601.
6. S. Patai, ed. *The Chemistry of the Amino Group.* New York: Interscience Publishers, 1968.
7. C. C. Price and S. Oae. *Sulfur Bonding.* New York: Ronald Press, 1962.
8. R. O. C. Norman and R. Taylor. *Electrophilic Substitution in Benzenoid Compounds.* New York: Elsevier Publishing Co., 1965.
9. G. M. Badger. *Aromatic Character and Aromaticity.* London: Cambridge University Press, 1969.
10. D. F. Mowery, Jr. "Criteria for Optical Activity." *Journal of Chemical Education* 46 (1969): 269.

Section 15

Biochemistry

Section 15

Biochemistry

Labor: The Major Cost in Monster Building

It is not a simple matter to establish the cost of building a monster. If only elemental chemicals are used, basic monsters composed merely of skin, bone, and hanks of hair can be produced from materials costing as little as \$3.50. The cost of labor will naturally depend upon which model is built. A conservative estimate would be 6480 man-hours per unit at current union scale (Amalgamated Monster Assemblers). If much of the work is subcontracted, however, and luxury options such as type AB^+ blood are desired, the unit price per monster immediately skyrockets. (The current market price for type AB^+ blood is approximately \$270 per liter.)

Note. The reader may question why the literature on monster manufacturing procedures is limited almost exclusively to male monsters. The reason is that, although the cost of materials for producing female monsters is lower because of lower average body weight, once created their demands for optional modification often outstrip any savings accrued in the initial production. In all his many years, Dr. Frankenstein's monster was never concerned with his appearance. This is seldom the case for the more fashion-conscious female monster.

The assembly sheet used in the production of the 1976 GM-XMD (General Monsters Experimental Male Deluxe Model) has been reproduced with the permission of General Monsters. (Figure 15–1.) It is hoped that it will provide the student with a better grasp of what is involved in manufacturing a monster. The monster model specified is very similar to a real, live, honest-to-goodness human being. The student can therefore assume that with the exception of the capricious skin coloring option of XMD76 the biochemistry of this monster very closely parallels human biochemistry. Knowledge of monster specifications, and their options, can as a consequence be extremely useful to the understanding of the biochemistry of humans.

Specifications

Body type

Whatever skeletal body type is chosen, the bone should be constructed of a cartilaginous matrix, similar to collagen, impregnated

Assembly Sheet GM-XMD 76

1. Body type. . . . . Skeletal humanoid
   Size option – 6 ft.
2. Exterior covering material. . . . . Flesh – protein base
   Color – green
3. Body insulation. . . . . Lipids
4. Temperature, fuel, and hydraulic system. . . . . Blood type – A+
5. Electrical and communication system. . . . . Deluxe neurological network
   sensor systems – a. Visual
   b. Auditory
   c. Olfactory
   d. Tactile
   e. Taste

WARRANTY: This product is guaranteed against all manufacturing defects for 90 full moons. During this period, all defects not the fault of the purchaser will be rectified free of charge if the product is returned to the manufacturer. If anyone other than authorized GM personnel attempts repairs on the monster during the period of guarantee, this warranty is voided.

FIGURE 15–1 Assembly Sheet for Monster

with mineral matter to give it rigidity. Typically, 60 to 70 percent of the bone should be composed of minerals in the following approximate percentages:

Calcium	–	37%
Magnesium	–	1%
Phosphorus	–	17%
Carbonate	–	5%

The inorganic compounds represented by these percentages are not substances of fixed composition. The mineral content of living bone is in a state of constant flux. Carbonate ion may increase at the expense of phosphate ion. Magnesium may replace calcium. Either or both of these processes can also be reversed. For these reasons, the mineral content of bone is normally considered only in terms of an average value. This average can be represented by

$$[Ca_{8.5}Mg_{0.25}Na_{0.19}][(PO_4)_{5.97}(CO_3)_{1.24}](H_2O)_2$$

In many ways this composition is similar to the geological class of minerals known as apatites.

Exterior covering

A large percentage of skeletal covering is composed of proteins. Proteins are giant biological polymers of *amino acids* whose molecular weights may be as great as several millions. Their shapes may vary from coiled chains to hollow spheres. In addition to being the most important component for the manufacture and repair of body tissue, proteins are necessary for the syntheses of enzymes, hormones, immunological antibodies, and certain vitamins. At least twenty-two different amino acids are commonly found as subunits of proteins.

Most amino acids can be represented by the general formula

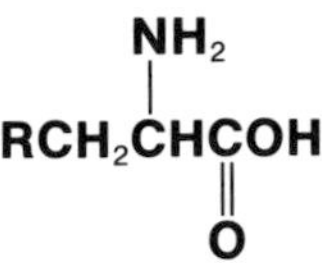

The **R** may be either an aliphatic, aromatic, or heterocyclic functional group. (A *heterocyclic* functional group is a ring whose skeleton contains atoms other than carbon; e.g., nitrogen or sulfur.) With the exception of glycine, all amino acids contain asymmetric carbon atoms and are therefore optically active. Additionally, all naturally occurring amino acids possess an (L) special configuration of atoms.

Amino acids are capable of ionizing either as acids or as bases. At times they may even do both simultaneously. When a double ionization does occur, the resulting product is called a *zwitterion* ("half-breed ion" in German).

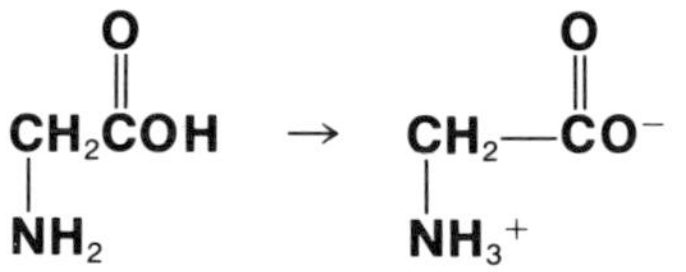

$$\underset{\text{Glycine}}{\underset{\underset{NH_2}{|}}{CH_2}\overset{\overset{O}{\|}}{C}OH} \rightarrow \underset{\text{Glycine zwitterion}}{\underset{\underset{NH_3^+}{|}}{CH_2}—\overset{\overset{O}{\|}}{C}O^-}$$

If an acid is added to a zwitterion solution, the protons will combine with the carboxyl group. If a base is added, protons will be removed from the amino group. The ability of amino acids to neutralize both acids and bases makes them extremely useful in maintaining the proper *p*H of blood and other body fluids.

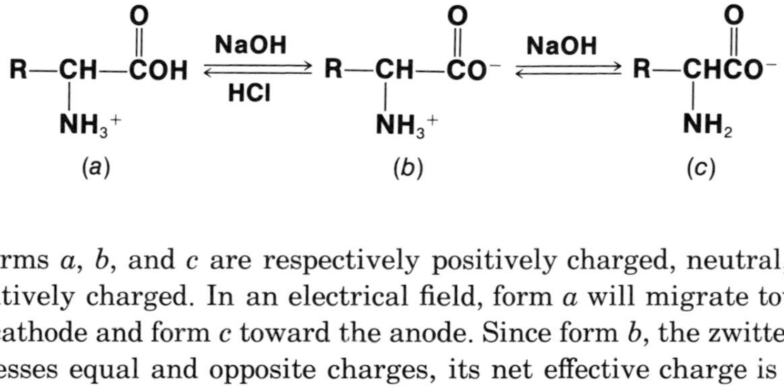

$$\underset{(a)}{R—\underset{\underset{NH_3^+}{|}}{CH}—\overset{\overset{O}{\|}}{C}OH} \underset{HCl}{\overset{NaOH}{\rightleftharpoons}} \underset{(b)}{R—\underset{\underset{NH_3^+}{|}}{CH}—\overset{\overset{O}{\|}}{C}O^-} \overset{NaOH}{\rightleftharpoons} \underset{(c)}{R—\underset{\underset{NH_2}{|}}{CH}\overset{\overset{O}{\|}}{C}O^-}$$

Forms *a*, *b*, and *c* are respectively positively charged, neutral, and negatively charged. In an electrical field, form *a* will migrate toward the cathode and form *c* toward the anode. Since form *b*, the zwitterion, possesses equal and opposite charges, its net effective charge is zero, making it immune to the electric field. The *p*H that converts an amino acid to its zwitterion is known as the isoelectric point of that acid. Each amino acid has its own characteristic isoelectric point. Because of this, mixtures of amino acids and proteins can be separated by judicious adjustment of *p*H and electric fields. The procedure is known as electrophoresis. When applied to body fluids, it is of great value in diagnosing pathologic conditions.

The classification of amino acids according to molecular structure, although useful, has limited application. Of greater significance is the division of acids into essential and nonessential groups. Of the approximately twenty amino acids found in human protein, most can through enzymatic action be synthesized by the body itself. Those that cannot are the *essential* amino acids. They must be supplied by the diet in

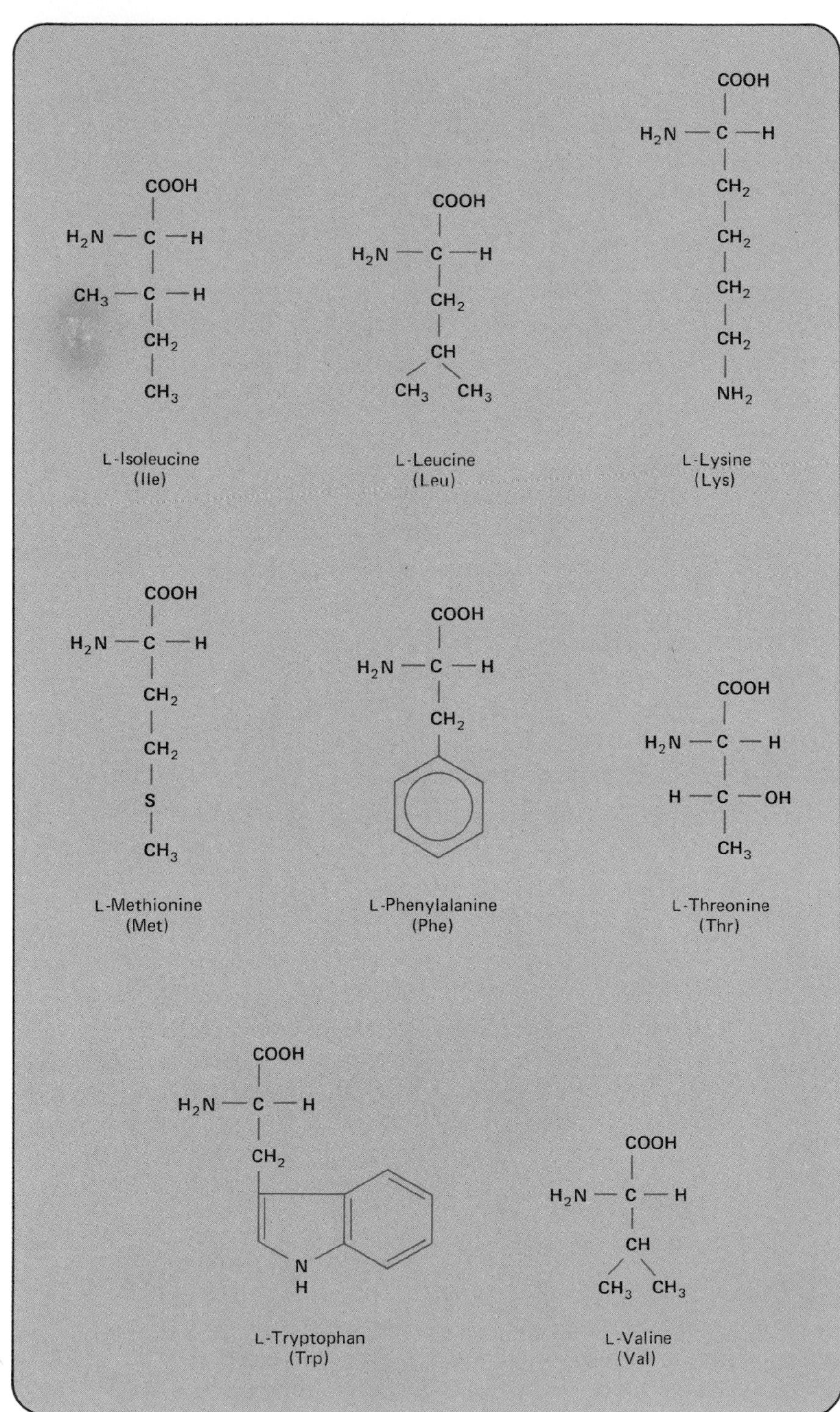

FIGURE

15–2 The Eight Amino Acids Essential to Man

amounts sufficiently large to prevent the organism from developing deficiency diseases. The number and selection of essential amino acids vary with the particular species. The eight amino acids essential to man are shown in Figure 15–2.

A daily diet containing in excess of two and a half ounces of meat, fish, cheese, or eggs will supply sufficient quantities of all eight essential amino acids to meet the average human's daily needs. The distribution of essential amino acids among plants is generally not up to the quality found in animal foods. Cereal proteins tend, as a rule, to be deficient in one or more of the essential amino acids. (Wheat is deficient in lysine. Rice is deficient in lysine and threonine. Corn is deficient in lysine and tryptophane.) In India, where the basic diet is vegetable and where meat is considered a luxury, the government has decreed that bread produced in its bakeries should be fortified with 0.2 percent lysine as a means of preventing deficiency diseases.

The major process for polymerizing amino acids into proteins is peptide bond formation. Peptide bonds are formed by neutralizing the basic amino group of one amino acid with the carboxy group of another. The process occurs with the splitting out of a molecule of water from between the two reactants.

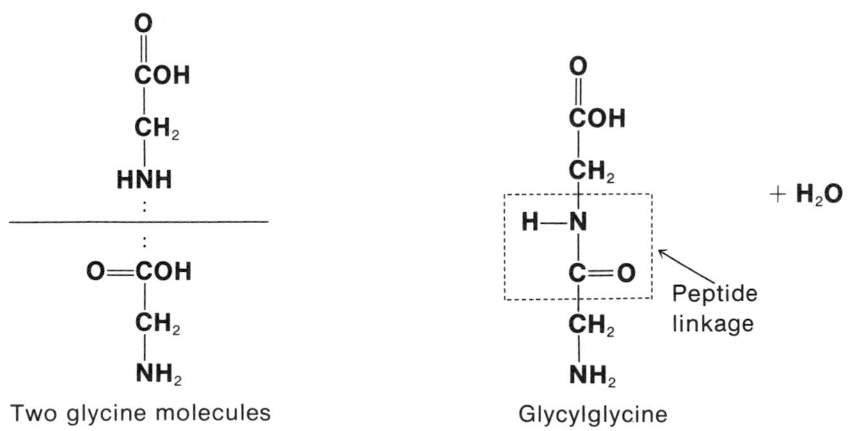

In this manner, although by a far more complicated process, the slightly more than twenty basic amino acids combine with each other to produce the more than one hundred thousand different kinds of protein molecules found in the human body. Figure 15–3 illustrates the composition of an enzyme protein composed of 120 amino acids. Each circle except those joined to an **S—S** bridge contains an amino

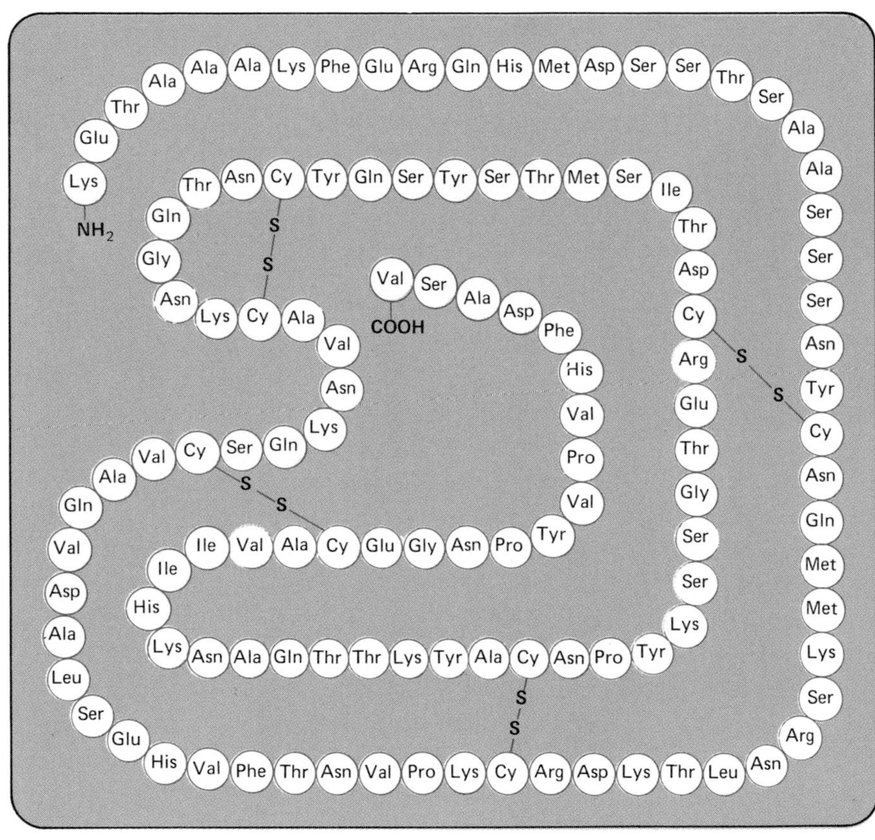

FIGURE

Ala—(L-alanine)	CyS-SCy—(L-cystine)	Ile—(L-iso leucine)	Ser—(L-serine)
Arg—(L-arginine)	Gln—(L-glutamine)	Leu—(L-leucine)	Thr—(L-threonine)
Asn—(L-asparigine)	Glu—(L-glutamic acid)	Lys—(L-lysine)	Tyr—(L-tyrosine)
Asp—(L-aspartic acid)	His—(L-histidine)	Phe—(L-phenylalanine)	Val—(L-valine)

15–3 Polymer Structure of the Enzyme Pancreatic Ribonuclease, Showing Its 120 Amino Acids

acid that is bonded to the two amino acids in adjacent circles through peptide linkages. The arrangement, type, and number of amino acids necessary to produce this enzyme or any other specific protein polymer in a living organism is governed by even more complex macromolecular compounds. These macromolecules (giant molecules), which sometimes have molecular weights as great as a hundred million, are the nucleic acids.

Ribonucleic acid and deoxyribonucleic acid, often abbreviated RNA and DNA, are the chemical substances that determine the characteristics of all living things. Besides guaranteeing that a brown-eyed gray elephant doesn't give birth to a pink-eyed white mouse, the nucleic acids specify that when tissue is damaged either as a consequence of accident or normal living, it will be replaced by new tissue of a similar type. If a child skins his knee, the new tissue, upon healing, will look like the original skin.

The *nucleic acids* are enormous polymers that are made up of nucleotides. (Figure 15–4.)

Nucleotides are three-part entities. They are composed of (1) a phos-

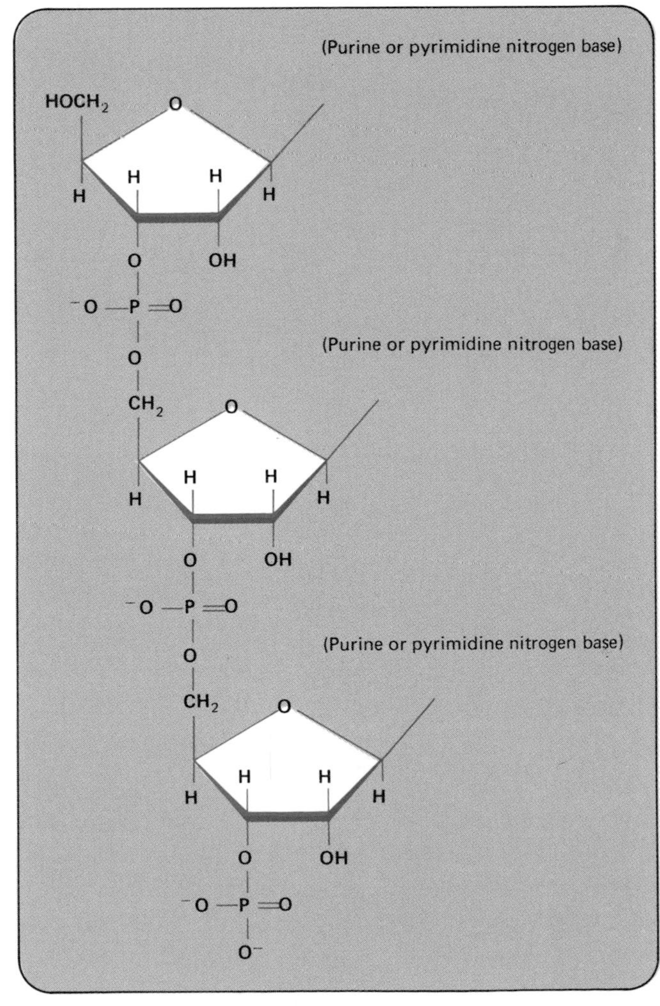

FIGURE 15–4 Nucleotide Composition of a Nucleic Acid Polymer

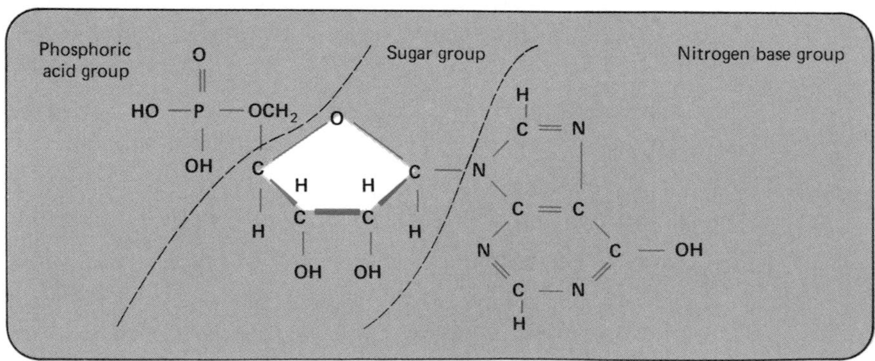

FIGURE 15–5 The Three-Part Composition of a Nucleotide

phate group that is attached to (2) one of two possible sugar groups that is attached to (3) one of a small number of possible heterocyclic nitrogen bases called purines and pyrimidines. (Figure 15–5.) The two possible sugar groups are D-ribose and 2-deoxy-D-ribose. (Figure 15–6.) Nitrogen bases commonly associated with RNA and DNA are adenine, guanine, cytosine, thymine, and uracil. (Figure 15–7.) Through phosphate-sugar linkages, nucleotides join together to form the gigantic polymers that are the nucleic acids. These acids are classified as either DNA or RNA by which one of the two sugars is incorporated into the polymer skeleton.

DNA is mostly found in the nucleus of living cells, where it serves as the prime source of direction for the chemical processes occurring in living organisms. The directions are coded into the DNA molecules by the arrangement of nitrogen bases. The system is very similar to the systems used to open combination locks. For a three-numbered

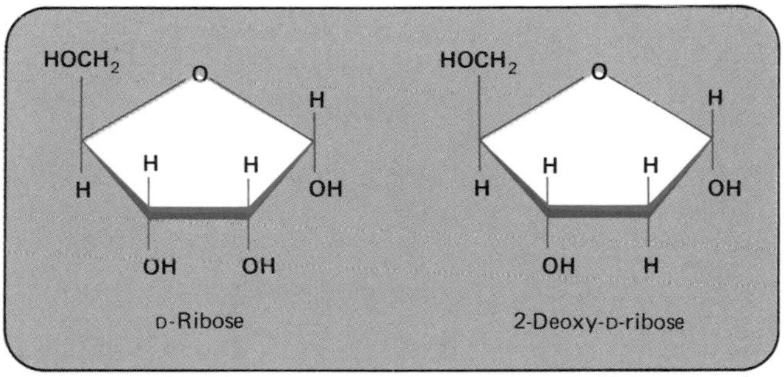

FIGURE 15–6 The Sugar Components of RNA and DNA

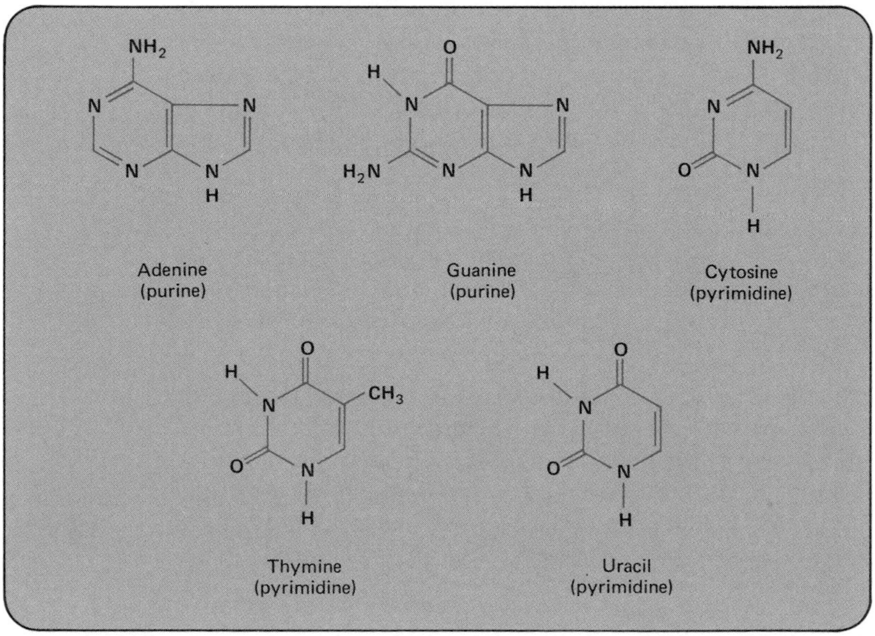

FIGURE 15–7 Nitrogen Bases That Are Often Components of RNA and DNA

combination lock to open, it is necessary to dial the first number correctly, reverse direction, dial the second number correctly, reverse direction again, and then properly dial the last number. If the selection of numbers on the dial run only from one to four, then there are only sixty-four combinations that the lock may have. (The first number may have any value from 1 through 4. The second number may have any value from 1 through 4. The same is true of the last. $4 \times 4 \times 4 = 64$.) Before a living organism can manufacture protein, it is first necessary that there be established an order in which the component amino acids are to be sequenced. This could be accomplished through the use of three nucleotide sections of nucleic acids. Since each nucleo-

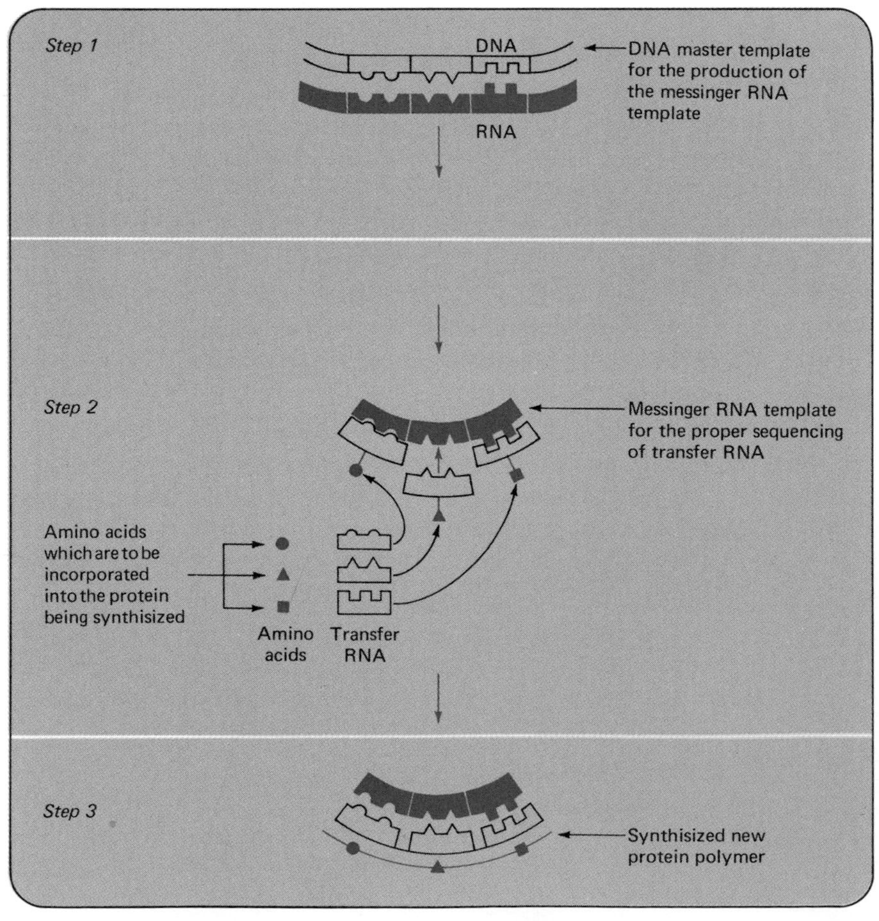

FIGURE 15–8 Process by Which DNA and RNA Direct the Synthesis of Proteins

tide may contain any of four possible nitrogen bases, a three-nucleotide section could be constructed in any of sixty-four possible combinations. Each of these can direct the addition of a specific amino acid to a particular position in a protein chain. A three-nucleotide section of a nucleic acid carries more than sufficient sequencing information for this process since there are only about twenty amino acids that are commonly incorporated into proteins. It is the arrangement of these nucleotide triads in the parent nucleic acid that determines the nature and sequence of the amino acids in the protein molecule under construction. The process, which is schematically depicted in Figure 15–8, can be subdivided into three steps.

Step 1. DNA, which is found in the nucleus of a cell, associates with monoribonucleotides to produce a strand of RNA having a nitrogen base sequence that is the complement of the nitrogen base sequence in the DNA. The result is very similar to what happens when a child fills a mold with wet sand. When the mold is turned out, the product is the complement of the mold impression. RNA produced in this manner is known as messenger RNA.

Step 2. The messenger RNA constructed through association with the nuclear DNA leaves the nucleus to become associated with the main body of the cell. In the cell's main body there also exists a second type of RNA, which has a lower molecular weight. This type of RNA is known as transfer RNA. Transfer RNA, because of its ability to form associations with particular amino acids, is the active component in protein synthesis.

Step 3. Transfer RNA molecules that are associated with specific amino acids align themselves to complement the sequencing in the messenger RNA. In so doing, they also sequence the amino acids with which they are associated. Once sequenced, the amino acids undergo peptide bond formation to produce proteins. The proteins produced are of a particular type because the DNA determined the sequencing of the messenger RNA, which in turn determined the sequencing of the transfer RNA and its associated amino acids. The remarkable part of this information storage system is that it is better than 99.997 percent accurate.

Body insulation

Fats are the single most important component of the oil soluble class of organic compounds called lipids. In deluxe model monsters as in many animals, lipids in general and fats in particular are utilized for

the purpose of insulation. Lipids not only shield the body against the stresses of reduced temperatures, they also prevent the electrical circuits of the nervous system from short circuiting by providing insulating sheaths around the individual nerves.

Stearin, a typical fat, is shown in Figure 15–9. It and all other fats are esters of glycerol, a trihydroxy alcohol.

$$
3\,C_{17}H_{35}COOH + \begin{array}{c} H \\ | \\ HO{-}C{-}H \\ | \\ HO{-}C{-}H \\ | \\ HO{-}C{-}H \\ | \\ H \end{array} \rightarrow \begin{array}{c} H \\ | \\ C_{17}H_{35}COO{-}C{-}H \\ | \\ C_{17}H_{35}COO{-}C{-}H \\ | \\ C_{17}H_{35}COO{-}C{-}H \\ | \\ H \end{array} + 3\,H_2O
$$

Stearic acid Glycerol Stearin

No naturally occurring fat is composed of pure tristearin. Most, depending on their source, are mixtures of stearin and other related esters.

$$
\begin{array}{lcl}
 & \begin{array}{c} O \quad H \\ \| \quad | \end{array} & \\
(1) & C_{17}H_{35}CO{-}C{-}H & \text{(Stearo)} \\
 & \begin{array}{c} O \\ \| \end{array} \quad | & \\
(2) & C_{17}H_{33}CO{-}C{-}H & \text{(Oleo)} \\
 & \begin{array}{c} O \\ \| \end{array} \quad | & \\
(3) & C_{15}H_{31}CO{-}C{-}H & \text{(Palmitin)} \\
 & \quad | & \\
 & \quad H &
\end{array}
$$

Mixed triglyceride

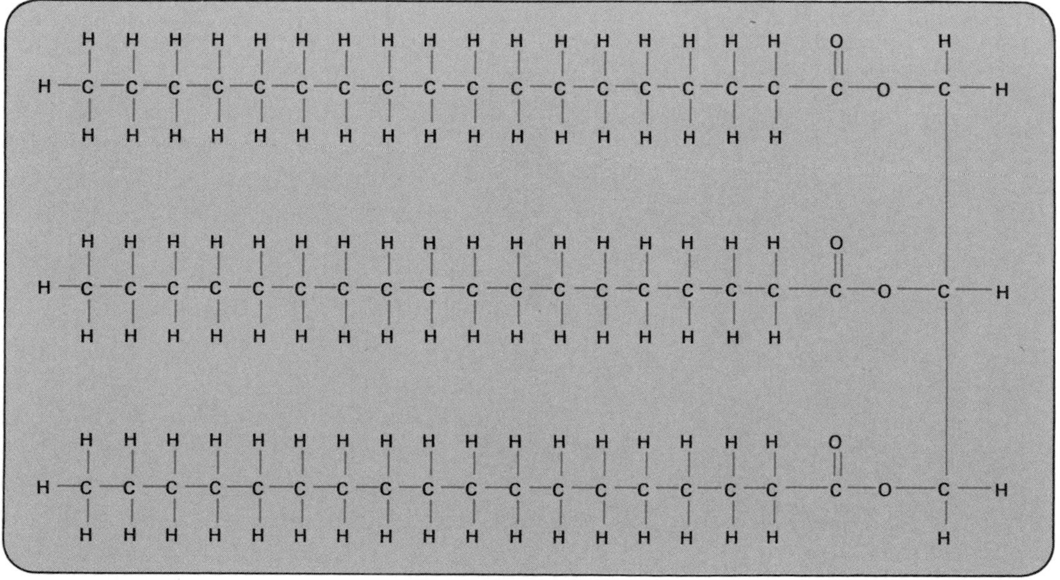

15–9 Stearin FIGURE

The long-chained fatty acids derived from natural fats such as beef tallow and lard are all strikingly similar. All (with one exception) have an even number of carbon atoms; all have long-chained structures with no branches; and all have their carboxyl group attached to the terminal carbon atom. The one exception is isovaleric acid, $(CH_3)_2CHCH_2COOH$, a component of dolphin and porpoise fats, which has a branched methyl group.

Fuel and hydraulic system

Any machine, whether living or mechanical, in order to function properly, must be supplied with fuel, a means of servicing worn-out parts, and a garbage disposal system. In animal machines, including the human body, these functions are accomplished by the flow of blood through a highly refined vascular system. During its transit, the blood performs all the functions of a quartermaster. It picks up hormones, which are chemical instructions for the operation of the body, at the ductless glands and delivers them to the appropriate operating region. It picks up antibodies at their point of manufacture and transports them to points of infection. Further, the body's hydraulic system is self-regulating, because the blood contains a special protein, called fibrinogen, that forms clots if the system should spring a leak for any reason.

Each liter of human blood also contains 150 grams of a composite protein material called hemoglobin. (In a 150-pound person this amounts to approximately 900 grams.) Vertebrate hemoglobins are usually composed of four heme molecules (Figure 15–10) attached to the protein globin. The molecular weight of this composite hemoglobin is approximately 68,000. Its function is to provide a connecting link for the transport of gases between the lungs and the rest of the body. It does this by combining reversibly with oxygen at the lungs and then allowing the blood to transport it to the cells, which contain fuel to be oxidized.

$$\underset{\text{Hemoglobin}}{4\,Hb} + 4\,O_2 \rightleftarrows \underset{\text{Oxyhemoglobin}}{Hb_4(O_2)_4}$$

The fuel was previously picked up at the digestive tract and delivered by the blood to the cell. Once the oxyhemoglobin arrives at the site where oxidation is to take place, it decomposes with the liberation of the oxygen necessary to consume the cellular fuel.

$$Hb_4(O_2)_4 \rightleftarrows 4\,Hb + 4\,O_2$$

It is important to realize that without the presence of hemoglobin

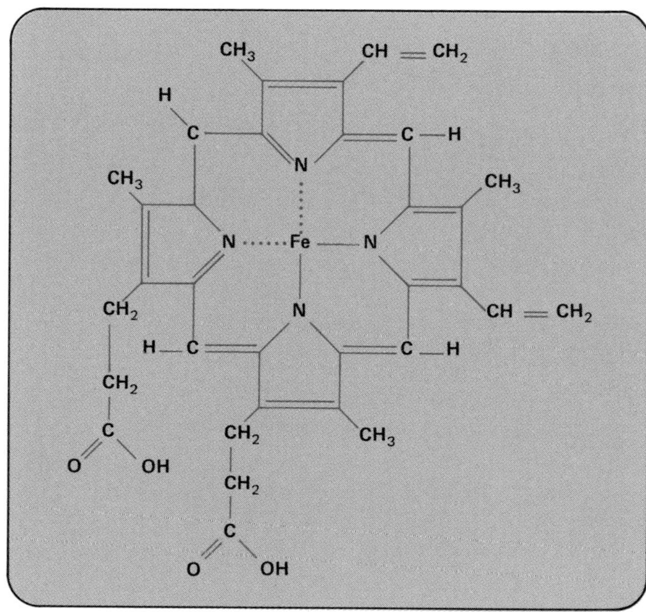

FIGURE

15–10 Heme Molecule

the blood would be capable of transporting only a small fraction of the amount of oxygen necessary to carry on the organism's life processes. The average man, instead of operating on six liters of blood, would need three hundred to perform the same functions. That is, he would have to transport within his body more than six hundred pounds of blood. Imagine the skeletal and muscular revisions that would be necessary to support this drastic change in weight.

The problem is immediately apparent when one begins constructing monsters. If for reasons of economy the amount of hemoglobin is drastically reduced or eliminated, the production of a humanoid-type monster weighing less than half a ton is not feasible. Although Dr. Frankenstein's complete papers were never found, it is quite possible that the inordinately large size of his monster was in part due to a hemoglobin economy measure.

Electrical and communication system

The nervous system provides a high-speed electrochemical means of transporting information from any portion of the body to the decision-making facilities of the brain. As with a human nervous system, a deluxe monster may be provided with sensors that are capable of detecting changes in light, sound, taste, smell, or touch. Although the complete chemistry involved in the operation of these transducers is in many cases unknown or beyond the scope of this text, each sensor in its own way converts a change in the particular variable that it monitors into an electrochemical impulse that can be interpreted by the brain. These impulses or signals are carried to the brain by elongated cells, called nerves, which function much like the electrical wires that couple a computer-driven machine.

The signals transmitted along this nerve network travel at high speeds and are very specific. They often cover the distance between sensor and brain in less than a thousandth of a second. Their transmission to a large part is due to the action of chemicals such as acetylcholine and acetylcholinesterase. Acetylcholine is synthesized in humans from phospholipids such as lecithins. (Lecithins are phosphoric acid esters of diglycerides and choline.)

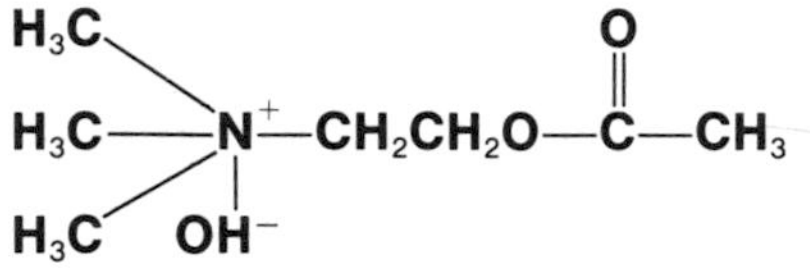

$$(H_3C)_3N^+(OH^-)-CH_2CH_2O-C(=O)-CH_3$$

Acetylcholine

It is capable of combining with receptor proteins to produce a change in a nerve's permeability (degree to which it permits diffusion) to the movement of ions. This change alters the nerve's electrical characteristics so that it can conduct the signals being sent by the sensor organ.

The duration of any signal is determined by the enzyme cholinesterase. A few millionths of a second after the signal has been transmitted via the nerve to the brain, cholinesterase destroys by hydrolysis the acetylcholine that made the transmission possible. The purpose of this destruction is to allow the brain to differentiate between a short-term excitation of a particular sensor and one that continues over an extended time. (If information is to be conveyed by impulse signals, it is an absolute necessity that these signals be switched on and off according to a prescribed recognition pattern. A message in Morse code can be neither sent nor understood if the sending key remains continually in either the depressed or the released position.) When a sensor is subjected to stimulation over an extended period of time, this information is conveyed to the brain in the form of a continual series of pulses during that period. The hydrolysis of acetylcholine at the end of each nerve impulse is a reversible process. In the presence of potassium and magnesium ions and enzymes such as acetylase, acetylcholine is resynthesized. The regeneration of acetylcholine serves to recock the nerve's "trigger" in preparation for the transmission of a new impulse.

$$CH_3\overset{O}{\overset{\|}{C}}OCH_2CH_2\underset{OH^-}{N^+}(CH_3)_3 + H_2O \underset{\text{Acetylase, K, Mg}}{\overset{\text{Cholinesterase}}{\rightleftharpoons}} CH_3\overset{O}{\overset{\|}{C}}—OH + HOCH_2CH_2\underset{OH^-}{N^+}(CH_3)_3$$

Acetylcholine — Acetic acid — Choline

Some of the most deadly poisons known to man are the neurotoxins. These poisons interfere with the acetylcholine synthesis and hydrolysis cycle. This they do in either of two ways. They may, if they are structurally similar to acetylcholine, bond to the enzyme cholinesterase and deactivate it, or they may, as in the case of atropine and curare, combine with receptor protein and in so doing prevent it from combining with acetylcholine in the normal fashion. The situation where the cholinesterase is deactivated by bonding is typical of the mode in which nerve gases and many insecticides operate. These agents cause death because of the overstimulation of nerves, glands, and muscles that results from the inability of bound cholinesterase to hydrolyze acetylcholine.

When atropine or curare occupy a receptor site, impulse transmission is prevented. Because of this, neurotoxins of this type are often useful in medicine. A number of neurotoxins and some of their medical applications are listed in Table 15–1.

It Lives. Give It Vitamins and Minerals.

One would hardly expect an automobile that had never received any maintenance to run as well after 50,000 miles as the day it was bought. Although the oil that is periodically applied to an engine does not normally act as a fuel to operate it, lack of it would in very short order result in sufficient friction to cause the engine to break down. Vitamins in a similar manner maintain living organisms at their peak operating condition. These powerful organic compounds, which are required daily in microgram (10^{-6} gm.) to milligram (10^{-3} gm.) amounts to prevent deficiency diseases, cannot for the most part be synthesized by the

TABLE 15–1 Neurotoxins (Acetylcholine Competitors)

Name	Use	Lethal Dose (150-lb. man)	Formula
Atropine	Dilation of pupil of the eye	0.1 g	CH₃, N, CH₂–CH₂, O, C, CH₂OH, C, H
Caffeine	Mild stimulant	—	O, CH₃–N, N–CH₃, O, N, N, CH₃, H
Cocaine	Anesthetic ointments	1 g	N–CH₃, O, C–OCH₃, O, O–C
Codeine	Pain killer	0.3 g	CH₃, N, CH₃O, O, OH
Curare	Muscle relaxant, treatment of mental disorders	—	OCH₃, CH₃, O, CH₃, +N, OH, CH₂, +N, CH₃, O, CH₂, CH₃, CH₃O, OH
Morphine	Pain killer	0.2 g	CH₃, N, HO, O, OH
Nicotine	Insecticide	0.3 g	N, N, CH₃

body. The vitamins and minerals essential for the health and well-being of humans are listed in Table 15–2.

Vitamins and minerals can be considered the fuel and oil additives that keep the human machine running smoothly. They are chemically discrete compounds that have no single functional group common to them all. The amounts listed are applicable to the average, medium-size, human male engine. Smaller female and child engines require as a rule proportionately less of each vitamin. It should be remembered that the quantities of vitamins listed are the absolute *minimum* amount a person must have on a daily basis in order to prevent the occurrence of deficiency diseases. The amounts necessary for the optimum operation of the human machine are often many times these amounts.

All vitamins can be classified as either water soluble (e.g., B complex and C) or oil soluble (e.g., A and D) compounds. Water-soluble vitamins must be ingested on a daily basis since the body has little storage capacity for these substances. Excess quantities of water-soluble vitamins that have not been utilized in a given day are normally excreted. This process presents the body with a need each day and only memories of the plenty that it had the previous day.

Unlike water-soluble vitamins, fat-soluble vitamins can be stored. It is not necessary that their daily requirement be consumed each and every day. If the amount consumed in one day is more than is used that day, the excess will be stored in the body's fatty and organ tissue until it is needed. On the days that insufficient quantities of these vitamins are available, the body can draw from its larder to meet its needs. This storage capacity has obvious advantages for wild animals that must forage for their food. It has a disadvantage that the body is storing extremely powerful compounds which are capable of poisoning it if their concentrations become larger than that which the body can assimilate.

A story is told about some Arctic explorers who, stranded without food, killed and ate a polar bear. They considered themselves fortunate to have had the luck to find food and thus save their lives. Unfortunately, however, the bear's liver, because of its high concentration of vitamin D, was poisonous to humans, and the explorers who ate it unluckily died.

Although the vitamin D did the explorers no good, it was a definite asset to the bear while he was alive. Bears in the normal course of events eat heavily in the spring and summer and grow fat. In the winter, when food is scarce, they hibernate and draw upon their fatty reserves to carry them through until spring. Because different species have different vitamin requirements, both in terms of amounts and

15–2a Some Vitamins Essential for Proper Human Metabolism

TABLE

Vitamin	Source	Adult daily requirements	Function in the body	Effects of a deficiency
A Retinol	Dairy foods, yellow and green vegetables, organ meats	5000 I.U.*	Visual pigments Maintenance of epithelial tissue	Nightblindness Softening of cornea
B complex (B_1) Thiamine	Whole grains, yeast, meat, legumes	1 to 2 mg	Carbohydrate metabolism	Disturbances of nervous system Fatigue Beriberi
(B_2) Riboflavin	Yeast, dairy foods	1.7 mg	Protein metabolism	Inflammation of the tongue, mouth, and skin
B_3 Pantothenic acid	Widely distributed liver, molasses, rice, bran, royal jelly	10 mg	Carbohydrate and fat metabolism	Nervous disorders Irritability Fatigue Poor skin tone
B_5 Niacin (nicotinic acid)	Yeast, meat	18 mg	Anti-pellagra vitamin	Pellagra
(B_6) Pyridoxine	Whole grains, egg yolk, yeast, meat, fish, legumes	1 to 2 mg	Associated with nerve impulse transmission, amino acids, nucleic acids, and protein metabolism	Convulsions Hyperirritability Nerve inflammation Edema
(B_7) Biotin	Egg yolk, organ meats	Unknown	Associated with carbon dioxide fixation	Loss of appetite Malaise Skin inflammation
(B_{12}) Cyanocobalamin	Eggs, meat, liver	3 to 5 mcg	Prevention of pernicious anemia	Pernicious anemia
C Ascorbic Acid	Fresh fruits and vegetables, rose hips	70 mg	Associated with oxidation-reduction processes Maintenance of connective tissue	Scurvy Anemia Bleeding gums
D Calciferol	Fish liver oils, dairy foods	400 I.U.	Absorption and utilization of calcium	Rickets in children Bone softening in adults
E Tocopherol	Wheat germ, egg yolk, liver, leafy vegetables	30 I.U.	Essential for normal red blood	Hemolytic anemia
K Phyllo Quinone	Spinach, cabbage	Unknown	Essential for proper blood clotting	Slow blood clotting

*I.U. = International units

15–2b Minerals Essential for Proper Human Metabolism

TABLE

Mineral	Source	Adult daily requirements	Function in the body	Effects of a deficiency
Sodium	Table salt, fruits	0.5 to 1.0 g	Control of body water and electropotentials of nerves, muscles and glands Regulation of pH	Weariness, vomiting, cramps, low blood pressure
Potassium	Present in all foods	2.0 g	Regulation of pH, and the electropotentials of cell membranes	Mental disorders, hallucinations Poor muscle tone
Calcium	Dairy foods, grains	0.8 g	Calcification of bones and teeth, blood clotting	Muscle spasms Bone softening
Magnesium	Fruit, legumes, nuts, grains	0.4 g	Enzyme reactions, regulation of the body's electrical potentials	Neuromuscular disorders
Iron	Meat, fruits, vegetables, grains	12 mg	Component of hemoglobin	Anemia
Copper	Shell fish, fruits, organ meats	1 to 2 mg	Associated with synthesis of hemoglobin	Anemia

specific types, the quantity of oil-soluble vitamins necessary to sustain the bear's life proved too high a concentration for the explorers and resulted in their untimely demise.

REVIEW

1. The shape that most humans are in is, in the main, due to their distribution of proteins, lipids, and inorganic chemicals.
2. Bones are composed of mixtures of inorganic phosphate compounds and proteins.
3. Proteins are long-chained polymer molecules composed of smaller amino acid subunits.
4. All the amino acids—with the exception of glycine, the simplest—contain asymmetric carbon atoms.
5. Amino acids are capable of ionizing as either acids or bases.

6. The *p*H at which both the carboxyl and the amine group of an amino acid are ionized is its iso-electric point.
7. "Essential" amino acids are those that cannot be synthesized in sufficient quantity by the living organism and therefore must be acquired in the form of food.
8. Peptide bond formation is the major process for polymerizing amino acids into protein polymers.
9. The number and arrangement of amino acids within a protein polymer is determined by the organism's nucleic acids.
10. Nucleic acids are polymers of nucleotides.
11. A nucleotide is a three-part molecule composed of a phosphate group, a sugar group, and a heterocyclic nitrogen base group.
12. DNA and RNA possess different sugar and nitrogen base components.
13. Fats are esters of glycerol and long-chained fatty acids.
14. Hemoglobin is the blood component that combines reversibly with respiratory gases.
15. Acetylcholine is the chemical substance that permits the transmission of nerve impulses.
16. Vitamins are compounds needed regularly in small quantities by living organisms in order to function at optimum efficiency.

REFLEXIVE QUIZ

1. The major inorganic component of bones and teeth is ________________.
2. The lowercase letters d and l describe the ________________________ of a molecule.
3. An amino acid whose carboxyl and amino groups are both ionized is known as a ______________.
4. Amino acids that an organism needs to function properly but cannot synthesize are known as ______________ amino acids.
5. The bond produced between an amine group of one amino acid and the carboxyl group of another is known as a ____________ linkage.
6. Polymers of nucleotides are known as ____________ acids.
7. RNA is a nucleic acid that contains the sugar ____________.
8. In humans ____________ are used for both thermal and nerve insulation.

9. Neurotoxins interfere with the proper functioning of ________.

10. ________ are powerful compounds essential daily in milligram or microgram quantities for the proper functioning of living organisms.

Answers: (1) Calcium phosphate [306–308, 327]. (2) Spatial configuration [308, 327]. (3) Zwitterion [309, 328]. (4) Essential [309–311, 328]. (5) Peptide [311, 328]. (6) Nucleic [313, 328]. (7) Ribose [312–315]. (8) Lipids [317–320]. (9) Acetylcholine [321–323]. (10) Vitamins [323–328].

SUPPLEMENTARY READINGS

1. B. Commoner. "DNA and the Chemistry of Inheritance." *American Scientist* 52 (1964): 365.
2. E. Zuckerkandl. "Evolution of Hemoglobin." *Scientific American* (May 1965), p. 110.
3. F. D. Gunstone. *The Chemistry of Fats and Fatty Acids.* New York: John Wiley & Sons, 1958.
4. A. Meister. *Biochemistry of Amino Acids,* 2nd ed. New York: Academic Press, 1965.
5. A. F. Wagner and K. Folkers. *Vitamins and Coenzymes.* New York: Interscience Publishers, 1964.
6. R. F. Steiner. *The Chemical Foundations of Molecular Biology.* Princeton, N.J.: Van Nostrand Co., 1965.
7. J. D. Watson. *The Double Helix.* New York: Atheneum Publishers, 1968.

Section 16

Consumer Chemistry

Section 16

Consumer Chemistry

The Borgias Were Pioneers in Food Additives.

Back in fifteenth-century Italy, the Borgia family found that the addition of small but significant amounts of poisons to the dietary intake of their enemies could effectively promote their plans for Italy and their own well-being. The additives of today are as a rule designed with less drastic purposes in mind. Usually they are meant to retard spoilage or to produce a more appetizing taste, color, or consistency. A listing of some of the over three hundred chemicals approved by either the U.S. Food and Drug Administration or the U.S. Department of Agriculture for addition to foods suitable for human consumption appears in Table 16–1. The American populace consumes approximately three-quarters of a billion pounds of these materials each year. There is hardly a processed food available that does not contain at least one additive.

The motives of the Borgias for employing their additives are easily imagined: revenge, jealousy, fear, and other such basic human emotions. Cynics might say that the motives of modern food processors for using additives are similar: greed and ambition. The more fair-minded, however, recognize that modern processors are acting in the interest of improved business, aiming to increase the saleability of their products and thereby increase their profits. Their motives are thus more subtle than the Borgias', and the effects of their additives are, happily, less lethal.

Two qualities are involved in increasing a product's saleability: initial product appeal and long-term product stability. With additives, a processor can cause his product to look, taste, smell, or feel better than an identical product that does not have the benefit of additives. It does him little good, however, to produce a tasty, eye-appealing product if it is likely to go rancid, change color, or separate while sitting on the supermarket shelf. Saleable products must therefore be both attractive and durable.

Product Appeal

Milk and its associated products provide examples of the first of these two qualities. The packaged milk sold in supermarkets is usually quite different from the product that Farmer Grey's friendly cow originally produced. It may be skimmed to reduce the cream, homogenized to disperse the cream, or fortified to increase its vitamin or protein content. The fortification of milk by the addition of vitamin D serves to make the milk calcium content more utilizable. This is of significant benefit to growing children since calcium is essential for the proper formation of bones and teeth.

Though you may lead a child to milk, you can't necessarily make him

16–1 A Partial List of Food Additives Currently Recognized as Safe by the U.S. Food and Drug Administration (1974) TABLE

Alcohols	Pimenta leaf oil (allspice)	Fumaric acid
Glycerol	Saccharin (sweet)	Lactic acid
Sorbitol	Vanillin (vanilla)	Phosphates
Mannitol	Wintergreen oil (wintergreen)	Potassium acid tartrate
Propylene glycol		Sorbic acid
		Tartaric acid
Colors	**Nutrients**	
Annatto (yellow)	Amino acids	**Sequestering Agents**
Carbon black (black)	Minerals	Citric acid
Carotene (yellow-orange)	Vitamins	EDTA
Cochineal (red)		Pyrophosphate
Titanium dioxide (white)		Sodium tartrate
	Preservatives	Sorbitol
Flavors	Ascorbic acid	Tartaric acid
Acetanisole (slight haylike)	Benzoic acid	
Allyl caproate (pineapple)	Butylated hydroxyanisole (BHA)	**Surface Active Agents**
Amyl acetate (banana)	Butylated hydroxytoluene (BHT)	Cholic acid
Amyl butyrate (pear-like)	Calcium propionate	Glycerides: mono- and diglycerides of fatty acids
Carvone (spearmint)	Lecithin	Polyoxyethylene (20) sorbitan monopalmitate
Cinnamaldehyde (cinnamon)	Methylparaben	Sorbitan monostearate
Citral (lemon)	Oxytetracycline	
Ethyl cinnamate (spicy)	Propionic acid	**Viscosity | Determining Agents**
Ethyl formate (rum)	Propyl gallate	Agar-agar
Ethyl propionate (fruity)	Propylparaben	Algins
Ethyl vanillin (vanilla)	Sodium benzoate	Carrageenin
Eucalyptus oil (bittersweet)	Sorbic acid	Gelatin
Eugenol (spice, clove)	Sulfites	Gum acacia
Geraniol (rose)		Gum tragacanth
Geranyl acetate (geranium)	***p*H Determining Agents**	Sodium carboxymethyl cellulose
Ginger oil (ginger)	Acetic acid	
Menthol (peppermint)	Acetates	
Methyl anthranilate (grape)	Calcium lactate	
Methyl salicylate (wintergreen)	Citric acid	
Monosodium glutamate (MSG) (flavor enhancer)	Citrates	
Orange oil (orange)		
Peppermint oil (peppermint)		

drink. To aid in this noble cause, chocolate flavoring may be added. Flavoring milk with chocolate, however, presents a problem. Chocolate derives its taste from cocoa, a substance that is insoluble in milk. If cocoa were added to the milk at the dairy, most of it would settle to the bottom by the time the container reached the consumer. This problem is solved by the addition of a hydrocolloid, a substance capable of producing a colloidal solution in water, called carrageenin. Strangely enough, it is derived from edible seaweed. When added to milk, carrageenin combines with the protein constituent casein to produce a thixotropic solution, which is capable of keeping the cocoa particles in suspension. (Thixotropy is a property of certain gels that causes them to liquify on shaking and to reform on standing.)

Margarine is a synthetic product specifically formulated to mimic the taste and appearance of the milk by-product butter. It is usually produced by emulsifying refined vegetable oils in cultured skim milk. Without the addition of a dye to color it, it looks just like lard. During the war years of the 1940s, a dye known as butter yellow was used to put the "sunshine" in the "low-priced spread."

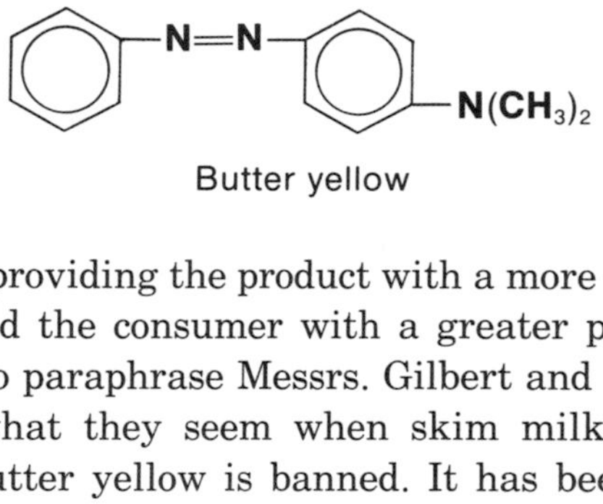

Butter yellow

This dye, besides providing the product with a more palatable appearance, also provided the consumer with a greater probability of contracting cancer. To paraphrase Messrs. Gilbert and Sullivan: "Things are indeed not what they seem when skim milk masquerades as cream." Today, butter yellow is banned. It has been replaced by β-carotene, the orange dye that gives carrots their color.

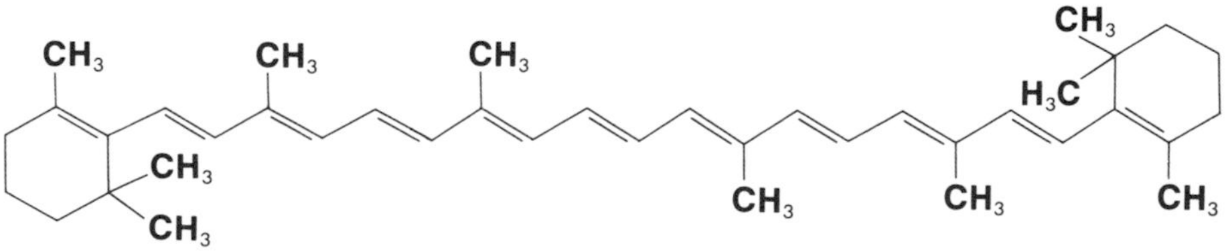

β-Carotene

What was learned from the butter-yellow experience was that any substance added to a food stuff must be evaluated not only for its desired effect but also for all other effects that it might have on the consumer.

Product Stability

When a shopper removes a food product from a supermarket shelf, he does so with the expectation that it will be as good as the glowing claims of its producer. If, however, the product has been shelf-bound for any length of time, the chance is much better than even that, unless the producer has taken steps to prevent it, the product has deteriorated in quality. The two main factors involved in the deterioration of foods are *oxidation* and *microbes*.

Going rancid is just a slow burn.

Atmospheric oxidation is the chief factor in causing unsaturated fats and oils to go rancid. Oxygen reacts with these fats to produce hydroperoxides.

$$\sim\sim\overset{\displaystyle H}{\overset{|}{C}}=\overset{\displaystyle H}{\overset{|}{C}}-\underset{\displaystyle H}{\underset{|}{\overset{\displaystyle H}{\overset{|}{C}}}}\sim\sim + O_2 \rightarrow \sim\sim\overset{\displaystyle H}{\overset{|}{C}}=\overset{\displaystyle H}{\overset{|}{C}}-\underset{\displaystyle OOH}{\underset{|}{\overset{\displaystyle H}{\overset{|}{C}}}}\sim\sim$$

Unsaturated portion of a fat molecule — Hydroperoxide of an unsaturated fat molecule

The reaction proceeds according to a free radical mechanism. In the first step atmospheric oxygen reacts with the fat to produce two free radicals.

$$\underset{\text{Fat}}{RH} + O_2 \rightarrow \underset{\text{Organic free radical}}{R\cdot} + \underset{\text{Peroxide free radical}}{HO_2\cdot}$$

The organic free radical that results then reacts with atmospheric oxygen to produce an organic peroxy free radical.

$$R\cdot + O_2 \rightarrow \underset{\text{Organic peroxy free radical}}{ROO\cdot}$$

It is this organic peroxy free radical which reacts with other fat molecules to produce a hydroperoxide and another free radical to carry on the chain.

$$ROO\cdot \quad + \quad RH \rightarrow \underset{\text{Hydroperoxide}}{ROOH} \quad + \quad \underset{\text{Organic free radical}}{R\cdot}$$

The hydroperoxy group, because it is capable of attacking double bonds, can produce a complex mixture of volatile aldehydes, ketones, and acids that will give fats or oils rancid tastes and odors.

To prevent the formation of these hydroperoxides, antioxidants are often added to fatty components of foods or to the packaging materials in which they are wrapped. Formulas of three of the more commonly used antioxidants appear in Figure 16–1. It is no accident that when these phenolic compounds are listed on packages of food such as breakfast cereals, they are named in a very unconventional manner. Food packagers feel that the public might have established adverse psychological associations between carbolic acid and phenols in general. Their use of unorthodox names is an attempt to remove the phenolic stigma from these compounds.

These phenols inhibit the oxidation of fats by terminating the free

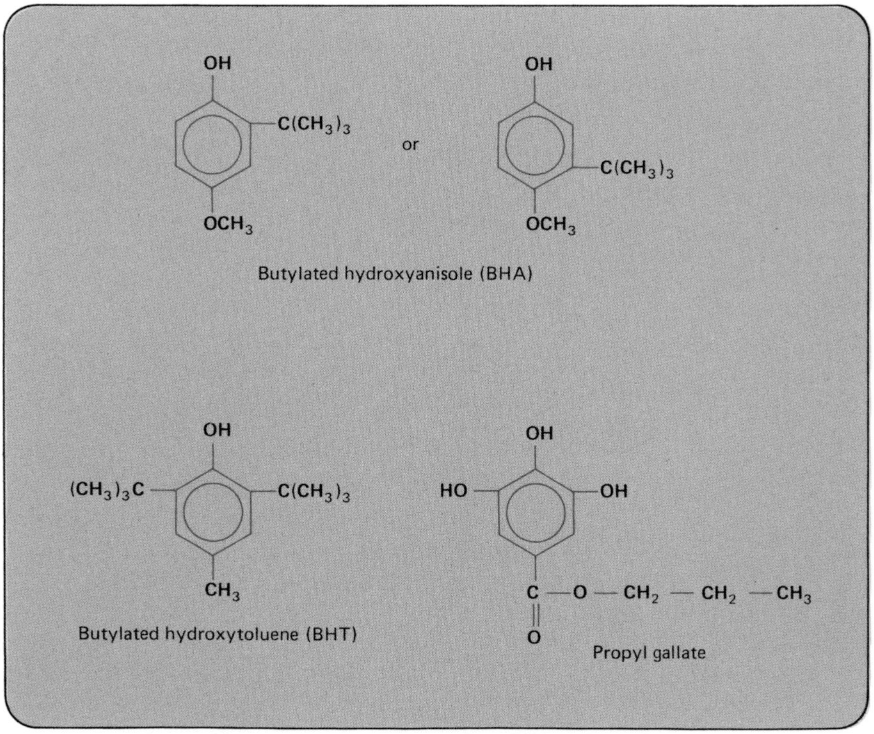

OH
C(CH3)3
OCH3
or
OH
C(CH3)3
OCH3
Butylated hydroxyanisole (BHA)

OH
(CH3)3C
C(CH3)3
CH3
Butylated hydroxytoluene (BHT)

OH
HO
OH
C—O—CH2—CH2—CH3
O
Propyl gallate

FIGURE 16–1 Phenolic Compounds Commonly Used as Antioxidants

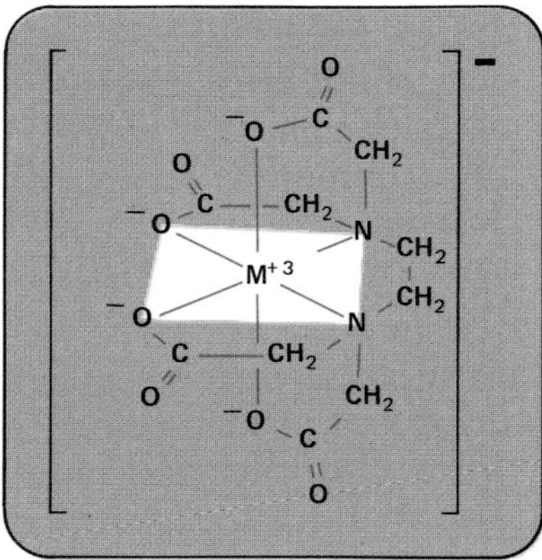

FIGURE 16–2 Ethylene Diaminetetraacetic Acid (EDTA) Chelated to a Metal

radical chain. This they do by forming a relatively stable, nonreactive radical when their phenolic hydrogen atom is abstracted by one of the free radicals derived from the fat.

$$R\cdot + (CH_3)_3C\text{–}C_6H_2(OH)(CH_3)\text{–}C(CH_3)_3 \rightarrow RH + (CH_3)_3C\text{–}C_6H_2(\dot{O})(CH_3)\text{–}C(CH_3)_3$$

Antioxidant — Stable antioxidant radical

Many oxidation reactions are catalyzed by the presence of metallic ions. These as a rule are a result of the reaction of food with the machinery used to process it. When catalytic ions such as iron and copper are removed, antioxidants can perform much more effectively. Their removal is usually accomplished through the process of *chelation* ("to be held in a claw" in Greek). Chelation involves the capture of the metallic ion by the "claws" of a complex organic molecule. Once captured, the metallic ion is so tightly held that it is no longer capable of acting as a catalyst. Figure 16–2 illustrates the structural formula and method of action of one of the chelating sequesterants, ethylene diaminetetraacetic acid (EDTA). This compound may be found in commercially prepared salad dressings, beverages, margarines, and crabmeat. When chelated, the metal ion is bound in each of the five five-membered rings that form the EDTA metal complex.

Don't bug me.

All natural foods that have not undergone processing contain live microbes. These little creatures, which are probably at least as hungry as the person who buys the product, account for much food spoilage. If the food is tasty, they immediately set about to consume it. Lacking proper sanitary conditions, however, they are forced to excrete where they eat. The toxins produced by these excrements are a major contributor to the spoilage of the food. Propionates, benzoates, nitrates, nitrites, and other chemical preservatives are added to foods to inhibit the growth of these organisms. Although the exact method of action of each of these is not completely known, most postulated mechanisms can be grouped into three possible categories.

1. Interference with cellular permeability. Propionic, benzoic, and salicylic acid are believed to coat the surface of the bacterial cell walls, producing a reduction in their permeability.
2. Interference with intracellular enzyme activity. Sorbic acid is known to interfere with the enzymes that allow molds to derive nourishment from the fatty acids present in cheese.
3. Interference with mold genetics. Chloramphenicol and streptomycin are known to combine with ribosomes to inhibit protein syntheses. As a result, it is felt that antibiotics of this type may be useful in disrupting mold growth.

Before leaving the topic of food additives, mention should be made of DDT, arsenic, and strontium 90. The presence of these components in food is neither intentional nor desirable. Their addition could more properly be described as Borgiaesque.

A Drug for All Reasons

Unable to sleep, eat, snore, or scratch? Take Mother Murphy's pink pills for blue ladies and purple guys. Suffering from lumbago, impetigo, chilblains, or fallen arches? Panacea Plus, the little effervescent tablet that acts twice as fast as the tablets of any competitor, will rid you of your ills. Feeling fine? There is a pill for that too.

The American Pharmaceutical Association, in its *Handbook of Non-Prescription Drugs*, lists almost one thousand drug preparations available for the asking price at any of the many pharmacies across the country. Television, radio, newspapers, and your back-fence neighbor are all eager to convince you that the use of these preparations will bring you to a state of fulfillment exceeded only by nirvana.

As late as the 1930s, preparations such as tonics, snake oils, and liniments that were good for either man or beast were the common form of medicine. Most were shotgun preparations in which the formulator compounded in one container as many diverse noxious materials as he could without seriously risking the life of the user. The logic behind these preparations was simple: the greater the number of ingredients, the greater the chance that one, by accident, would be of benefit to the afflicted customer. In recent years these all-purpose preparations have for the most part been replaced by particular drugs for particular needs. The consumer must now purchase a different drug for each condition. Lacking the formal medical training that would allow him to write prescriptions, his choice is restricted to "over-the-counter" drugs. This restriction does not, however, prevent him from attempting to cure

1. Pains, irritation, and fever
2. Coughs and colds
3. Disorders of the digestive tract
4. Restlessness and sleeplessness

"Oh, my achin' back!": Analgesics

In 1968 the American public spent over half a billion dollars on over-the-counter, internal analgesic (pain-relieving) products. More than 90 percent of this money was spent on aspirin or aspirin-containing drug mixtures. The applications varied from fever to hangovers and included a variety of other inflammatory conditions. The structures of aspirin and some of the more common analgesics used in conjunction with it are shown in Figure 16–3. The quantities of these compounds that are employed in the more common over-the-counter preparations are listed in Table 16–2. As formulated they are useful in the reduction of fever as well as the alleviation of pain.

The reason for the presence of caffeine in analgesic preparations is not entirely clear. Caffeine in itself is not an analgesic. Although claims have been made that it is helpful in certain types of migraine headaches, clinical evidence indicates that its presence can either increase or decrease the analgesic effect of individual drugs. Because of its effect on the central nervous system, cardiovascular system, and gastric secretions, its blanket inclusion in analgesics compounded for general application is less than a wise procedure.

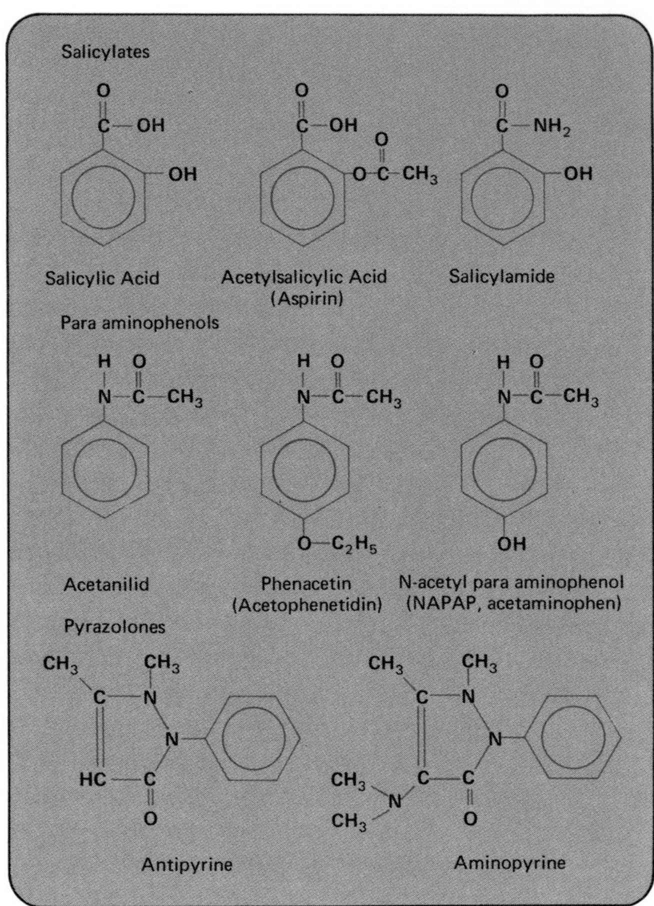

FIGURE 16–3 Analgesic Drugs

TABLE 16–2 Internal Analgesics Sold Without Prescription

Product	Aspirin	Phenacetin	Caffeine	Salicylamide	NAPAP (acetaminophen)
A	190 mg.	—	60 mg.	130 mg.	90 mg.
B	227 mg.	162 mg.	32 mg.	—	65 mg.
C	230 mg.	—	30 mg.	—	230 mg.
D	233 mg.	166 mg.	30 mg.	—	—
E	300 mg.	—	—	—	—
F	—	—	30 mg.	225 mg.	325 mg.
G	—	—	75 mg.	150 mg.	150 mg.
H	—	—	—	—	325 mg.

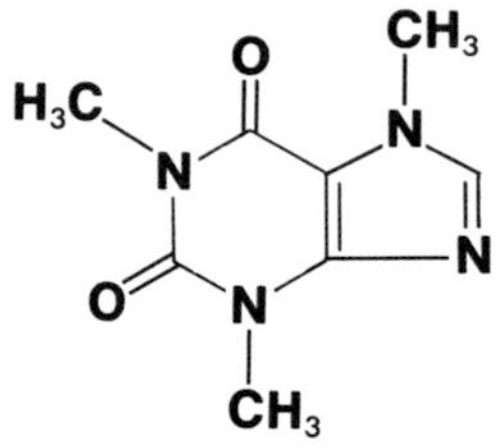

Caffeine

Topical analgesics usually depend upon methyl salicylate (oil of wintergreen), benzocaine, or camphor for their therapeutic actions. In addition to these components they may, depending on their intended application, also contain counterirritants and aromatic oils.

"Achoo!" "Gesundheit!": Cold remedies

The common cold and its associated symptoms have been a cause of misery to humanity since time began. Great-grandpa tried curing it with cupping. Grandma used dried raspberry tea. Uncle Edgar often resorted to the potent properties of "fire water" to ease his ailing. Few if any achieved success in less than a week. (Uncle Edgar really didn't care.) In recent years, modern chemotherapy has caused a shift away from these old-fashioned remedies. They have been replaced by decongestants to unplug stuffed noses and sinus cavities, antitussives to relieve coughing, analgesics for the relief of general discomforts, and antihistamines, which work on the assumption that a cold is not really a cold but an allergy. These medicants have allowed modern man to establish an average of seven days as the cure time for the common cold. Although these chemical agents have limited value as cures, they do to some degree eliminate the symptoms of a cold. Once the symptoms have been reduced, the patient feels better. His cold is no better than before the medication; he is just less conscious of it.

When a cold is not really a cold but an allergic reaction, antihistamines may be of value. Histamine is a compound produced by the body in response to exposure to an allergen (a substance capable of producing an allergic reaction).

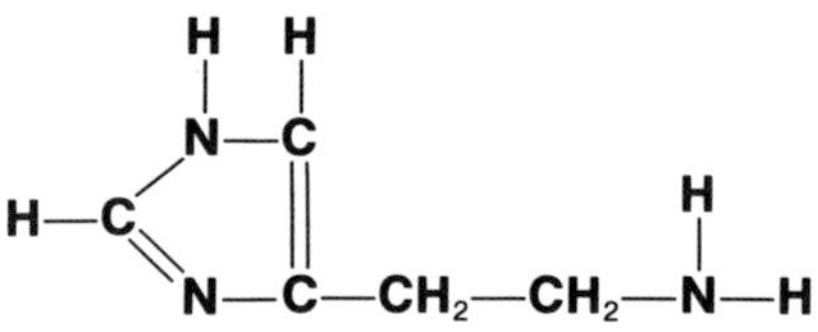

Histamine

Its presence produces a drop in blood pressure, breathing difficulties, headaches, and itching. Antihistamines are drugs that antagonize the action of histamine by competing with histamine for the receptor sites on nerve cells. Since histamine and antihistamines are all substituted ethyl amines, it is this portion of these molecules that is believed to be the active agent. The structures of three of the more common antihistamines are shown in Figure 16–4. Once the histamine has been neutralized by one of these drugs, allergic reactions often mistaken for colds are alleviated. This is essentially what happens when a person suffering from hay fever or some similar allergy takes an antihistamine preparation.

Congestion of the nasal passages is one of the more distressing symptoms of a cold. Its cause is probably the release of free histamine as a result of a virus infection. Decongestants are vasoconstrictors. They function by shrinking the blood vessels of the nose. Once these vessels are shrunk the nasal passages reopen to allow the air to flow again. The chemical formulas of decongestants are for the most part related to the hormone adrenaline (epinephrine), which allows mammals to react to stress situations. This is the hormone that gives a man the "get up and go" to run a four-minute mile, swim a record 100 meters, or chew out the driver who has just cut him off on the freeway. It, too,

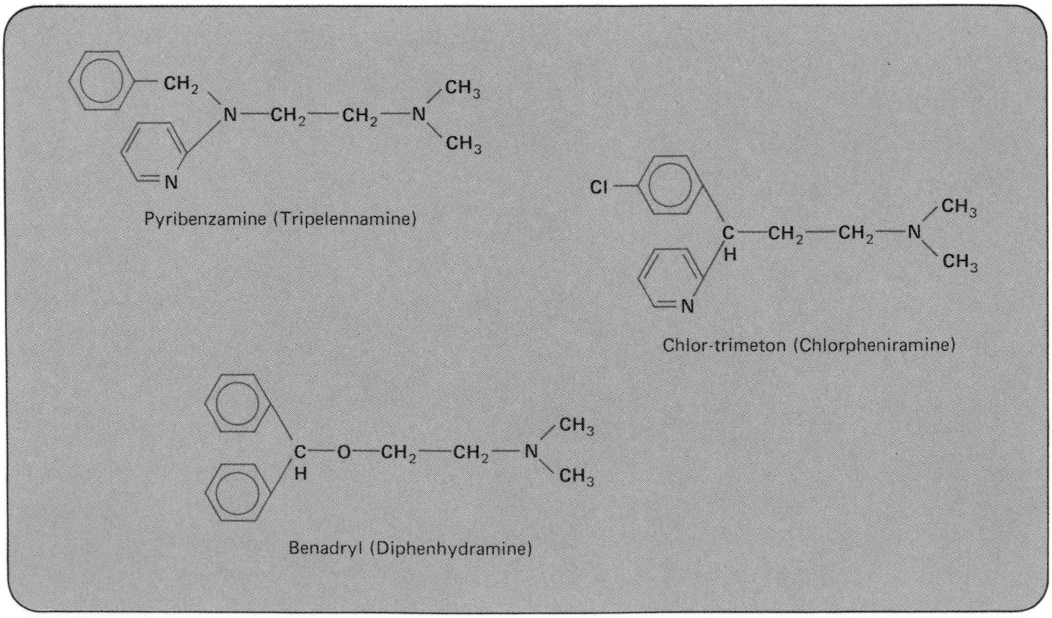

Pyribenzamine (Tripelennamine)

Chlor-trimeton (Chlorpheniramine)

Benadryl (Diphenhydramine)

FIGURE 16–4 Antihistamine Drugs

is a vasoconstrictor. Decongestants, because of their chemical similarity to this hormone, share many of its physiologic properties. Consequently, their use may produce a number of side effects above and beyond the clearing of one's nose.

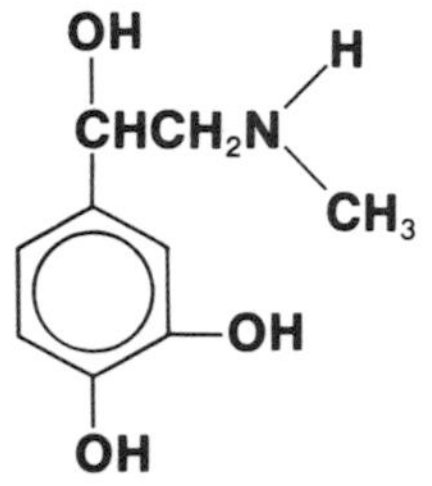

Adrenaline (Epinephrine)
(A vasoconstricting hormone)

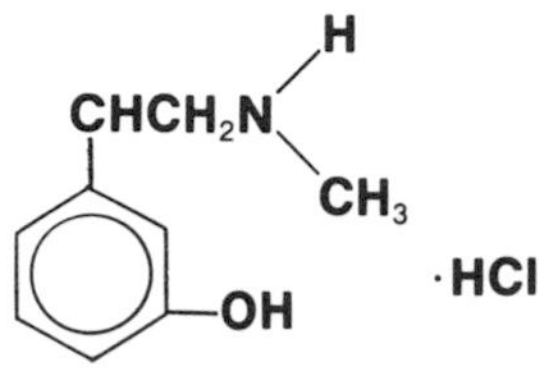

Phenylephrine hydrochloride
(A vasoconstricting decongestant)

Lemon and honey or hard candies for the control of coughs are for the most part passé. They have been replaced by drugs that raise the threshold of excitation of the nerves that control the cough reflex. These nerves are found in the medulla of the brain. The first drugs to be used for this purpose were the narcotics; however, because of their side effects, they have to a great degree been replaced by dextromethorphan (D-3-methoxy-n-methylmorphinan), a compound closely related to methadone. On a weight basis it is only half as potent as the narcotic codeine. Because of its minimal side effects, however, it—and to a lesser extent the naturally occurring, nonnarcotic alkaloid noscapine—has replaced codeine as the active ingredient in proprietary antitussives.

"If you want to grow big, eat!": Alkalizers

In affluent societies where overeating is common, heartburn and acid indigestion often result. The best cure for these disturbances is basic medicine. Basic medicine consists of neutralizing the excess acid of the stomach with one or more nontoxic alkalizers. Typical compounds employed for this purpose are sodium bicarbonate, $NaHCO_3$; calcium carbonate, $CaCO_3$; aluminum hydroxide, $Al(OH)_3$; magnesium oxide, MgO; magnesium hydroxide, $Mg(OH)_2$ (milk of magnesia); and magnesium trisilicate, $Mg_2Si_3O_8 \cdot 5\,H_2O$. Optimally, the use of these basic materials should increase the *p*H of the stomach to a value between 4 and 5. The adjustment of stomach *p*H to within this range is important for two reasons.

Below *p*H 4 the pepsinogen that is secreted by the cell walls of the

LYE
LYE
MAY BE A
POWERFUL
BASE BUT IT'S
A BAD NEWS
ANTACID

stomach is converted to the proteolytic enzyme pepsin. If the stomach has insufficient protein in it to digest the pepsin, the pepsin may just take a mind to digest the walls of the stomach, thus producing a peptic ulcer. This can be a very traumatic experience for any stomach.

Above *p*H 5 the stomach is stimulated to produce additional hydrochloric acid, which, if produced in sufficient quantity, may defeat the purpose for which the antacid was originally intended. The increased *p*H also results in a more rapid gastric emptying, which means that the antacid will remain in the stomach for a shorter time and will as a result be less effective.

Irrespective of which antacid is employed, its mode of action is always the neutralization of excess hydrochloric acid.

$$Mg(OH)_2 + HCl \rightarrow MgCl_2 + H_2O$$

(Milk of Magnesia)

In the arms of Morpheus: Sedatives and mind-boggling drugs

Most drugs, in addition to performing their intended function, also produce a number of side effects. Before beginning any drug treatment an appraisal must be made as to which is worse, the cure or the affliction. Similarly, the side effects of a drug used in one treatment may prove to be the desired effects in another.

The antihistamines, for example, besides combating the effects of histamine, have as a side effect the ability to cause drowsiness. For this reason they are dispensed with the precaution: "Do not drive or operate heavy machinery while under the influence of this drug." Because of their hypnotic effect, however, compounds related to the antihistamines benadryl and pyribenzamine are used as the active ingredients in a number of proprietary sleeping pills.

Before the discovery of these compounds, ethyl alcohol was the drug most commonly used. In the forms of wine, beer, and whiskey it has, without a doubt, relaxed more people than any other single drug. It accomplishes this by depressing the activity of the body's central nervous system. The degree of depression is dependent on the concentration of alcohol in the blood. It may vary from being "happy" at a concentration of 0.5 percent to being "totaled" at a concentration of 5.0 percent. (Figure 16–5.)

In an attempt to relieve his mind of the stresses put upon him by society, man has sometimes resorted to the use of psychedelic or hallucinogenic drugs. Delta-L-*trans*-tetrahydrocannabinol (the active ingredient in marijuana and hashish), mescaline (the major active ingredient in peyote), and D-lysergic acid diethylamide (LSD) are typical compounds of this type. (Figure 16–6.) The ability of these compounds to induce extraordinary sight and auditory experiences has caused many primitive cultures to assign mystical and religious qualities to them. Although each can be used as a hallucinogen, their histories are quite different.

Marijuana has been a valuable crop throughout history not merely because it is effective as a drug but more importantly because it is a source of hemp, the substance from which rope, burlap, and certain types of linen can be manufactured. The library of the Assyrian king Ashurbanipal, who reigned in the seventh century B.C., and other references confirm the knowledge and use of marijuana long before

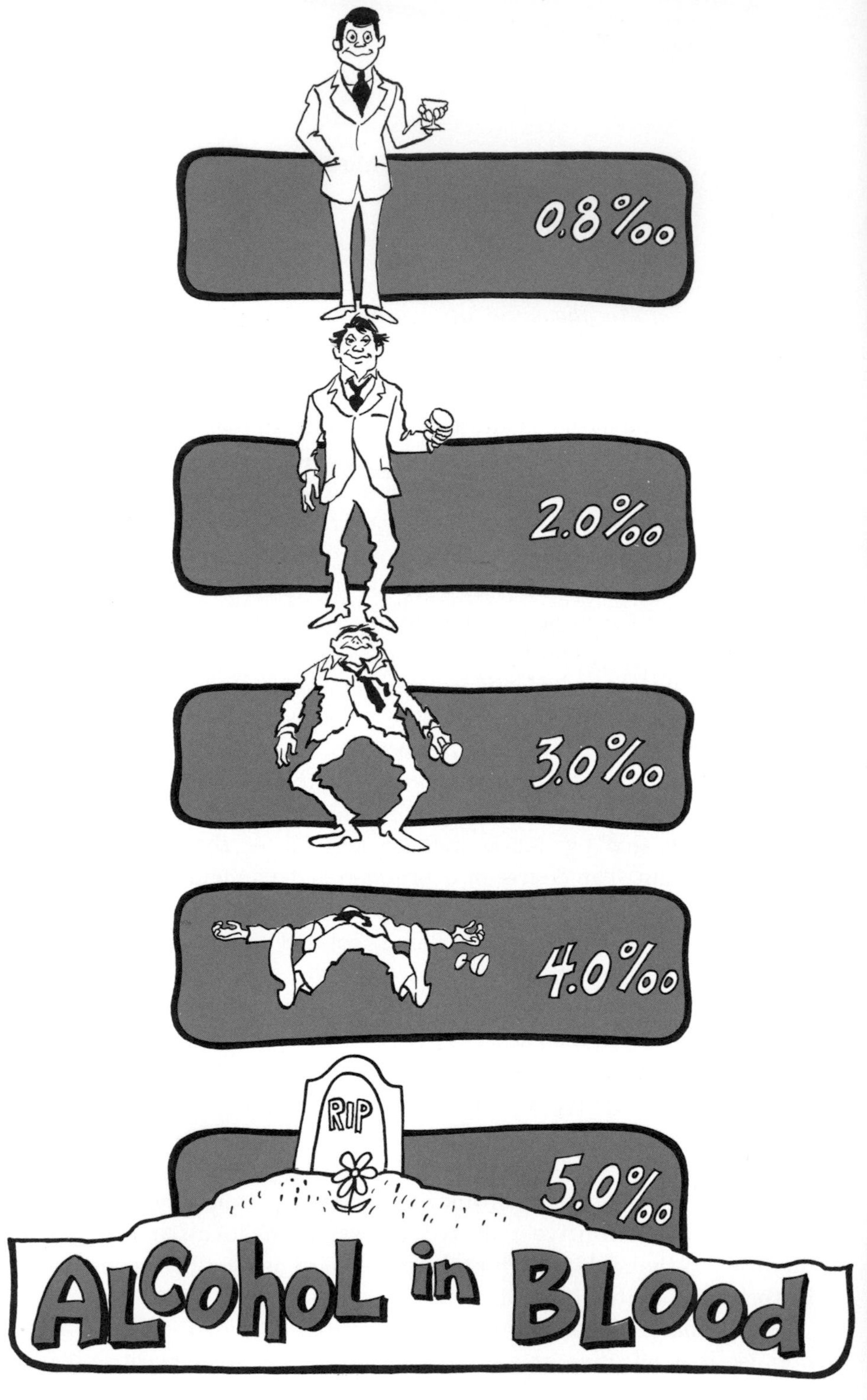

FIGURE

16–5 Effects of Concentrations of Alcohol in Blood

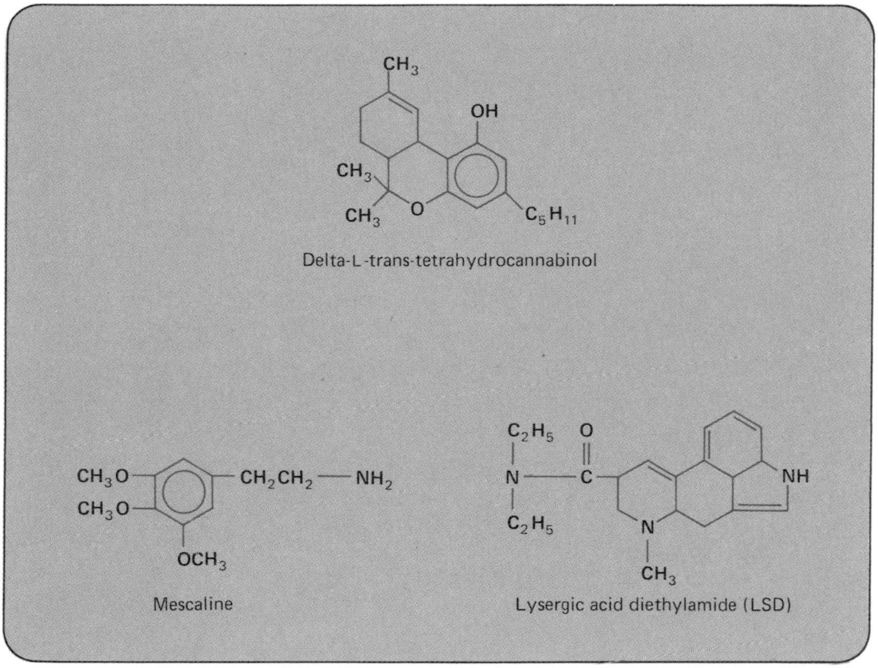

FIGURE 16–6 Hallucinogenic Drugs

the birth of Christ. Between 1850 and 1937 marijuana was widely used in the United States as a medicine. It was listed as "Extractum Cannabis" in the *U.S. Pharmacopia*, the *National Formulary*, and the *U.S. Dispensatory*. The U.S. Dispensatory touted its beneficial effects for all sorts of conditions ranging from gout to hydrophobia. (Present-day medical authorities would of course find its use for either of these purposes highly questionable.) To meet the demand for this "wonder drug," leading pharmaceutical houses such as Parke Davis, Squibb, Lilly, and Burroughs Wellcome marketed a liquid extract of marijuana for some eighty-odd years.

Marijuana is a sedating drug that when taken in sufficiently high quantities can cause intoxication. When it is taken at prescribed dosages for medicinal purposes, marijuana seldom produces hallucinations. As with alcohol, extrasensory experiences, when they are produced, are the result of overdosing.

Although in recent years newer drugs have for the most part supplanted marijuana's usefulness as a medicinal, many bird lovers insist that it would be difficult to find a bird seed component that would have as beneficial an effect on the quality of their bird's song.

The long-term effects of marijuana upon humans are not yet known.

Because of some similarities in molecular structure between its active ingredient and compounds that are part of the body's female-hormone metabolism, some chemists suggest that large dosages of marijuana might adversely affect these metabolic processes.

Mescaline can be extracted from a small, spineless cactus with a long, carrot-like root called peyote. The crown or button of the cactus has been used since pre-Columbian times as part of the religious practices of many Indian cultures, notably the Aztec. As a hallucinogen, it appears to alter the visual, auditory, and tactile senses so that they often become indistinguishable one from the other. Because LSD is usually considered superior to mescaline as a psychotherapeutic drug, the use of mescaline for medicinal purposes in the United States has for the most part been limited.

LSD was discovered for the second time on the afternoon of April 16, 1943, by the Swiss pharmaceutical chemist Dr. Albert Hoffmann. While working on a derivative of ergot, Dr. Hoffmann became dizzy and restless and experienced optical disturbances. The symptoms he displayed were caused by D-lysergic acid diethylamide, which he had synthesized that morning. Strangely enough, LSD had been discovered by Dr. Hoffmann and a collaborator, Dr. Stoll, five years previously, but because preliminary testing with experimental animals suggested no medical applications for the compound, it was shelved.

As a result of the ability of LSD to alter human perception even at extremely low concentrations, it was stockpiled by the U.S. military establishment as a possible weapon for disabling enemy forces. With the advent of more psychoactive chemicals, which are capable of even more bizarre effects, it was never put to this potential use. Its major licit use today is in the area of psychotherapy. A number of physicians claim that, when clinically administered, it is an effective agent in the treatment of many forms of advanced mental disorders.

Cosmetics and Toiletries

In the days of Adam and Eve, beauty may have been only skin deep. Today, however, its depth has been increased appreciably by layers of cosmetics. The application of most cosmetic and toiletry preparations can in general be classified into one of three categories:

1. Improved appearance of the hair
2. Improved appearance of the teeth

3. Improved appearance of the skin
4. Improved social acceptability

Of the three, the most time, effort, and expense is normally expended on increasing the attractiveness of hair where it is wanted and on removing it from where it is not.

The crowning glory

Hair can be shampooed, brushed, bleached, conditioned, colored, curled, and straightened to enhance its appearance. In 1969 alone the American consumer spent 1.5 billion dollars on preparations designed to improve the appearance of his hair. Today, through the marvels of chemistry, even people who are born redheads or brunettes can personally determine whether blonds have more fun. All that is needed is a bottle of hydrogen peroxide and a pinch of ammonia water to increase the *p*H and oxidizing power of the peroxide. This magic elixir when liberally applied to the hair will oxidize the brown-black melanin pigment to a stable colorless compound. In so doing the color of the hair will be changed from its natural red or brown to a less intensely colored blond.

The principles involved in dying hair are very similar to those employed in dying textiles. Two types of dye can be employed for this transformation: those that are water soluble only coat the hair and are thus temporary; those that are soluble in organic solvents penetrate the hair fibers and are permanent. Of the two, the latter is by far the more interesting. They are usually organometallic complexes (organic complexes containing metal ions) of chromium or cobalt and derivatives of phenylenediamine. Once these complexes have permeated the body of the hair, the amine groups are oxidized either by hydrogen peroxide or air to produce more intensely colored nitro compounds. It is these nitro compounds that darken the hair.

$$\underset{\text{Amine}}{-N\begin{matrix}H\\H\end{matrix}} + 3\,H_2O_2 \xrightarrow{\text{Oxidation}} \underset{\text{Nitro compound}}{-N\begin{matrix}O\\O\end{matrix}} + 4\,H_2O$$

The oxidized compounds are held permanently within the hair shaft either by hydrogen bonds or the decreased solubility of the nitro complex as compared to the original amine complex. They will not wash out. When applied skillfully they very effectively mimic natural hair coloration.

Just as the color of hair can be changed through chemistry so too can its configuration . The chemical setting of hair in a given configuration, whether curly or straight, is accomplished by reorienting the polymeric proteins that comprise the hair fibers. In the normal state these protein molecules are joined to one another by means of linking disulfide (**—S—S—**) bonds. The net effect of these bridges is to arrange adjacent component polymers into longitudinal bundles. (Figure 16–7.) If these bundles are attacked by basic thioglycollate or other suitable reducing solutions the sulfur bridges are broken, allowing one molecule to slip past the other. (Figure 16–8.) In this plastic state, the hair can be molded by rollers or what-have-you to the desired fashion arrangement. The molding process places **—SH** groups that were formerly joined together in a single disulfide bond opposite different **—SH** groups. (Figure 16–9.) To establish this new arrangement so that it does not change when the curlers are removed, the modified **—SH** group configuration is fixed by the addition of an oxidizing agent such as hydrogen peroxide. The hydrogen peroxide converts adjacent **—SH** groups into the disulfide bonds of the new configuration (Figure 16–10), and in addition neutralizes any excess reducing agent.

$$2\,HOOCCH_2SH + H_2O_2 \rightarrow HOOCCH_2S{-}SCH_2COOH + 2\,H_2O$$

Reducing agent (Thioglycollic acid) — Oxidizing agent — Disulfide compound

The complete neutralization of excess reducing agent is of more than passing interest. If it is not inactivated, it acts as a depilatory, causing the hair to become so plastic that it can be rubbed off the skin. Such was the unfortunate experience of a number of women who experimented with the novel "cold-wave, home-permanent" preparations when they first became consumer items during the 1940s.

Depilatories used for the removal of unwanted hair are in many ways identical to permanent-waving and hair-straightening preparations. The only important exception is the higher concentration of reducing agent in the depilatory. When the hair attains a plastic condition, it can easily be removed.

Besides controlling hair through chemical modification, sprays and coatings may be used to straightjacket each fiber in place. This is accomplished by applying some plastic film to the hair, which after

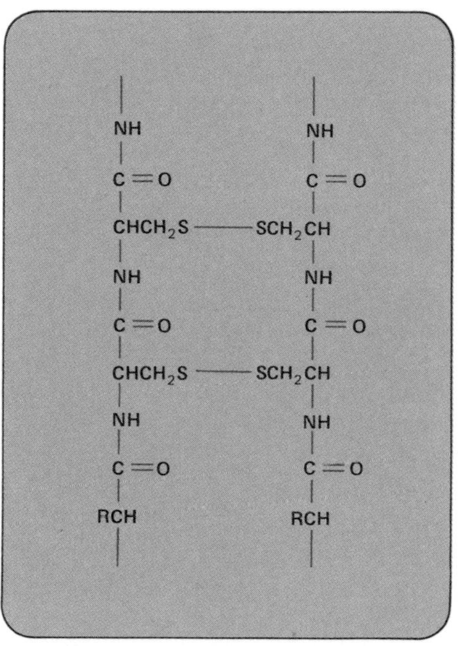

FIGURE 16–7 Longitudinal Bundle Configuration of Component Hair Polymers

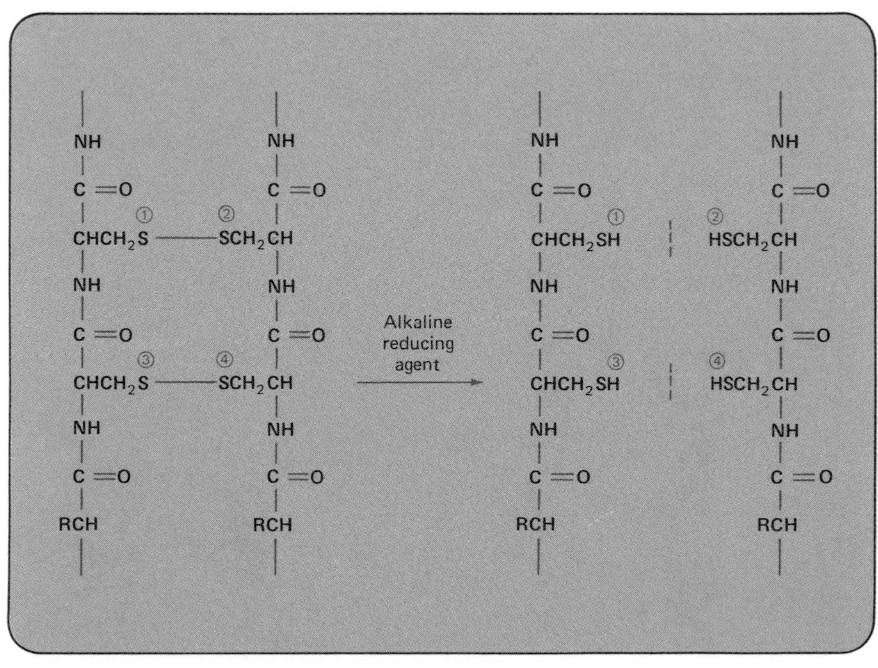

FIGURE 16–8 Reduction and Cleavage of —S—S— Bonds

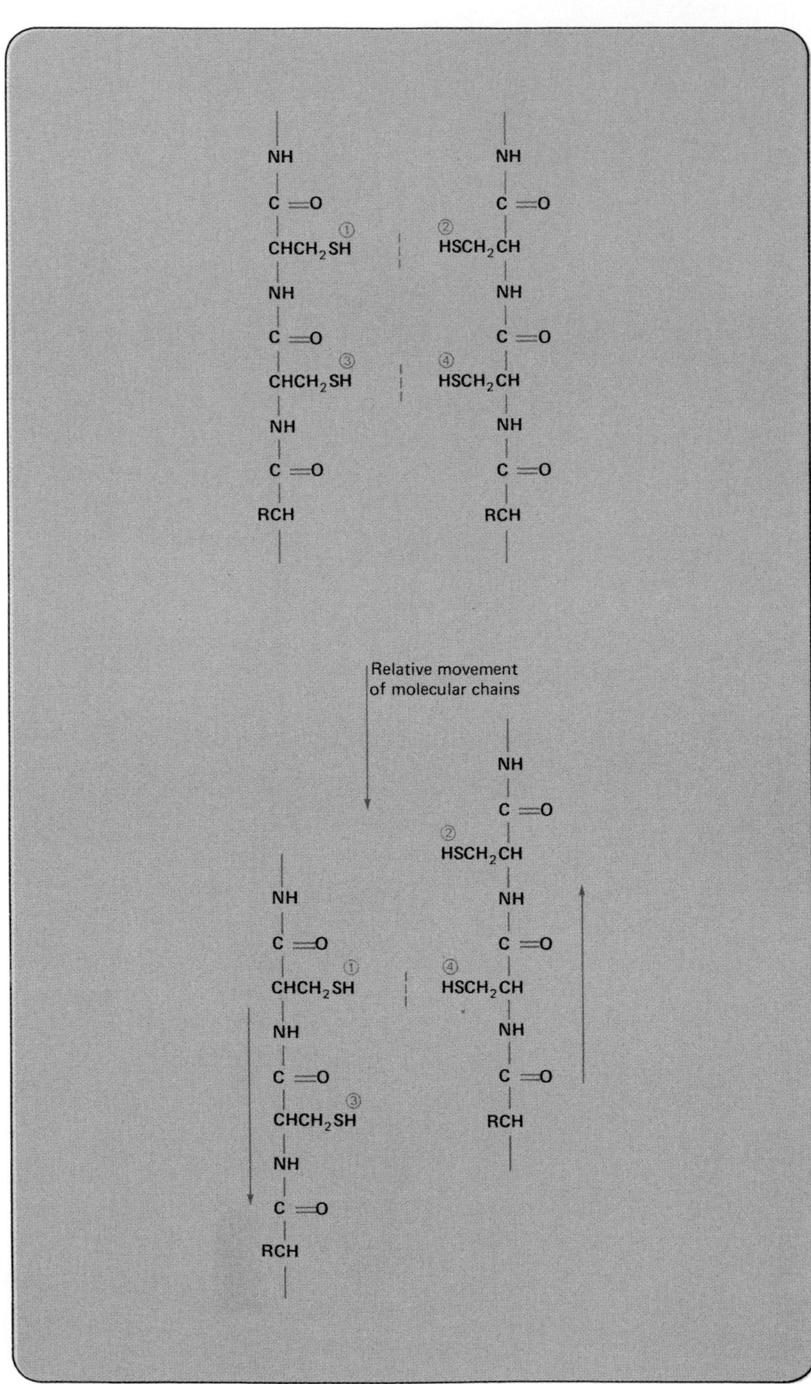

FIGURE

16–9 Relative Movement of Molecular Chains

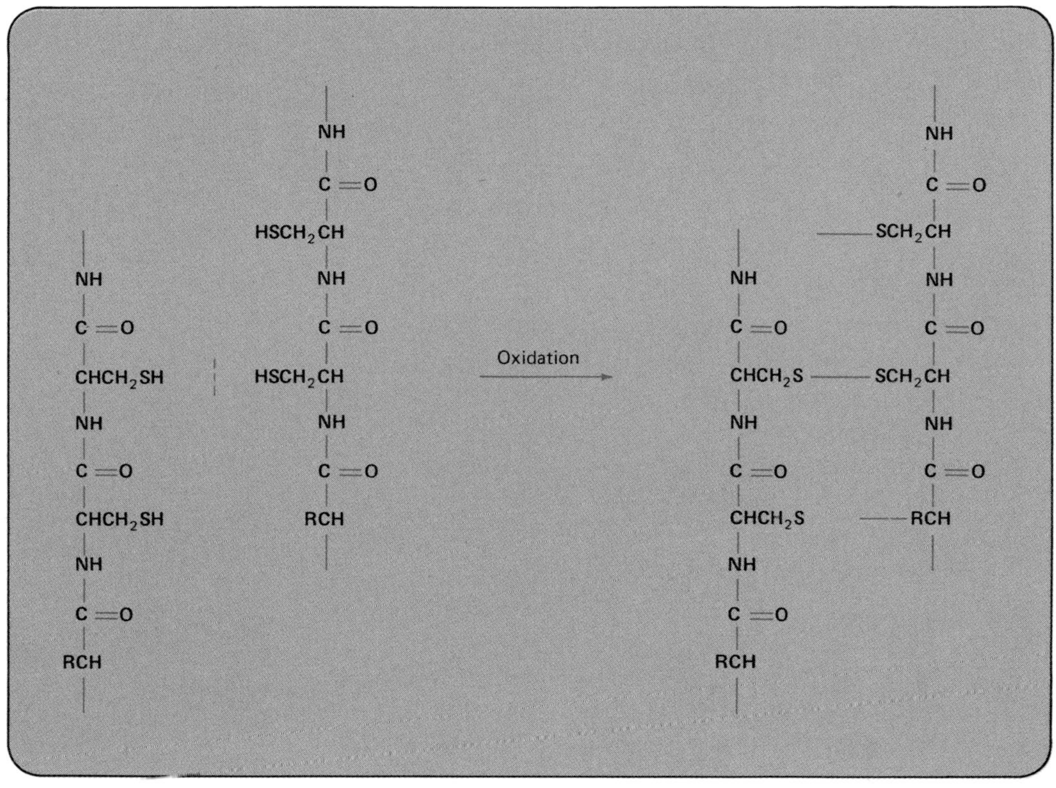

16–10 Oxidation of —SH Bonds to Establish New Configuration **FIGURE**

hardening maintains the proper orientation. Materials used as coating agents vary from acrylic acid resins and shellac, both of which can be used to paint walls, to copolymers (combined polymers) of polyvinyl pyrrolidone (PVP).

Polyvinylpyrrolidone (PVP)

PVP is a synthetic resin that forms a smooth transparent film. By itself, however, it becomes brittle when dry and forms flakes resembling dandruff. The formation of copolymers of PVP with vinyl alcohol or the addition of plasticizers such as isopropyl laurate can to a great degree control this problem.

After the hair has been dyed and molded, shampooing provides the crowning touch to the well-groomed head. Special shampoos are sold for dry hair, normal hair, oily hair, tinted hair, bleached hair, and so on. Each is formulated to have in varying degrees the following three properties:

1. Foaming action, to remove gross filth (and to give the user a psychological boost)
2. Cleansing action, to remove body oils and perspiration residues
3. Residual effects, to provide luster and body

Most modern shampoos contain a combination of soaps and synthetic detergents as cleansing agents. Because these agents are composed of long-chained molecules that possess both fat-soluble and water-soluble ends, they are able to remove oily residues that cling to the hair and that would normally be insoluble in the wash water. (Figure 16–11.)

Soaps are produced by hydrolyzing naturally occurring fats and oils in strongly basic solutions. The process is known as saponification.

$$\begin{array}{l} CH_3(CH_2)_{16}COO—CH_2 \\ \qquad\qquad\qquad\qquad\quad | \\ CH_3(CH_2)_{16}COO—CH \\ \qquad\qquad\qquad\qquad\quad | \\ CH_3(CH_2)_{16}COO—CH_2 \end{array} + 3\,NaOH \rightarrow 3\,CH_3(CH_2)_{16}COO^-Na^+ + \begin{array}{l} HO—CH_2 \\ \qquad\quad | \\ HO—CH \\ \qquad\quad | \\ HO—CH_2 \end{array}$$

Glycerol tristearate (Fat) | Sodium hydroxide (Base) | Sodium stearate (Soap) | Glycerol

If olive oil is used as the source of the fatty acid castile soaps are produced. Olive oil has the distinction of having a higher percentage of oleic acid, $CH_3(CH_2)CHCH(CH_2)_7COOH$, than most fat sources. Whether or not this should be a significant factor in formulating a shampoo is a moot point to say the least.

Detergents used in shampoos are usually condensation products of amines and fatty acids.

$$HN(CH_2CH_2OH)_2 + CH_3(CH_2)_{10}COOH \rightarrow CH_3(CH_2)_{10}\overset{\overset{\large O}{\|}}{C}—\overset{\overset{\large H}{|}}{N}(CH_2CH_2OH)_2$$

Diethanol amine (Amine) | Lauric acid (Fatty acid) | Lauric diethanol amide (Detergent)

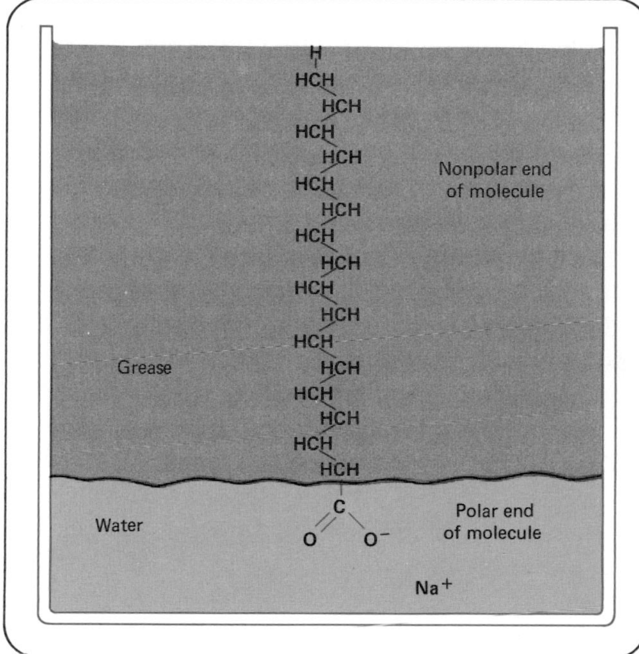

FIGURE 16–11 The Dual Solubility of a Detergent Molecule in Polar as Well as Nonpolar Substances

Although soaps alone perform well as shampoos in soft water, when used in hard water (water containing calcium and magnesium ions) they leave a dulling film of insoluble calcium and magnesium salts on the hair. To minimize this, EDTA and detergents, which do not produce insoluble calcium and magnesium salts, are often added to shampoos. The EDTA functions to tie up the free calcium and magnesium ions so that they cannot react with the soap to form insoluble salts.

Contrary to popular opinion, a shampoo that produces ultraclean hair is not a desirable product. Devoid of all its surface oils, hair becomes unmanageable. The complaint "I've just washed my hair, and I can't do a thing with it" is the result of such a condition. To counter this difficulty, manufacturers invariably formulate their preparations so that a conditioning residue remains on the hair after most of the shampoo is rinsed off. It is this quality that is modified when companies compound different shampoos for dry and oily hair. The function of shampoos is therefore twofold: to remove a coating of dirt from the hair and to replace it with a more desirable one.

Say cheese—smile!

Unlike most cosmetic preparations, those that are generally used to improve the appearance of the teeth also perform vital health functions. Regular brushing of teeth reduces the incidence of dental caries (tooth decay), helps to maintain healthy perdontium (tissue that supports the teeth), and aids in the reduction of mouth odors. The brushing of teeth removes fermenting foods. If not removed, the enzymatic bacterial decomposition of the carbohydrates in these foods produces organic acids that convert tooth enamel, the highly insoluble tooth hydroxyapatite, $3\,Ca_3(PO_4)_2 \cdot Ca(OH)_2$, into a soluble calcium salt. The action of bacterial-produced proteolytic enzymes (enzymes that hasten the hydrolysis of proteins) also destroys the organic matrix of tooth enamel. Net result: tooth decay.

The interest in the maintenance and appearance of teeth is not a modern one. Ancient Greeks, Romans, Hebrews, and Buddhist monks all employed toothpicks, chewing sticks, or some similar device. The Aztecs even used toothpaste. In the Aztec Herbal the following advice is given:

"Rough teeth are to be diligently made smooth; the dirt being removed, they are to be rubbed with white ashes in white honey, using a small cloth, whereby elegant cleanness and real brightness will stay."

The compositions of modern toothpastes are a bit more complex. In the main, they can best be described by the mnemonic BASHFUL.

Binding agents are used in toothpastes to prevent the solids from separating from the liquids. For the most part they are gums, hydrocolloids, or clays. Typical examples are karaya gum, tragacanth, carragheen, methylcellulose, bentonite.

Abrasives are finely divided solid materials with sufficiently high degrees of hardness that they can be used to polish away the bacterial plaque surrounding the teeth. It is essential in this polishing that the abrasive not be too course lest it scratch the tooth enamel and do more harm than good. Typical polishing agents are calcium carbonate, $CaCO_3$; dicalcium phosphate dihydrate, $CaHPO_4 \cdot 2\,H_2O$; hydrated alumina, $Al(OH)_3$; calcium pyrophosphate, $Ca_2P_2O_7$.

Sudsers are soaps or detergents that function as surface active agents (allow the toothpaste to "wet" the tooth). Aside from low toxicity, they do not differ appreciably from laundry products. Typical compounds are sodium lauryl sulfate, $CH_3(CH_2)_{10}CH_2OSO_3Na$; sodium lauryl sulfoacetate, $CH_3(CH_2)_{11}OCOCH_2SO_3Na$; sodium lauroyl sarcosinate, $CH_3(CH_2)_{10}CON(CH_3)CH_2COONa$.

Humectants function to prevent the dentifrices from drying out and hardening when exposed to air. Compounds suitable for this purpose are glycerine, $CH_2OHCHOHCH_2OH$; propylene glycol, $CH_3CHOHCH_2OH$; sorbitol, $CH_2OH—CHOH—CHOH—CHOH—CHOH—CH_2OH$.

Flavors are added both for the obvious reason and for the psychological effect of a "clean-mouth, tingling sensation." Spearmint, peppermint, wintergreen, sassafras, and cinnamon are among the more popular flavors.

Unique additives is a term that encompasses a diversity of chemicals. Sodium dehydroacetate, sodium salt of 3-acetyl-6-methyl-1,2-pyran-2,4(3H)-dione, may be added to reduce enzyme action; stannous fluoride, SnF_2, may be added to reduce hydroxyapatite solubility; sodium perborate, $NaBO_3 \cdot 4\,H_2O$, may be added to enhance whitening.

Liquid component of toothpastes is for the most part water, although mineral oil is sometimes added to minimize hardening and to assist in extrusion from the tube.

Body paint

In the category of preparations designed to beautify the skin, there are a wide variety of relatively nontoxic paints and coating products. They vary from pleasant smelling powders composed mainly of talc, chalk, zinc oxide, and a bit of coloring to lipsticks, eye makeup, and nail lacquers having high-intensity coloration.

Lipsticks and eye makeups are essentially mixtures of oils, fats, and waxes to which finely dispersed dyes and pigments of appropriate colors have been added. Brom fluorescein dyes such as tetrabromofluorescein are often the coloring matter in lipsticks. Lamp black and oxides of iron and chromium are used for mascara and eye liners and shadow.

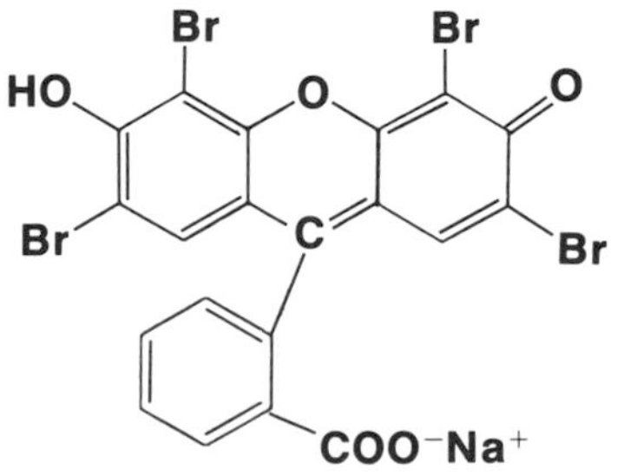

Tetrabromofluorescein
(Sodium salt)

Nail lacquers are varnishes containing nitrocellulose or a similar polymer to which coloring matter, plasticizers, and resins have been added. The plasticizers prevent the nail polish from becoming too brittle and cracking when the nail is flexed. The resin provides adhering properties so the polish doesn't flake off.

Most skunks suffer from body odor.

Deodorants, antiperspirants, and perfumes are products designed to make the user socially acceptable. Depending upon the conditions of temperature, humidity, and physical and mental activity, the average adult produces between one and three pints of perspiration per day. These secretions do not of themselves normally have an objectionable odor. What is objectionable, though, are the odors resulting from the chemical and microbial degradation of these secretions. Decomposed sweat owes its aroma to organic acids of low molecular weight (caproic, caprylic, isovaleric, and butyric), mercaptans, indoles, amines, ammonia and hydrogen sulfide. The formation of these compounds can be prevented if the perspiration is kept bacteria free. Commercial deodorants operate by one or a combination of any of the following three mechanisms.

1. They may, by acting as antiperspirants, reduce the volume of perspiration and as a consequence also the amount of perspiration decomposition products on the skin. Antiperspirants are usually salts of either zinc or aluminum that restrict the flow of perspiration from the sweat glands when they are applied to the skin.

2. They may, by means of their antibacterial activity, reduce the population of bacteria on the skin so that the amount of noxious products produced is significantly reduced. Among the compounds that have been used for this purpose are hexachlorophene, quaternary ammonium compounds, and antibiotics. Each of these has a history of undesirable side effects.

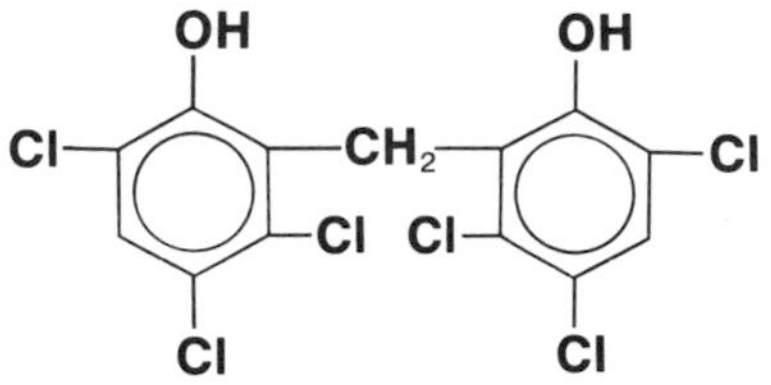

2,2′-Dihydroxy-3,3′,5,5′,6,6′-hexachlorodiphenylmethane
(Hexachlorophene)

Hexachlorophene has been known to produce lesions of the nervous system. Quaternary ammonium compounds produce skin irritations, and antibiotics cause sensitizing reactions (sensitizing reactions do not produce adverse effects on initial exposure but may after subsequent exposure. The initial exposure makes the subject sensitive.) With the ban against the use of hexachlorophene in 1972, other chemicals have been sought to replace it. Some of the substitutes that have been tried are tetramethylthiuramdisulfide, 8-hydroxyquinoline, 3,4,4-trichlorocarbanalide. Unfortunately, these compounds, too, have histories of sensitization.

3. They may because of their own pleasant odor mask the odor of the perspiration decomposition products. There are over five thousand commercially available natural and synthetic materials with sufficiently pleasant odors and low volatility to classify as possible perfume compounds. In choosing a perfume it is more important that it provide the wearer with a pleasant fragrance than that it smell good in the bottle. Because of individual body chemistries the same perfume on two different people may smell entirely different. Depending upon the intended fragrance, perfumes may be formulated from plant and flower extracts, ketones derived from Asian musk deers, secretions of Ethiopian civet cats, or ambergris from sick whales. These fragrances are all complex mixtures of a number of compounds. Their source is unimportant as long as the final product smells good.

During the Middle Ages, when baths in Northern Europe were extremely rare, social intercourse at the royal courts was possible thanks mainly to perfumes. So highly valued were they that many an explorer traveled thousands of miles in the hope that he might return with a fortune in fragrances from the East.

REVIEW

1. Additives are used to increase the saleable value of food products. They may be designed either to improve the original product or to prevent it from spoiling.
2. Antioxidants impede the decomposition of a product by terminating free radical oxidation chains.
3. Most antioxidants are phenolic compounds.
4. Chelating agents are useful in controlling reactions that are catalyzed by metallic ions.
5. Bacterial decomposition of foods can be controlled by chemical additives that interfere with the normal life processes of bacteria.

6. Allergic discomforts are caused by reaction to histamine. They may be alleviated through the use of antihistaminic drugs.
7. Antacids are mild bases that operate by neutralizing excess stomach acids.
8. Compounds capable of reducing disulfide bonds are used either to style or to remove hair.
9. Soaps are produced by the saponification of fats.
10. Soaps and detergents are useful as cleansing agents because the ends of these molecules are soluble in polar and nonpolar materials.
11. Toothpastes are complex mixtures of chemicals designed to improve the appearance and decrease the decay of teeth.
12. Deodorants prevent offensive body odor by reducing the skin concentration of perspiration breakdown products and masking existing odors with more pleasant odors.

REFLEXIVE QUIZ

1. The production of hydroperoxides in unsaturated fats occurs by a ________________ mechanism.
2. EDTA sequesters catalytic metal ions by the process of ____________.
3. The most popular drug for the relief of pain and fever is ____________.
4. Because of their structural similarity to histamine, the antihistamines function by competing with ________________ for receptor sites on nerve cells.
5. Nasal decongestants, like the hormone ________________, function by acting as vasoconstrictors.
6. Some ____________________ are examples of the use of the side effects of one drug to alleviate the symptoms of an entirely different condition.
7. Permanent-wave solutions may be neutralized by ____________________.
8. Oxidation of the ____________ structure in penetrating hair dyes produces more intensely colored nitro compounds.
9. The process of hydrolyzing fats with strongly basic solutions is known as ____________________.
10. The three processes by which commercial deodorants function are are (*a*) odor masking, (*b*) ________________, and (*c*) ________________.

Answers: (1) Free radical [335–338, 359]. (2) Chelation [337, 359]. (3) Aspirin [338–341]. (4) Histamine [341–343]. (5) Adrenalin [342–343]. (6) Sleeping pills [345]. (7) Oxidizing agents (hydrogen peroxide) [349–353]. (8) Amine [349]. (9) Saponification [354, 360]. (10) Inhibition of bacterial growth, reduction of perspiration [358–360].

SUPPLEMENTARY READINGS

1. "Consumerism: A Special Report on the Drug Industry." *Chemical and Engineering News* (26 July 1971), p. 24.
2. L. Grinspoon. "Marijuana." *Scientific American* (December 1969), p. 17.
3. R. G. Harry. *Modern Cosmetology.* New York: Chemical Publishing Co., 1947.
4. T. H. Mikuriya. "Marijuana in Medicine, Past, Present, and Future." *California Medicine* (January 1969).
5. J. S. Slotkin. *The Peyote Religion.* Glencoe, Ill.: The Free Press, 1956.
6. H. O. J. Collier. "Aspirin." *Scientific American* (May 1963), p. 96.
7. J. Y. Bogue. "Drugs of the Future." *Journal of Chemical Education* 46 (1969): 468.
8. G. O. Kermode. "Food Additives." *Scientific American* (March 1972), p. 15.
9. M. Sasman. "*Cannabis indica* in Pharmaceuticals." *Journal of the Medical Society of New Jersey* 35 (1938): 51.

Section 17

Chemical Pollution

Section 17

Chemical Pollution

Life and Breath

In order to avoid the unpleasantries of suffocation, a full-grown man requires about 8500 liters of "pure air" each day. This is roughly half the volume of an 8 × 10-foot room. If this requirement is not met, one may be assured that his life expectancy will be markedly affected. "Pure air" is composed mainly of nitrogen (78 percent), oxygen (21 percent), and, in decreasing amounts, traces of helium, krypton, xenon, methane, hydrogen, oxides of nitrogen, and ozone. In addition, it may, depending on the relative humidity, contain anywhere from 0.1 to 5 percent water vapor. Water vapor, however, because of its high variability, is not normally included in the basic listing of component percentages.

In this day and age, "pure air" is a luxury item. Air that is available to support life processes is, at best, minimally polluted. Pollutants are substances that, by virtue of either their nature or abnormally high concentrations, are less than desirable additions to the basic mixture of gases known as "pure air." They may be the products of geologic cataclysms, natural life processes, or the result of man's industriousness.

The eruption of the volcano Krakatoa near Java in 1883 set free into the atmosphere such a gigantic quantity of particulate pollutants that for several years after the explosion spectacular sunsets were seen around the world. In England, on the other side of the globe, artists were so impressed by the splendor of these sunsets that they dubbed them "Chelsea Sunsets."

The Great Smoky Mountains dividing North Carolina and Tennessee appear smoky not because of an inordinately large number of forest fires but because the terpenes released into the atmosphere by local plant life create a haze. (Terpenes are organic compounds whose molecular formula is based on a $\mathbf{C_{10}H_{16}}$ unit.) A typical terpene found in pine needles and citrus fruit is limonene.

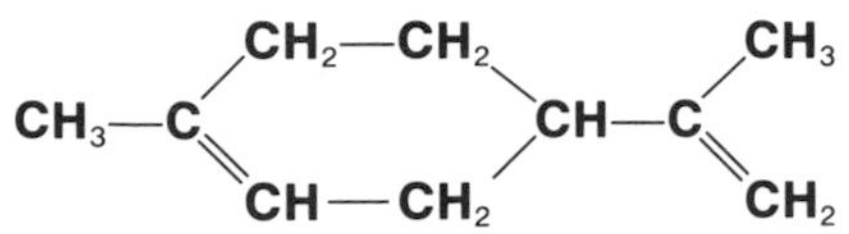

```
          CH2—CH2              CH3
         /       \            /
   CH3—C          CH—C
        \\       /     \\
          CH—CH2          CH2
```

Limonene

Fumes from the fires used to warm and feed man as well as from the machines he has developed to produce a "better world" are the third major type of pollution. Of these three sources of pollution only the man-made one is man-controllable. Dissuading a volcano intent upon blowing its top is an extremely difficult task. Similarly, a pine forest doing its own thing cannot be easily made to cease and desist.

Unlike natural pollutants, those produced by man create a disproportionately large environmental disturbance. This is because (1) man-made pollutants are produced at maximum concentrations in populated areas; (2) man-made pollutants have been produced for only a very short period in the long history of the earth. As a result, insufficient time has elapsed to allow the earth to generate adaptive mechanisms. When such adaptive detoxifying mechanisms do in fact exist, they more often than not are overtaxed by the concentrations at which these artificial pollutants are produced.

Reducing the impact of man-made pollutants to simplest terms, it may be said that before man began to put natural forces to his own uses, the natural polluting and detoxifying mechanisms of this planet had reached a state of relative equilibrium. Through his activities, man disturbed this natural balance, both by the production of inordinately high concentrations of pollutants and by the introduction of compounds for which no place exists in the *natural* scheme of things. Freons, for example, the compounds used as propellants in aerosol preparations, were unknown in nature before man. Their release into the atmosphere presents no hazard at this time; however, because there is no natural system for their degradation, they could with increasing time and concentration become objectionable pollutants.

Note. The classification of a substance as a pollutant is usually the result of the negative effects it may have upon man. Such classification is admittedly arbitrary, since one creature's pollutants may be another's meat. Substances that reduce the effectiveness of air to act as a respiratory gas are presently considered pollutants. However, this has not always been the case.

When the Earth Was Young

Before primitive life appeared, the atmosphere surrounding the earth was composed of methane, ammonia, water, and hydrogen. It contained little, if any, oxygen. The high-energy radiation of the sun and the natural radioactivity of the earth at that time reacted with molecules of these atmospheric substances to break and rearrange their

bonds, causing some of them to be converted to sugars and amino acids.

$$\mathbf{CH_3} + \mathbf{NH_3} + \mathbf{H_2O} + \mathbf{H_2} \xrightarrow[\text{energy}]{\text{Radiation}} \text{Sugars} + \text{Amino acids}$$

Fermentation of the resulting sugars by primitive life forms provided them with the energy necessary to sustain their life processes.

$$\text{Sugar} \xrightarrow{\text{Fermentation}} \text{Alcohol} + \mathbf{CO_2} + \text{Energy}$$

As the number of life forms increased, the amount of available fermentable materials was depleted. If the new phenomenon called life was to survive, a means for supplying additional fermentable material had to be developed. Mustering all their resources in an effort to alleviate their food crisis, primitive organisms developed the process of *photosynthesis*. Using the energy of sunlight to break the interatomic bonds of carbon dioxide and water, they caused the atoms to recombine to form sugar molecules. This sugar was then fermented to provide the energy the creatures needed.

$$6\,\mathbf{CO_2} + 6\,\mathbf{H_2O} \xrightarrow{\text{Sunlight}} \underset{\substack{\text{Sugar} \\ \text{(Glucose)}}}{\mathbf{C_6H_{12}O_6}} + 6\,\mathbf{O_2}$$

The process of photosynthesis that was developed three billion years ago to store the energy of the sun until it could be used by living creatures is the very same process that is employed by the more complex plants of today. With the advent of photosynthesis came the first example of air pollution by living beings. The pollutant in this instance was oxygen. The air at that time contained no oxygen. The creatures that produced it had no need for it. Since no natural processes existed to dispose of it, it simply accumulated. Eventually, it became so prevalent and so concentrated that it destroyed the organisms that created it.

A record of the initial production of oxygen 1.8 billion years ago can be found on the walls of the Grand Canyon. Sedimentary rocks containing iron, which are older than 1.8 billion years, lack the red coloration of iron oxide that could be formed only in an oxygen-containing atmosphere. It was the atmospheric pollution with oxygen by primitive organisms that caused the iron in the sedimentary rocks to be oxidized to ferric oxide ($2\,Fe + 3\,O_2 \rightarrow Fe_2O_3$), allowed the development of the aerobic (oxygen-breathing) forms of life popular on earth today, and resulted in the decline of the anaerobic (non-air-breathing) forms that started it all. The sad tale of these primordial creatures is a useful point of departure for the evaluation of the current pollution crisis.

Dirty Air

Man, like the primitive life forms, has always been faced with a need for energy. Energy is required to cook his food, warm his home, and fabricate his tools. Early man satisfied most of these needs by building campfires. These isolated fires of themselves had little environmental impact; the small amount of smoke produced by each soon dispersed to an unobjectionable level. Only with the rise in population and the concentration of people into urban and industrial centers has the accumulation of the by-products of burning become a problem. When natural processes for the disposal of pollutants are insufficient to cope with their rate of production, the passage of time can only increase their concentration. If the pollution is allowed to continue unchecked, eventually it will produce a toxic atmosphere. This is what happened to the primordial life forms. They polluted the atmosphere with oxygen to a concentration that for them was lethal. It is true that in so doing they made the atmosphere suitable for life forms such as humans, but there is small consolation in this knowledge if you happen to be an anaerobic organism.

Hittite records as early as 900 B.C. describe air pollution in areas where asphalt was mined. These areas, however, were geographically localized, and the objectionable air conditions were only partially

man-made. Man's coming of age as an air polluter arrived with the building of large cities. Urban man, through combined efforts with his neighbors, was able to produce a level of air pollution unattainable from isolated camp fires.

The city of London has long been plagued by air pollution. During the reign of Edward II in the fourteenth century the "pestilential odor" of London air became so acute that people who were caught burning sea coal (a significant cause of this problem) were tortured and put to death. Under Henry V in the early fifteenth century, these penalties were relaxed. Taxation was substituted for torture. Pollution as a result prospered. By the nineteenth century, the poet Shelley considered it fairly accurate to describe Hell as a city much like London, "populous and smoky." Little did he know how prophetic his words were. The "Black Fog" that enshrouded London in December 1952 was an extraordinary pea-souper, so dense that bus conductors walked in front of their buses to guide the drivers, flying ducks crashed into buildings, and singers at Sadlers Wells were forced to cancel *La Traviata* after the first act because the "fog" that seeped into the theater was so heavy that they could not see the conductor. In four days four thousand people died of the smog, creating the most tragic air-pollution disaster in recorded history. To prevent the reoccurrence of such a diaster, additional restrictions on the use of soft coal were instituted. These restrictions, although they carried penalties far less severe than those of Edward II, were nevertheless sufficient to increase the visibility by a factor of 3 and the number of sunny days by 70 percent.

The agents that caused the "Black Fog" and air pollution in general can be classified into six groups of substances.

1. Oxides of Nitrogen ($\mathbf{NO}$, $\mathbf{NO_2}$)
2. Oxide of Oxygen ($\mathbf{O_3}$)
3. Oxides of Sulfur ($\mathbf{SO_2}$, $\mathbf{SO_3}$)
4. Oxides of Carbon ($\mathbf{CO}$, $\mathbf{CO_2}$)
5. Organics (alkanes, alkenes, aldehydes, aromatics, etc.)
6. Particulate matter (dust, smoke, etc.)

Any of these substances present at an objectionably high concentration produces air pollution. If these components of air pollution are compared with the components of pure air, one sees that the oxides of nitrogen and oxygen as well as carbon dioxide appear in both categories. Oxides of nitrogen and oxygen are produced in the atmosphere as a result of the action of sunlight and lightning on air. These compounds are sometimes considered natural pollutants. However, at their natural abundances, they produce minimal objectionable effects. It is only when their concentrations are markedly increased through the activities of man that they become serious threats to health and life.

Man produces fifty million tons of nitrogen oxides per year. Although this is only one-tenth the amount produced by natural processes, its potential for destruction is markedly greater because it is produced at higher concentrations. The concentration of $\mathbf{NO_2}$ when generated in nature seldom exceeds five parts per billion. When produced by man in crowded urban areas, however, values as high as five hundred parts per billion are not uncommon. This is one hundred times greater than the natural concentration. Table 17–1 lists some typical U.S. cities and their maximum nitrogen oxide levels during the period 1962–1968.

Atmospheric **NO** can be produced by any one of three processes: (1) bacterial action, (2) the action of ultraviolet light on atmospheric gases, or (3) the heating of mixtures of nitrogen and oxygen to temperatures above 1210°C (a common occurrence in internal combustion engines, i.e., automobiles). **NO** is a colorless, odorless gas that is only one-quarter as toxic as the reddish-brown, choking, higher oxide $\mathbf{NO_2}$. Ultraviolet radiations and heat can provide sufficient energy to split the normally stable oxygen ($\mathbf{O_2}$) and nitrogen ($\mathbf{N_2}$) molecules, which then recombine to form oxides of nitrogen. Unless sufficient energy is provided to make these reactions go, atmospheric mixtures of nitrogen

TABLE 17–1 Maximum Nitrogen Oxide Levels, 1962–1968

City	Maximum NO Concentration (ppm) (In parts per million)	Maximum NO_2 Concentration (ppm)
Chicago	0.91	0.47
Cincinnati	1.38	0.56
Denver	0.54	0.33
Los Angeles	1.42	0.68
Philadelphia	1.87	0.32
St. Louis	0.75	0.22
San Francisco	1.30	0.41
Washington, D.C.	1.14	0.30

Data selected from U.S. Dept. of Health, Education, and Welfare, *Air Quality Criteria for Nitrogen Oxides*, pp. 6-10–6-13.

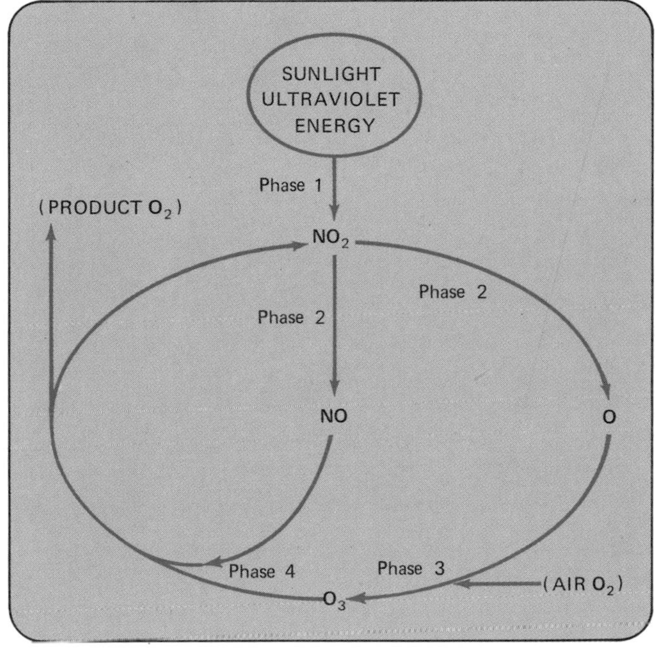

FIGURE 17–1 NO_2 Cycle (Photolytic)

and oxygen show practically no tendency to react. This energy, whether provided thermally or through radiation, is known as the energy of activation. It is the minimum energy barrier that the reactants must overcome before a reaction can occur.

$$N_2 + O_2 \rightleftarrows 2\,NO$$

$$2\,NO + O_2 \rightleftarrows 2\,NO_2$$

When this energy is provided through the sun's radiation, the NO_2 produced does not present a pollution problem. This is because the NO_2, when generated, enters into a series of reactions known as the "photolytic NO_2 cycle." (Figure 17–1.)

Phase 1. Energy from the sun is absorbed by NO_2 molecules.

$$\text{(Light energy)} + NO_2 \rightarrow \underset{\text{(Activated)}}{NO_2^*}$$

Phase 2. Absorbed energy causes one of the nitrogen-oxygen bonds to rupture, producing **NO** and an oxygen atom (**O**).

$$NO_2^* \rightarrow NO + O$$

Phase 3. Highly reactive oxygen atoms react with atmospheric oxygen (O_2) to produce ozone (O_3)

$$O + O_2 \rightarrow O_3$$

Phase 4. Ozone reacts with **NO** to produce $NO_2 + O_2$.

$$O_3 + NO \rightarrow NO_2 + O_2$$

The net effect of these reactions is merely to cycle the component molecules. The cycle produces no change in the ambient concentrations of **NO** or NO_2 since O_3 and **NO** are formed and destroyed in equal amounts. If, however, this cycle is interrupted through the introduction of organic compounds, a variety of ketones and aldehydes may be produced.

Such compounds are the result of the action of oxygen atoms and ozone molecules, produced by the photochemical decompositions of NO_2, on organic pollutants arising from the incomplete combustion of automobile and industrial fuels. It is the ketones and aldehydes plus the organic lachrymators (substances that cause eye tearing) produced by the action of **NO** and NO_2 on unsaturated organics that gives Los Angeles smog its characteristic properties. Figure 17–2 shows the formulas of four typical aldehydes and nitrates found in photochemical smog.

Table 17–2 lists six reactions characteristic of those involved in the mechanism of photochemical smog production. The scheme is simplified and accounts for only a small number of the reactions at play in the production of photochemical smog. It does show, however, the interplay of the oxides of nitrogen, oxygen, and organic compounds in the

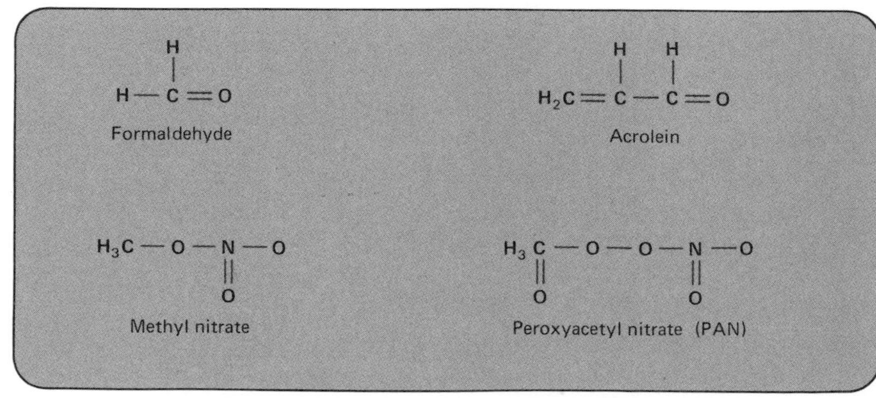

FIGURE 17–2 Aldehydes and Nitrates Found in Photochemical Smog

17–2 Characteristic Reactions of Photochemical Smog Production

TABLE

1. $O\cdot + (Or) \rightarrow (Or)O\cdot$
   Organics
2. $(Or)O\cdot + O_2 \rightarrow (Or)O_3\cdot$
3. $(Or)O_3\cdot + (Or) \rightarrow RCHO$ *or* $R{-}C({=}O){-}R$
   Aldehydes — Ketones
4. $(Or)O_3\cdot + NO \rightarrow (Or)O_2\cdot + NO_2$
5. $(Or)O_3\cdot + O_2 \rightarrow O_3 + (Or)O_2\cdot$
6. $(Or)O_3\cdot + NO_2 \rightarrow R{-}C({=}O){-}O{-}O{-}N(O)O$ + Other products
   (PAN)

production of California smog. If approximately 0.5 parts per million (ppm) of SO_2 is added to this mixture, the extremely destructive sulfuric acid aerosols typical of London smog are produced.

Most of the sulfur dioxide (SO_2) in the atmosphere comes from power plants burning sulfur-containing fuel, from smelters roasting metallic sulfide ores, and from chemical plants producing sulfuric acid.

$$S + O_2 \rightarrow SO_2$$

SO_2 has a relatively short atmospheric life. It and the sulfuric acid formed from it dissolve readily in streams and lakes, producing a lowering of the waters' *p*H. If the *p*H of a body of water drops below 5.5, its ability to support fish is seriously affected.

$$H_2O + SO_2 \rightleftarrows \underset{\text{Weak acid}}{H_2SO_3} \rightleftarrows H^+ + HSO_3^-$$

$$H_2SO_4 \rightarrow H^+ + HSO_4^- \rightleftarrows H^+ + SO_4^=$$

The effects of oxy-sulfur compounds on plants are equally serious. Vegetation growing as far as fifty miles from SO_2-producing power plants can be found bleached and stunted from the stack fumes.

Besides causing harm to living things, atmospheric sulfur dioxide and its derivatives take their toll of inanimate objects. They can corrode metals, age fabrics, stiffen leather and paper, and convert limestone and marble ($CaCO_3$) to friable calcium sulfate ($CaSO_4$). Cleo-

patra's Needle (an obelisk), for example, has deteriorated more in the single century that it has stood in London than in the thirty preceding centuries when it stood in Egypt.

Available technology can markedly reduce the concentration of sulfur oxide pollutants. This is usually accomplished by converting $\mathbf{SO_2}$ into a nonvolatile form. One such process involves the reaction of $\mathbf{SO_2}$ with **CaO**, produced by heating limestone. The calcium sulfate resulting from this reaction is then separated from the other stack emissions by electrostatic precipitation.

$$\mathbf{CaCO_3} \xrightarrow{\text{Heat}} \mathbf{CaO} + \mathbf{CO_2}$$

$$2\,\mathbf{CaO} + 2\,\mathbf{SO_2} + \mathbf{O_2} \longrightarrow \underset{\text{Calcium sulfate}}{2\,\mathbf{CaSO_4}}$$

Such methods are, however, economically feasible only when they are employed on an industrial basis. The best procedure for reducing $\mathbf{SO_2}$ emissions from consumer sources is the utilization of low sulfur fuels.

The oxides of carbon pollute in two distinctly different ways. Carbon monoxide (**CO**), the lower of the two oxides, is a colorless, odorless, tasteless, toxic compound that is not appreciably soluble in water. It is produced industrially by the incomplete combustion of carbon and carbon-containing compounds and naturally by certain bacteria, marine plants, and jellyfish.

$$2\,C + O_2 \rightarrow 2\,CO$$

The toxic effects of **CO** result from its ability to combine with blood hemoglobin to form a compound that is two hundred times more stable than oxyhemoglobin. When it does, the oxygen-carrying ability of the blood is reduced by an amount equivalent to the carboxyhemoglobin that is formed.

$$CO + \text{Hemoglobin} \rightarrow \text{Carboxyhemoglobin}$$

To counteract this oxygen depletion, the heart must pump at a greater rate. When the rate necessary to provide the proper amount of oxygen is beyond the pumping capabilities of the heart, the animal dies. Because of its high affinity for hemoglobin, concentrations of **CO** in air as low as 0.40 percent can prove fatal.

Man releases approximately 2×10^{11} kilograms of **CO** into the atmosphere each year, which is enough to double the atmospheric concentration in five years. Each gallon of gas burned in an automobile, for example, liberates approximately one kilogram of **CO**. In spite of this alarming rate of production, however, there has been no significant actual increase in **CO** concentration.

The natural mechanism by which **CO** is removed from the air is not well understood. It is believed, however, that it is accomplished by soil microorganisms and fungi. Authoritative estimates put the **CO**-removal capabilities of the United States soil sink at five hundred million tons per year, five times greater than the current rate at which we discharge **CO** into the atmosphere. These figures, although they may appear comforting at first glance, are misleading. Most man-made **CO** is produced in areas of high population where industrialization has produced a deficiency of the soil organisms necessary for the removal of **CO**. As a result, the concentration of **CO** over these areas can in fact reach dangerous proportions. It is little comfort to a man

on a street of downtown Manhattan to know that sufficient organisms exist in an Oregon forest to detoxify the **CO** in the automobile exhaust fumes that envelop him. For such reasons, the use of broad-based statistics to describe the dangers of a pollutant is an unsatisfactory procedure.

The fact that a mechanism exists for the detoxification of **CO** points up nature's ability to react and adapt to pollutants. Because **CO** has been produced for eons by natural processes, nature has had sufficient time to develop these mechanisms. The relatively constant atmospheric **CO** concentration is merely an indication that at present the combined production of **CO** from natural and man-made sources does not as yet exceed the capacity of this naturally developed detoxification mechanism.

Carbon dioxide (CO_2), the higher oxide of carbon, is colorless, slightly soluble in water, and by comparison to other atmospheric pollutants relatively nontoxic. Its main impact as an environmental pollutant is dependent more on its physical rather than its chemical properties. Chemically, it is a well-adjusted molecule firmly established in the photosynthetic cycle. Physically, however, it can react at increased concentration to cause an appreciable rise in the temperature of the atmosphere. Between the years 1860 and 1970, the average concentration of CO_2 in the atmosphere increased from 292 to 322 parts per million. Although in itself this might not be considered a large increase, it has resulted in an increase of 0.16°C in the average atmospheric temperature of the earth. To put this figure in its proper perspective one must realize that a change of 2.3°C in the average atmospheric temperature could, by melting the polar ice caps, easily raise the levels of the oceans some 200 feet and drown entire countries such as Denmark and the Netherlands. The ability of CO_2 to produce this temperature rise is commonly known as the "greenhouse effect."

The last major group of atmosphere pollutants is the *particulates*. These particles, such as soot, clays, and dust, reflect the sun's radiation and, in so doing, tend to reduce the average atmospheric temperature. It has jokingly been proposed that if man could pollute the atmosphere with the proper proportions of CO_2 and particulates, he could pollute to his heart's content without appreciably affecting the mean atmospheric temperature. This, however, is only fanciful thinking, since it is not possible to alter two components of the atmosphere without disturbing the delicate balance that has been established by nature over millions of years.

Dark Waters

"Water, water, everywhere, nor any drop to drink." From the wettest jellyfish to the driest desert lizard the maintenance of life as we know it could not be accomplished without water. The problem with water today, however, is that with few exceptions its formula is no longer H_2O but $H_2O \cdot n$ garbage (where n is a significantly large number). Our water, like the air we breathe, is polluted. Rain water and the driven snow are no longer pure; they have been tainted by the atmosphere. Ground water, which results from the settling out of this precipitation, is even more contaminated. It contains, in addition to the pollutants washed from the air, decaying plant and animal materials, infectious agents, dissolved chemicals, and suspended silt.

Over 1.5 million billion gallons of precipitation fall on the United States each year. This may at first seem like a fantastically large amount of water, but the shocking fact of the matter is that only about 10 percent of this precipitation is currently available for use by man, and he is already using the major portion of it.

Two factors make this information particularly alarming. (1) If the personal and commercial use of water in the United States continues to grow at the current rate, within a short amount of time it will exceed the available water supply. (2) Because of man's industriousness, the quality of available water has decreased with the passage of time.

Before the industrial revolution, the major pollutants in water were waste products of living systems. Other living systems (bacteria) detoxified these pollutants by converting them to carbon dioxide, water, and nitrates that could be utilized by plants as nutrients. Man through his activity has altered in both a quantitative and a qualitative fashion the nature of these pollutants.

Long before man, flowing rivers washed out minerals from the soil and carried them to the sea. Man has, however, changed the quantities of these minerals through mining operations and their quality through chemical processing. In addition, the rivers that serve the major cities of the world are constantly being assaulted with inordinately large quantities of biological and industrial sewage. Their volume staggers the natural detoxification capabilities that the rivers have taken hundreds of thousands of years to develop. The net result is that waste materials, rather than being destroyed, increase in concentration with time, eventually destroying the natural detoxification processes and those that depend on it.

The problem of water pollution has no one simple solution. Its complexity can perhaps be better understood in terms of a common everyday pollutant such as insecticides. Nearly all insecticides in use today have been introduced into the environment by man. They are sprayed in large quantities to protect crops from insects, and humans and animals against insect-borne diseases.

Prior to World War II the major components of insecticides were lead arsenate, nicotine sulfate and pyrethrins. The pyrethrins are esters that are extracted from chrysanthemum flowers native to Asia. With the advent of World War II and the Japanese occupation of Southeast Asia and Indonesia, the supply of pyrethrins was cut off. Lead, arsenic, and copper were needed for military purposes. A new type of insecticide had to be devised to protect the fighting troops from the scourge of diseases transmitted by insects (yellow fever, malaria, typhus, and plague, among others).

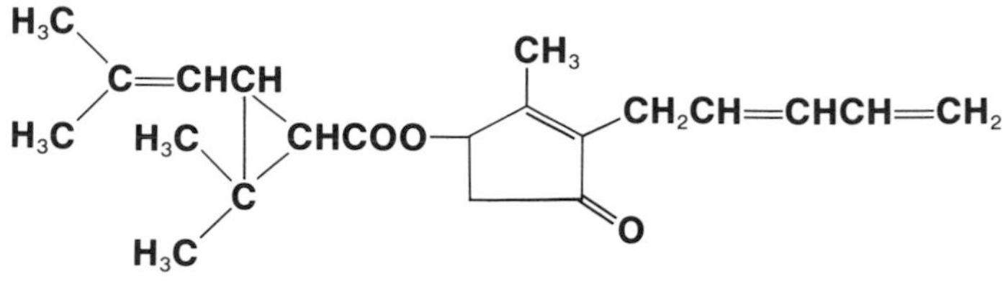

$(H_3C)_2C{=}CHCH$ — $C(CH_3)_2$ — $CHCOO$ — (cyclopentenone ring with CH_3, $=O$) — $CH_2CH{=}CHCH{=}CH_2$

Pyrethrin I

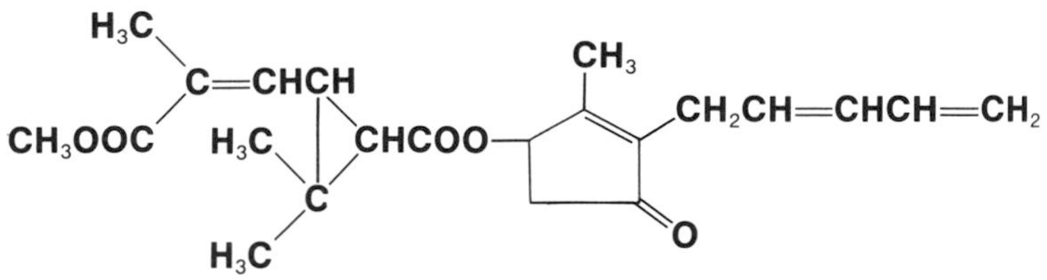

Pyrethrin II

At the time, the logical solution seemed to be DDT. (Its full name is dichlorodiphenyltrichloroethane, and the IUPAC calls it 1,1,1-trichloro-2,2-*cis-p*-chlorophenyl ethane; the abbreviation is understandably the more commonly used.) This compound, which was first synthesized in 1874, was rediscovered in 1939 by Paul Müller, a Swiss scientist, who found it to have excellent insecticidal properties. (He received a Nobel Prize for this discovery in 1948.) DDT was cheap, easily synthesized from chlorobenzene and chlorohydrate, and seemed to have little deleterious effect on man or large animals.

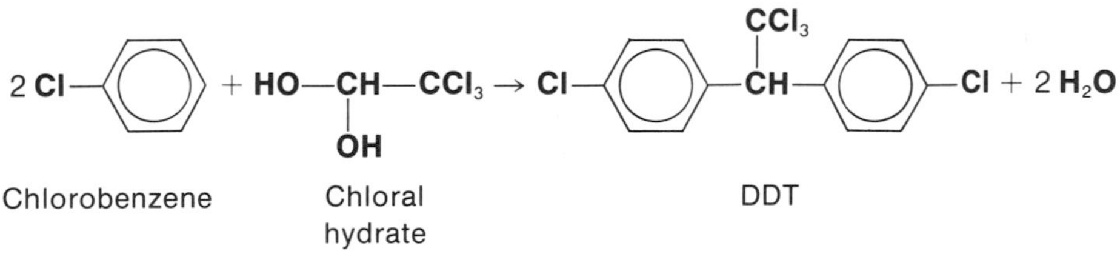

$$2\,Cl{-}C_6H_4 + HO{-}CH(OH){-}CCl_3 \rightarrow Cl{-}C_6H_4{-}CH(CCl_3){-}C_6H_4{-}Cl + 2\,H_2O$$

Chlorobenzene — Chloral hydrate — DDT

It kept Allied soldiers virtually free from lice-borne typhus, and it significantly reduced the incidence of mosquito-borne malaria. It seemed to be all that one could ask for in an insecticide. During the Allied invasions of Naples its effectiveness was dramatically demonstrated. "Softening up" bombings had destroyed the city's water supply. The only available water was from the sewers. The lack of general sanitation led to large insect and rat populations, which produced a massive typhus epidemic. The conquering Allies responded by spraying anything and everywhere they could with DDT. In short order the epidemic was brought under control. It has been estimated that during World War II alone the use of DDT resulted in saving the lives of five million people who would otherwise have died from insect-borne diseases.

Crowned with such apparent success, DDT was quickly incorporated into the postwar economy. It was used to spray crops, homes, insect

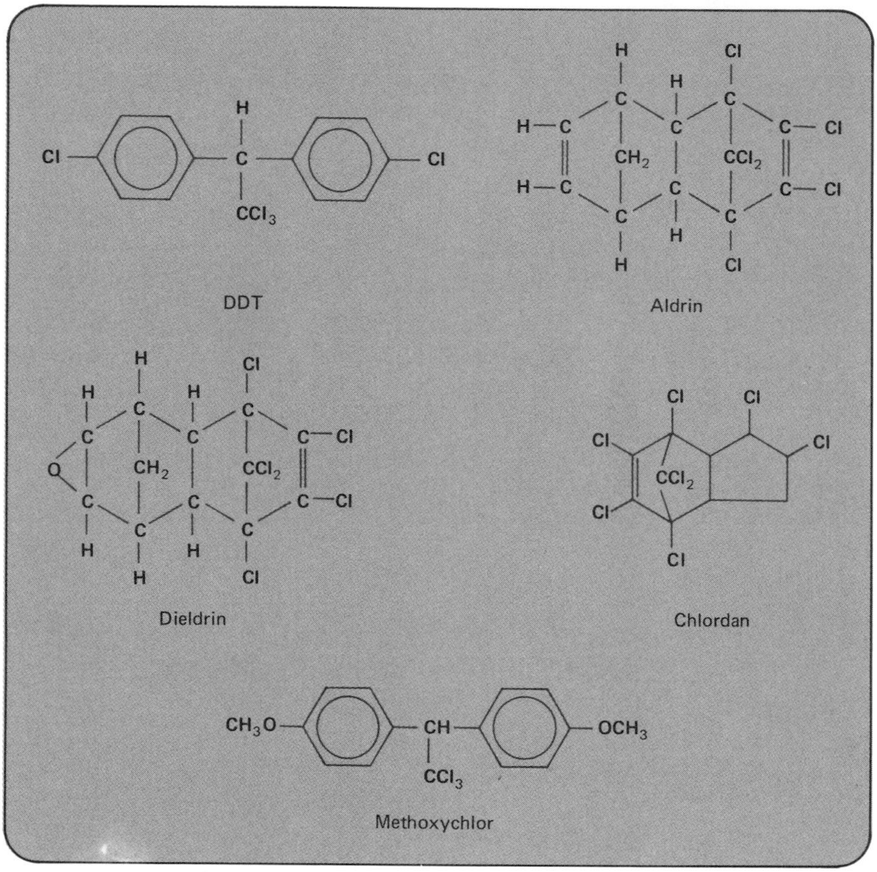

FIGURE 17–3 Chlorinated Hydrocarbon Insecticides

breeding grounds, and any place else an unwanted bug was thought to be. Man was soon to be freed of his insect afflictions. No such luck. The insects, which made their appearance on earth many eons before man and which will probably remain long after man has disappeared, were not to be done in that easily. As early as the late 1940s, flies and mosquitos with genetic resistance to DDT were discovered.

New compounds to fill the need for insecticides were synthesized. Like DDT, its successors were fat-soluble molecules of chlorinated hydrocarbons. (Figure 17–3.) Fat solubility is a desirable characteristic for a good contact insecticide because it enables the molecule to penetrate the insect's cuticle to kill him. It is, however, an undesirable characteristic for other life forms, which, although they do not succumb to small dosages, tend to deposit them in their fatty tissue. The stability of DDT and other chlorinated insecticides to natural degradation makes this property of fat storage a particularly serious problem. If an organism that is repeatedly exposed to small quantities of insecticides stores rather than detoxifies them, the level within its body will constantly increase. If such an organism is eaten by another form, the process of insecticide concentration is carried one step further. As the concentration increases, the level eventually becomes sufficiently high to kill life forms that normally would not be seriously affected by the level of insecticide necessary to kill insects.

Two examples point up the dangers of the concentrating effects of stable, fat-soluble insecticides. The first concerns the spraying of a California lake in 1957 to control the gnat population. Immediately after the spraying, the concentration of DDT in the water was 0.02 parts per million. The microscopic plant and animal life in the lake, however, contained two hundred and fifty times this concentration (5 ppm). The fish that ate the microscopic plankton contained 2000 parts per million. The birds that ate these fish died because the concentration of DDT was one hundred thousand times greater than the amount originally sprayed on the water and was thus sufficiently high to kill them.

The second example concerns an extremely obese woman who died quite suddenly and strangely after going on a crash diet. The coroner who investigated the cause of her death found that because of her sudden loss of fat, her body could no longer store the large amounts of insecticides it had accumulated from the foods she had previously eaten. Consequently, they were released into her bloodstream in a concentration sufficiently great to kill her. Had her loss of weight not been so drastic, she probably would have survived to display chronic symptoms of insecticide poisoning.

Insecticides have been discussed as only one of a large number of materials that pollute our environment. What must not be forgotten is they, like all other pollutants that now afflict us, were once part of some process or product that was wanted and that in some way was beneficial. The problem is that man, for all his advancement, does not in fact differ so radically from apes who throw today's banana peels on the ground to be slipped on tomorrow. Pollution is yesterday's discarded peels. Its solution is not abstinence from bananas, but beneficial utilization of the peels.

REVIEW

1. Pollution is the presence in the environment of substances that, by virtue of their nature or concentration, are considered undesirable.
2. Pollution can be produced both by natural processes and by man.
3. Because most natural pollution processes have occurred for a comparatively long time, the environment has, in many cases, had sufficient time to develop adaptive mechanisms to neutralize their effects.
4. The pollutants produced by man have appeared in the environment relatively recently.
5. Because of the newness of man-made pollutants, the environment has had comparatively little time to adapt to them. The lack of existing natural detoxification processes and the inherent stability of many of these compounds make their presence in the environment long-lived. They therefore often have a disproportionate effect on the environment.
6. Man often produces pollutants over small areas (cities) at much higher concentrations than might be formed naturally.
7. Most man-made pollutants are the by-products of some process that is beneficial to man. As in all processes, the benefit of the product must be weighed against its associated disadvantages.
8. The solution to man-made pollution is the more complete integration of the by-products of man's activities.

REFLEXIVE QUIZ

1. Two natural processes by which air pollutants can be generated are ________________ and ________________.
2. Large amounts of naturally produced pollutants are often less injurious to the environment than smaller amounts of the same pollutant when it is artificially produced because the ________________ at which they are produced are lower.
3. When evaluating the impact of a substance as a pollutant it is not sufficient to be concerned with its rate of production. Its rate of ____________ must also be considered.
4. Photochemical smog is mainly produced by a ________________ reaction mechanism.
5. The burning of sulfur-containing fuels in power plants results in the unwanted production of ______.
6. The toxic effects of ______ from auto exhausts are due to its ability to combine with hemoglobin to produce a more stable product than oxygen can.

7. It is feared that excessive production of _____ will produce a "green house effect" and raise the temperature of the earth's atmosphere.

8. ___________ provide a pollution threat to man and higher animals because they are comparatively stable compounds that tend to ________ in the fatty tissue of animals at concentrations that are often higher than those at the time of their application.

9. Three sources for the production of nitrogen oxide pollutants are (*a*) bacterial action, (*b*) _______________, and (*c*) _______________.

10. The organic compounds that give Los Angeles smog its characteristic properties are mostly __________, __________, and __________.

Answers: (1) Forest fires, electrical storms, decaying organic matter, volcanic eruptions [364–365]. (2) Concentrations [365, 370]. (3) Disappearance [364–367, 379–381]. (4) Free radical [370–373]. (5) $\mathbf{SO_2}$ [373]. (6) **CO** [375]. (7) $\mathbf{CO_2}$ [376–377]. (8) Insecticides, concentrate [379–381]. (9) Ultraviolet light on atmospheric gases, internal combustion engine [367–373]. (10) Aldehydes, ketones, organic nitrates [372–373].

SUPPLEMENTARY READINGS

1. M. Jacobson. *Insect Sex Attractants.* New York: Interscience Publishers, 1965.

2. G. M. Woodwell. "Toxic Substances and Ecological Cycles." *Scientific American* (June 1967), p. 24.

3. J. R. Clark. "Thermal Pollution and Aquatic Life." *Scientific American* (March 1969), p. 18.

4. W. McDermott. "Air Pollution and Public Health." *Scientific American* (October 1961), p. 49.

5. G. N. Plass. "Carbon Dioxide and Climate." *Scientific American* (June 1959), p. 41.

6. P. A. Leighton. *The Photochemistry of Air Pollution.* New York: Academic Press, 1961.

7. R. Carson. *Silent Spring.* Boston: Houghton Mifflin Co., 1970.

8. F. Graham, Jr. *Since Silent Spring.* Boston: Houghton Mifflin Co., 1970.

Appendixes

Appendix I

Measurement

Advantages of the Metric System

The British Imperial System of measurement, in which the U.S. Customary System has its origins, is not particularly suitable to scientific measurement. The units of measurement of length, for example, are based on the size of a dead king. An inch is the length from the end to the first joint of his thumb. A foot is the length of his foot, and a yard the length between his nose and the thumb on his outstretched arm. The difficulty in using this system is that there is no uniform progression between successively smaller units. Each unit—not only of length but of weight, volume, and area—is defined in terms of the other units. One arm equals three feet or thirty-six knuckles. Everything is relative.

A convenient progression does exist, however, in the International (Metric) System of measurement. This system is based on absolute standards of measurement, the meter and the kilogram, not on relative units. Because it is a decimal system, each unit is ten times greater than the next smaller one. One meter equals 10 decimeters or 100 centimeters or 1,000 millimeters, and so on. (A meter is the length of a large and very precise number of wavelengths of the orange-red radiation which is produced when krypton 86 is excited. In our inconvenient but familiar terms, that is about 39.37 inches.) One kilogram (2.2 pounds) equals 1,000 grams, and one liter (1.057 quarts) equals 1,000 milliliters (ml); one milliliter is the same as one cubic centimeter (cc) of space. Furthermore, on the Celsius (and Kelvin) temperature scales, 100 degrees separate the freezing point of water from its boiling point. (See the back end papers for a chart of metric measurements.)

Exponential Numbers

The advantages of the metric system become apparent when we are dealing with the extremely large or extremely small numbers of the subatomic world. 1,000,000,000,000 protons placed side by side in a straight line, for example, would occupy about a centimeter; 100,000,000,000 protons would occupy about a millimeter. In writing such measurements, an abbreviated notation can be adopted to eliminate the inconvenience of writing strings of zeros. Rather than writing a one with twelve zeros after it, we can express the same number by writing 10^{12}. This notation states that one followed by twelve zeros is equivalent to the number that would be obtained if 10 were used 12 times as a factor in a multiplication process. It is referred to as "ten to the twelfth power." The notation for $\frac{1}{10^{12}}$ is 10^{-12} and is referred to as "ten to the negative twelfth power."

This system of *exponential numbers* expresses the magnitude of a number at the expense of its precision. To get an idea of the relative importance of magnitude and precision, consider the following. A man desires to purchase a house for $30,000. He asks his bank to send him a statement of his assets. The statement he receives reads $779480, with no indication of a decimal point. His ability to purchase the house depends on whether a decimal point exists between the 4 and the 8. The decimal point indicates the magnitude of his assets. It makes no difference what any of the five integers are if the amount is hundreds rather than thousands of dollars. If his bank statement were written in exponential numbers, the statement would have read either 7×10^5 or 7×10^3 dollars. It is true that neither of these figures expresses the precise amount of money the man has on deposit. However, his decision does not require this information.

Exponential numbers are based on the products derived from the multiplications of 10 by itself. (Table I–1.) A positive superscript indicates the number of 10's that must be used in the multiplication to get the desired quantity. It also corresponds to the number of zeros following the 1 to the right of the decimal point in the standard num-

TABLE I–1 Bases of Exponential Numbers

Exponential Numbers	Standard Number	Product Form
10^3	1000	$10 \times 10 \times 10$
10^2	100	10×10
10^1	10	10
10^0	1	See any elementary algebra text for a discussion of the zero exponent.
10^{-1}	.1	$\frac{1}{10}$
10^{-2}	.01	$\frac{1}{10 \times 10}$
10^{-3}	.001	$\frac{1}{10 \times 10 \times 10}$

ber. A negative superscript indicates the reciprocal of the positive number. Extremely small numbers are represented by large negative superscripts. When representing numbers that are not simple multiples of 10, the exponential is multiplied by the appropriate integer. The number 4000 would therefore be written as 4×10^3, and .007 would be written as 7×10^{-3}.

Logarithms

Just as 10^9 is a more convenient way to express a billion than 1,000,000,000, there also exists a more convenient way to describe 10^9. The system is known as *logarithms*. The logarithm of a number is merely its exponential value. Thus, the logarithm (to "base 10") of 10^9 is 9; the logarithm of 10 is 1; and the logarithm of 1 is 0. The conversion of numbers such as 3×10^9 into logarithms, however, is a bit more difficult. (There are "natural" logarithms to a base other than 10, but they do not concern us here.) Since 3 is a number smaller than 10, it will have a fractional logarithmic value. Its logarithmic value as well as the logarithmic values of all integers less than 10 are listed in Table I-2 (the logarithms are correct to two decimal places).

TABLE I–2 Logarithms of Numbers Less Than 10

Number	1	2	3	4	5	6	7	8	9
Logarithm	0	.30	.48	.60	.70	.78	.84	.90	.95

Once the logarithm of 3 is determined, the logarithm of the number 3×10^9 can be expressed as the sum of its logarithmic parts.

$$\begin{aligned}\text{logarithm of } 3 \times 10^9 &= \text{logarithm of } 3 + \text{logarithm of } 10^9 \\ &= \quad .48 \quad + \quad 9 \\ &= 9.48\end{aligned}$$

Logarithms are often very useful in expressing extremely small concentrations of ions, such as hydrogen ion and hydroxide ion, that may be found in solutions. Hydrogen ion concentration is frequently measured in *p*H units, which are merely the negative logarithms of their concentrations. A solution having a hydrogen ion concentration of 4×10^{-13} moles per liter would have a *p*H of 12.4.

$$
\begin{aligned}
p\text{H} &= -(\text{logarithm of hydrogen ion concentration})\\
&= -(\text{logarithm of } 4 + \text{logarithm of } 10^{-13})\\
&= -(\quad .60 \quad + \quad -13 \quad)\\
&= -(-12.4)\\
&= 12.4
\end{aligned}
$$

You can find complete tables of logarithms in many general mathematics and calculus texts.

Appendix II

Stoichiometry

Stoichiometry is the study of the numerical proportions that govern the manner in which chemicals react. It is useful in determining the quantities of reactants that are necessary to produce a desired amount of product.

The first step in such a process is to balance the chemical equation of a reaction so that there are an equal number of atoms of each type on each side of the equation. Such an equation, however, describes only what happens when one or more molecules of compound A react with one or more molecules of compound B. This information, although valuable, is not so practical as knowing how much of compound B will react with 5 grams of compound A to produce compounds C and D.

To derive such information it is necessary to convert the more abstract units of moles and equivalents into units of mass or volume.

$$(1) \qquad H_2SO_4 + Ba(NO_3)_2 \rightarrow BaSO_4\downarrow + 2\,HNO_3$$

Equation 1 is a balanced equation describing the reactions between one molecule of sulfuric acid and one molecule of barium nitrate. To convert this equation to one expressed in terms of mass, the molecular weights of each of these compounds must be utilized.

(2) H_2SO_4	+	$Ba(NO_3)_2$	→	$BaSO_4\downarrow$	+	$2\,HNO_3$
2 H = 2		Ba = 137		Ba = 137		2 H = 2
S = 32		2 N = 28		S = 32		2 N = 28
4 O = 64		6 O = 96		4 O = 64		6 O = 96
mol. wt. = 98		mol. wt. = 261		mol. wt. = 233		mol. wt. = 126

Equation 2 indicates that 98 grams, kilograms, ounces, or pounds of sulfuric acid will react respectively with 261 grams, kilograms, ounces, or pounds of barium nitrate to produce respectively 233 grams, kilograms, ounces, or pounds of barium sulfate and 126 grams, kilograms, ounces, or pounds of nitric acid. In making this conversion, it cannot be emphasized too strongly that the same units must be consistently utilized for the reactants as well as the products. That is, if one substance is measured in grams, all the other substances must also be measured in grams.

Equation 2 shows that if one started with either 98 grams of sulfuric acid or 261 grams of barium nitrate, the maximum amount of barium sulfate and nitric acid that could be produced would be respectively

233 grams and 126 grams. Similarly, if the equation is viewed from the vantage point of the products, in order for 233 grams of barium sulfate or 126 grams of nitric acid to be produced, the minimum amounts of sulfuric acid and barium nitrate necessary must be respectively 98 grams and 261 grams. If the quantity of either of these two reactants is less than these amounts, then the amounts of both products will be proportionally reduced. That is, as soon as one reactant is used up, it is of no consequence how much of the second reactant is available, since no further production of this product is possible.

If, however, all the reactants except one are in excess (that is, in amounts that are more than sufficient to use up all of the reactant under consideration), the quantities of products will be proportional to the amount of that one reactant. In a reaction between sulfuric acid and barium nitrate, for example, if the barium nitrate is in excess, the amount of sulfuric acid necessary to produce 378 grams of nitric acid can be determined by a simple proportion.

$$\underset{(98)}{\overset{(x)}{H_2SO_4}} + Ba(NO_3)_2 \rightarrow BaSO_4\downarrow + \underset{(126)}{\overset{(378)}{2\,HNO_3}}$$

$$\frac{x}{98} = \frac{378}{126}$$

$$x = \frac{378}{126} \times 98 = 3 \times 98$$

$$x = 98\text{ g}$$

In a similar manner, the amount of barium sulfate that can be produced from 294 grams of sulfuric acid can also be determined by using a simple proportion.

$$\underset{(98)}{\overset{(294)}{H_2SO_4}} + Ba(NO_3)_2 \rightarrow \underset{(233)}{\overset{(x)}{BaSO_4\downarrow}} + 2\,HNO_3$$

$$\frac{294}{98} = \frac{x}{233}$$

$$x = \frac{294}{98} \times 233 = 3 \times 233$$

$$x = 699\text{ g}$$

Note. These proportional relationships exist only when all the reactants except the one under consideration are in excess.

PROBLEMS

In the reaction, the molecular weights of the reactants and products are given.

$$\begin{array}{ccccccc} \mathbf{Na_2S} & + & \mathbf{Pb(NO_3)_2} & \rightarrow & \mathbf{PbS}\downarrow & + & \mathbf{2\,NaNO_3} \\ \mathrm{MW} = 78 & & \mathrm{MW} = 331 & & \mathrm{MW} = 239 & & \mathrm{MW} = 85 \end{array}$$

1. ______________ is the number of grams of Na_2S necessary to convert 993 grams of $Pb(NO_3)_2$ to **PbS.**
2. The maximum number of grams of **PbS** which can be produced from 993 grams of $Pb(NO_3)_2$ is ______________.
3. Irrespective of the amount of reactants the above reaction will always produce ______________ as many moles of $NaNO_3$ as compared to moles of **PbS.**

Answers: 1. 234 [391]; 2. 717 [391]; 3. Twice [391].

Glossary

Glossary

Absolute temperature (Kelvin). Celsius temperature plus 273°.

Acetylcholine. A compound necessary for the transmittance of impulses along a nerve.

Acid. A compound capable of either producing hydrogen ions or accepting an electron pair.

Acid-base reactions. Reactions involving the recombination of atoms to form salts and a change in hydrogen ion concentration where applicable.

Actinide elements. Elements 89–103. They illustrate the result of the filling of the $5f$ electron orbitals.

Activation analysis. The procedure of subjecting unknown substances to neutron bombardment in order that they may be analyzed for trace components. The capture of neutrons by the trace materials often produces radioactive products that can be identified by their characteristic emissions.

Activation energy. The amount of energy which reactants must acquire before a reaction will take place.

Aerobic organisms. Oxygen-breathing organisms.

Alcohol. An organic compound having the structure **R—OH**, where **R** is an aliphatic group.

Aldehyde. An organic compound having the general formula **RCHO**, where **R** may be either an aliphatic or aromatic group.

Aliphatic compound. A straight-chained or cyclic compound that does not possess a "benzene-like" structure or properties.

Alkane. An organic compound containing no multiple bonds.

Alkene. An organic compound containing one or more double bonds.

Alkyne. An organic compound containing one or more triple bonds.

Allotrope. One of the possible forms in which an element in a given state may exist, e.g., the three solid states of sulfur.

Alpha particle. A particle that may be emitted during the nuclear decomposition of certain radioactive elements. It is composed of two neutrons and two protons.

Amine. A compound having the general formula $\mathbf{RNH_2}$, where **R** may be either an aliphatic or an aromatic group.

Amino acid. A carboxylic acid in which an amine nitrogen is attached to the carbon atom immediately adjacent to the carboxyl carbon atom.

Amphoteric compound. A compound capable of reacting with either acids or bases to form salts.

Anaerobic organisms. Non-oxygen-breathing organisms.

Analgesic. A substance capable of alleviating pain.

Angstrom. 1×10^{-10} meters $= 1 \times 10^{-8}$ centimeters. (See Appendix I for discussion of exponential numbers.)

Anion. A negatively charged ion.

Anode. The electrode of an electrochemical cell at which oxidation takes place.

Antihistamine. A substance that reduces the histamine level of the blood and the allergic reactions it may produce.

Antioxidant. A substance capable of terminating free radical oxidation chains.

Antitussive. A substance capable of alleviating a cough.

Aromatic compound. A compound possessing a "benzene-like" structure and properties.

Asymmetric carbon atom. A carbon atom to which four dissimilar groups are bonded.

Atmosphere (Pressure). 760 mm. of mercury.

Atom. The smallest quantity of an element possessing the properties of the bulk element.

Atomic fission. The splitting of atoms into fragments of roughly equal mass.

Atomic mass unit (AMU). The fundamental subatomic unit of mass. The mass of a single proton or neutron.

Atomic number. A number equal to the number of protons in an atom of a given element.

Aufbau Principle. Guidelines for determining the placement of electrons in atoms.

Avogadro's number. The number of atoms in a gram atomic weight. The number of molecules in a gram molecular weight (mole). 6×10^{23}

Azimuthal quantum number (l). One of the four quantum numbers necessary to describe the location and properties of an electron in an atom. It may have integral values from zero to **n** − 1.

Balanced equation. An equation in which the number of charges and atoms of a given type on one side of the arrow equals exactly the number on the other side.

Base. A compound capable of either accepting hydrogen ions or donating an electron pair.

Battery. An electrochemical cell capable of pumping electrons.

Beta particle. A particle that may be emitted from radioactive materials. It has the mass and charge of an electron.

Boiling point. The temperature at which the vapor pressure of a liquid is equal to atmospheric pressure (760 mm. **Hg**).

Bonds. See *chemical bonds.*

Boyle's Law. The volume of a gas is inversely proportional to its pressure.

Carboxylic acid. A compound having the general formula **RCOOH**, where **R** may be either an aliphatic or an aromatic group.

Catalyst. A substance that shortens the time it takes for a reaction to come to equilibrium, without the substance itself being altered.

Cathode. The electrode of an electrochemical cell at which reduction takes place.

Cation. A positively charged ion.

Celsius. A temperature scale in which the freezing point and boiling point of water are respectively 0°C and 100°C. It was formerly called the centigrade scale.

Chain initiation reaction. The initial reaction that results in the production of a specie capable of initiating a series of reactions known as a "chain reaction."

Chain propagation reactions. The series of reactions necessary to maintain a chain mechanism.

Chain reactions. A series of reactions in which the products of earlier reactions initiate future reactions.

Chain termination reactions. Reactions that result in the reduction in concentration of the chain carrier specie.

Charge (Subatomic particles). The basic units of positive and negative electrical characteristics from which the electrical properties of matter are derived. The proton is the unit of positive charge; the electron is the unit of negative charge.

Charging (Battery). The expenditure of energy by an external circuit in order to produce an electron flow within a battery.

Charles's Law. The volume of a gas is directly proportional to its absolute temperature.

Chelating agent. A substance capable of sequestering metallic ions.

Chemical Bonds. The forces that join atoms together to form molecules.

***Cis* isomers.** Isomers in which both of the referenced groups protrude from the same side of the molecule.

Collagen. A protein that comprises the major fibrous constituent of skin, tendon, and bone. It can be converted to gelatin by boiling in water.

Compound. A substance composed of more than one element that possesses unique composition and chemical properties.

Congeners. Elements in the same periodic group.

Copolymer. A polymer made by polymerizing different compounds together.

Covalent bond. The bond produced as a result of two atoms sharing the same electron pair.

Cracking. Thermal degradation of large organic molecules to produce smaller fragments.

Critical mass. The minimum mass a substance must have before it can explode spontaneously.

Cyclo. A prefix used to describe organic compounds that contain ring structures.

Dalton's Law of partial pressures. The total pressure of a gas mixture is equal to the sum of its component partial pressures.

DDT. Dichlorodiphenyltrichloroethane, a powerful insecticide.

Dehydrating agent. A substance such as concentrated sulfuric acid that is capable of abstracting water from a reaction mixture.

Dehydrogenation reactions. Reactions such as cracking that reduce the number of hydrogen to produce unsaturated compounds.

Dextro isomer (+). The isomeric form of an organic molecule containing an asymmetric carbon atom that rotates the plane of polarized light in a positive or clockwise direction.

Depilatory. A substance used for the removal of unwanted hair.

Deuterium. Hydrogen with an isotopic mass of 2.

Di. A prefix indicating two or dual.

Dimerizing. Forming molecules of two identical atoms or molecules.

Dipole. See *polar molecule.*

Diprotic acid. An acid containing two ionizable protons.

Discharging (Battery). The expenditure of energy by a battery in order to produce an electron flow in an external circuit.

Disproportionation. When a chemical specie reacts to produce two new species in which the oxidation numbers of one of the component atoms in the new species are above and below that found in the original specie, i.e.,

$$3\,\overset{+5}{ClO_3^-} \rightarrow 2\,\overset{+4}{ClO_2} + \overset{+7}{ClO_4^-}$$

The chlorine oxidation number has disproportionated from +5 to +4 and +7.

Dithio acids. Organic compounds with carboxylic-acid-like structures in which the oxygen atoms have been replaced by sulfur atoms, e.g., RCSSH.

DNA. Deoxyribose nucleic acid. A nucleic acid in which the sugar group is deoxyribose.

Double bond. A bond composed of one sigma and one pi bond.

Electrolytic solution. The solution surrounding the electrodes in an electroplating cell.

Electron. The smallest unit of negative charge.

Electron affinity. A measure of the attraction of an atom for electrons.

Electron orbit. The volume of space within an atom most likely to contain the electron of interest.

Electromotive force. Force that causes the movement of electricity in an electrical circuit.

Electronegativity scale. A quantitative scale for the measurement of the attraction of atoms in a molecule for electrons.

Electronegative. Tends to form negative ions.

Electrophilic substitution. An organic substitution reaction in which the electron pair necessary for covalent bond formation is supplied by the molecule undergoing substitution.

Electroplating. The process by which materials are made the cathode in an electrolytic cell in order that a layer of metal can be deposited on their surface.

Electropositive. Tendency to form positive ions.

Element. One of the 105 chemically different atoms.

Enantiomer. Optical isomers that are mirror images of each other.

Enzyme. A protein molecule produced by living cells that acts as a catalyst for a chemical reaction associated with life processes.

Equation. A chemical sentence that describes what occurs during a chemical reaction.

Equilibrium. The point in a reversible reaction where the rate that reactants are converted to products is equal to the rate that products are converted to reactants. At this point no net chemical change is observable.

Equilibrium constant. A number that expresses the distribution of reactants and products in an equilibrium mixture. It is equal to the product of the product concentrations divided by the product of the reactant concentrations at equilibrium, each concentration being raised to the power dictated by the coefficients of the reaction equation.

Essential amino acids. Amino acids that cannot be synthesized in sufficiently large quantities by an organism to allow it to function properly. It therefore must acquire these amino acids from its diet.

Ethers. Compounds having the general formula **ROR′**, where **R** and **R′** may both be either aliphatic or aromatic groups.

Exponential number. A number expressed as a power of 10.

Fats. Esters of glycerol and long-chained acids.

Flux. A material added to a substance to increase its fusibility.

Free radical. An atom, molecule, or ion that contains an odd number of electrons.

Gamma ray. A high-energy ray that has neither charge nor mass and that may be emitted from radioactive materials.

Gay-Lussac's Law. The volume of a gas increases 1/273 of its volume at 0°C for every degree Celsius that its temperature is raised.

Geometrical isomers. Molecules that differ only in terms of the spatial orientation of their groups.

Graham's Law of diffusion. The velocity of a gas molecule at a given temperature is inversely proportional to the square root of its mass.

Greenhouse effect. The raising of the earth's atmospheric temperature as a result of air pollution. The "effect" is usually ascribed to CO_2. However, substances such as ozone can produce similar temperature changes.

Grignard reagent. A compound used in the syntheses of other organic compounds. It can be represented by the general formula **RMgX** where **R** is an organic group and **X** is a halide, usually chlorine or bromine.

Groups. Vertical columns of the periodic chart that contain elements with similar chemical properties.

Half-life ($T_{\frac{1}{2}}$). The time necessary to reduce the amount of reactant in a decomposition reaction to one-half its initial amount.

Halides. Compounds having the general formula **MX,** where **M** may be either an organic or an inorganic group and **X** may represent any of the group VII elements.

Halogen. An element found in group VII of the periodic chart (**F**, **Cl**, **Br**, **I**, **At**).

Hemoglobin. A complex organic molecule used to transport the respiratory gases through the bloodstream.

Heat of fusion. The amount of heat necessary to convert a substance from a solid to a liquid at its melting point.

Heat of vaporization. The amount of heat necessary to convert a substance from a liquid to a gas at its boiling point.

Hetero-atom compounds. Organic compounds containing atoms in addition to carbon and hydrogen, i.e., sulfur, nitrogen, oxygen, and so on.

Hybridization. The combining of *s* and *p* orbitals to form new orbitals with unique properties distinctly different from either of the parent orbitals of which they are composed (e.g., *sp*, sp^2, and sp^3 orbitals of carbon).

Hydrocolloid. A substance capable of producing a colloidal solution in water.

Hydrogen. Element 1, composed of one proton and one electron. It can occur in one of three isotopic forms: protium, deuterium, or tritium, which contain 0, 1, and 2 neutrons respectively.

Hydrogen bond. A weak electrostatic bond between molecules in which a component hydrogen atom serves as the bonding group between electronegative atoms such as oxygen, nitrogen, and fluorine.

Ion. A charged chemical specie.

Ionic bond. The bond produced between oppositely charged ions.

Ionization constant (K). A constant that expresses the ratio of concentrations at equilibrium between a compound and its ions. It equals the ratio of the product of the ion concentrations divided by the compound concentration, each concentration being raised to the power dictated by the coefficients of the reaction equation.

Iso isomer. An isomer containing a branch chain or group on its second carbon atom.

Isoelectric point. The *p*H at which the amino and the carboxyl groups of an amino acid are both ionized.

Isomer. One of the possible forms in which compounds possessing the same molecular formula may exist. Molecules have the same type and number of atoms bound together in different arrangements.

Isotope. One of the possible forms in which an

element may exist. The forms differ from each other only with respect to the number of neutrons in the nucleus.

K capture. The capture by the nucleus of the innermost or K electron. The process produces an atom with the same atomic weight but an atomic number one less than the original.

Ketones. Compounds having the general formula **RR′CO**, where both **R** and **R′** may be either aliphatic or aromatic groups.

Kinetic energy. Energy of motion.

Lachrymator. A substance that induces tearing of the eyes.

Lanthanide elements (Elements 57–71). They illustrate the filling of the $4f$ electron orbitals.

Levo isomer (−). The isomeric form of an organic molecule containing an asymmetric carbon atom that rotates the plane of polarized light in a negative, or counterclockwise, direction.

Lipids. A broad classification of compounds of which fats are the most common example.

Logarithms. A system of expressing numbers as powers of 10.

Magnetic quantum number (m). One of the four quantum numbers necessary to describe an electron. It may have integral values from $(-l)$ to $(+l)$, including zero.

Markownikoff's Rule. When an asymmetrical reagent adds to a double bond, the more negative portion of the reagent will bond to the carbon with the smaller number of hydrogen atoms.

Mass. A measure of the quantity of matter of a substance in terms of its heaviness.

Mechanism. The "nitty-gritty" steps involved in the production of a chemical reaction.

Melting point. The temperature at which a solid is converted to a liquid. It is also the freezing point of the liquid.

Mercaptans. Organic compounds having the general formula **RSH**, where **R** may be either an aliphatic or an aromatic group.

Meso compound. A nonoptically active compound composed of two asymmetric groups that are mirror images of each other.

Meta isomer. Disubstituted isomers of benzene in which the substituted groups are separated from each other by one carbon atom.

Metal. An element that forms ions through the loss of electrons.

Metric system. A system of measurement that utilizes meters, kilograms, and seconds as the units of length, mass, and time, respectively.

Molarity. Concentration expressed in gram molecular weights per liter.

Molecular orbital theory. A theory that describes molecular orbitals in terms of component atomic orbitals.

Molecule. An uncharged grouping of atoms that displays properties unique to the group and different from those of the individual component atoms.

Mono. A prefix indicating one or single.

Neutralization. The result of reacting an acid with a base.

Neutron. An uncharged particle having a mass of one AMU.

Nonmetal. An element that forms ions through the acquisition of electrons.

Normal isomer. A straight, nonbranched chained isomer.

Normality. Concentration expressed in gram equivalent weights per liter.

Nucleic acids. Macromolecular polymers of nucleotide units.

Nucleophilic substitution. An organic substitution reaction in which the group being substituted is able to supply the electron pair necessary for formation of a covalent bond.

Nucleotides. The basic units from which nucleic acids are constructed. They are composed of a phosphoric acid, sugar, and nitrogen base group.

Nucleus. The central portion of an atom containing the protons and neutrons, if any.

Nucleon. A nuclear particle, i.e., a proton or neutron.

Olefin. A compound that contains double-bonded carbon atoms.

Optical isomer. A molecule possessing one or more asymmetric carbon atoms which rotates the plane of polarized light.

Orbital. A volume of space, denoted by *s*, *p*, *d*, or *f*, that describes the location of electrons of a particular energy in an atom.

Organo-metallic complex. An organic complex containing a metallic ion.

Ortho isomer. Disubstituted isomers of benzene in which the substituted groups are on adjacent carbon atoms.

Oxidant. The substance in an oxidation-reduction reaction that causes the oxidation. It, as a consequence, is reduced.

Oxidation. The loss of electrons.

Oxidation number. A number describing the degree of oxidation of an atom in terms of the number of electrons necessary to convert it to its elemental state. For example, Na^+ has an oxidation number of +1; the addition of one electron to a sodium ion is necessary to convert it to a sodium atom. Similarly, F^- has an oxidation number of −1; the removal of one electron is necessary to convert fluoride ion to an elemental fluorine atom.

Oxidation-reduction (redox) reaction. Reactions involving a change in oxidation state of atoms in addition to their recombination.

Pairing (electron spin). The coupling of two electrons that differ in the sign of their spin quantum number.

Para isomer. Disubstituted isomers of benzene in which the substituted groups are separated from each other by two ring carbon atoms.

Pauli Exclusion Principle. No two electrons in an atom may have the same set of four quantum numbers.

Period. A horizontal row in the periodic chart.

Periodic chart. A system of organizing the elements according to quantum numbers and their chemical properties.

Permeability. The degree to which a substance permits diffusion.

***p*H.** The negative logarithm of the hydrogen ion concentration.

Phenol. A compound represented by the general formula **ROH**, where **R** is an aromatic group.

Photosynthesis. The process by which plants utilize carbon dioxide, water, and sunlight to manufacture carbohydrates.

Pi (π) bond. A secondary bond formed above and below the midline that joins two atoms in a sigma bond.

***p*OH.** The negative logarithm of the hydroxide ion concentration.

Polar molecule. A molecule that possesses a dipole (positive and negative regions).

Pollutant. A substance whose presence is undesirable because of the adverse effects it may have on either life or its surroundings.

Polymer. A large molecule composed of smaller repeating molecular units.

Positron. A particle with the mass of an electron and the charge of a proton.

PPB (parts per billion). A unit of concentration equal to ten million times its percentage. $PPB = \% \times 10^7$.

Precipitate. A nonsoluble substance.

Pressure. A measure of force per unit area.

Primary compound. A compound in which a first degree replacement has occurred. For example, a primary amine (RNH_2) is a compound where one of the three hydrogens of ammonia has been replaced by an alkyl group.

Principal quantum number (n). One of the four quantum numbers that determine the location and properties of an electron within an atom. When describing a valence electron of an atom, it is equal to the period in which the element is found in the periodic chart.

Products. The chemical substances produced by a chemical reaction. Traditionally they are written to the right of the arrow when writing a chemical equation.

Proteolytic enzymes. Enzymes that hasten the hydrolysis of proteins.

Protium. An isotope of hydrogen containing one proton, one electron, and no neutrons.

Proton. A positively charged particle having a mass of 1 AMU.

Quantum numbers. A set of four numbers used to describe the location and properties of electrons in atoms and molecules.

Racemic mixture. A mixture containing equal

amounts of an enantiomeric pair of compounds. It does not rotate the plane of polarized light.

Radioisotope. An unstable isotope capable of decomposing with radioactive emission.

Rare gases. The elements found in group VIII of the periodic chart.

Reactants. The chemical substances that initiate a chemical reaction. Traditionally they are written to the left of the arrow when writing a chemical equation.

Redox. A contraction of the words *reduction* and *oxidation*. Also see *oxidation-reduction reaction*.

Reductant. A substance in an oxidation-reduction reaction that causes reduction. It, as a consequence, is oxidized.

Reduction. The gain of electrons.

Reversible reaction. A reaction whose products can readily convert back to reactants.

Ring activating groups (Electrophilic substitution). Functional groups already substituted onto the benzene ring that aid in the further substitution of other electrophilic groups onto the ring (e.g., alkyl, amino hydroxy, etc.).

Ring deactivating groups (Electrophilic substitution). Functional groups already substituted onto the benzene ring that retard further substitution of other electrophilic groups onto the ring (e.g., nitro, sulfonic acid, carboxylic acid, etc.).

RNA. Ribonucleic acid. A nucleic acid in which the sugar group is ribose.

Saponification. The process of hydrolyzing fats in a strongly basic medium.

Saturated compound. A compound containing no multiple carbon-to-carbon bonds.

Satzev Rule. If more than one compound may be produced by a given reaction, such as dehydration, the product usually produced in greatest abundance is the one with the greater degree of substitution.

Schiff base. The condensation product of a primary amine and an aldehyde.

Scintillation counter. An instrument that determines the level of radioactivity by measuring the amount of light emitted by an indicator substance as a result of being exposed to radiation.

Secondary compound. A compound in which a second degree replacement has occurred. For example, a secondary amine (**RR′NH**) is a compound where two of the three hydrogens of ammonia have been replaced by alkyl groups.

Shell. A term used to describe the quantum or energy levels of electrons in an atom.

Sigma (σ) **bond.** A bond formed along the midline between two atoms.

Smog. A term originally coined to signify a mixture of smoke and fog. In popular usage, it has come to mean any form of air pollution.

Soap. A salt of a long-chained fatty acid that possesses both a water-soluble and an oil-soluble end.

Spatial configuration. The spatial arrangement of branched groups in an organic molecule. Glyceraldehyde is usually employed as the reference standard for this purpose. The spatial arrangement of the atoms of a molecule is usually denoted by the letters (D) or (L). These letters may or may not correspond to the optical activity of the molecule. If the direction in which a molecule rotates the plane of polarized light is to be specified, either the letters (*d*) or (*l*), or the symbols (+) or (−) are utilized.

Spectroscope. An instrument capable of analyzing substances from the way they interact with light.

Spin quantum number (s). One of the four quantum numbers necessary to describe an electron. It may have values of either plus or minus one-half.

Stoichiometry. The study of the numerical proportions that govern the manner in which chemicals react.

Structural isomers. Molecules that contain the same number and type of atoms but differ in the manner in which these atoms are bonded to each other.

Sublimation. A physical process by which a solid is converted directly to a gas without the formation of an intermediate liquid.

Substitution reactions. Reactions that occur as a result of the interchange of positive and negative ions. There is no change in oxidation state in a substitution reaction.

Suborbital. A volume of space that can contain as many as *2 electrons*. (For example, there can be as many as one *s* and three *p* suborbitals in the second principal quantum level of an atom. In other words, the *p* orbital can contain three *p* suborbitals.)

Temperature. A measure of heat intensity.

Tertiary compound. A compound in which a third-degree replacement has occurred. For example, a tertiary amine (**RR′R″N**) is a compound where all three hydrogens of ammonia have been replaced by alkyl groups.

Tetra. A prefix indicating four or quadruple.

Thioesters. Organic compounds with ester-like structures in which the oxygen atoms have been replaced by sulfur atoms (i.e., **RCSSR′**).

Thioethers. Organic compounds with ether-like structures in which oxygen atoms have been replaced by sulfur atoms (i.e., **RSR′**).

Thioketones. Organic compounds with ketone-like structures in which oxygen atoms have been replaced by sulfur atoms (i.e., **RR′C═S**).

Thixotropy. A property of certain gels that causes them to liquefy on shaking and reform on standing.

***Trans* isomers.** Isomers in which both of the referenced groups protrude from opposite sides of the molecule.

Transition complex. The intimate grouping that is produced when chemical species are mixed. As the name indicates, the complex serves as a transition stage between reactants and products. In a reversible reaction this complex may break up to produce either reactants or products.

Transition elements. Metals possessing incompletely filled inner *d* orbitals. They are noted for the variability of their permissible oxidation state. For example, manganese has two electrons in its 4*s* suborbital and five in its 3*d* suborbital. It exhibits oxidation states of +1, +2, +3, +4, +5, +6, and +7.

Tri. A prefix indicating three or triple.

Triple bond. A bond composed of one sigma and two pi bonds.

Tritium. An isotope of hydrogen possessing one proton, one electron, and two neutrons, and having an atomic mass of 3.

Ultraviolet light (UV). Radiation whose wavelength is slightly shorter than that of visible light.

Unsaturated compound. A compound containing one or more multiple carbon-to-carbon bonds.

Valence. An integral number that describes the combining power of an atom.

Valence shell. The outermost electronic shell of an atom. The shell that is involved in chemical reactions.

Vasoconstrictor. A substance that constricts blood vessels.

Vitamins. Powerful organic compounds necessary in small amounts for the proper functioning of living organisms.

Voltage. A measure of the electrical pressure that causes electrons to flow.

Walden inversion. The geometrical inversion of three groups attached to a carbon atom as a result of a nucleophilic substitution of that atom.

Zwitterion. The product that results when the amine and the carboxyl groups in an amino acid both ionize.

Index

Index